Radioisotopes: Applied Principles in Biomedical Science

Radioisotopes: Applied Principles in Biomedical Science

Edited by **Peggy Sparks**

NY RESEARCH
P R E S S

New York

Published by NY Research Press,
23 West, 55th Street, Suite 816,
New York, NY 10019, USA
www.nyresearchpress.com

Radioisotopes: Applied Principles in Biomedical Science
Edited by Peggy Sparks

International Standard Book Number: 978-1-63238-385-3 (Hardback)

Contents

Preface   IX

Part 1   Radioisotopes and Radiations in Bioscience   1

Chapter 1   **Radioisotopes and Nanomedicine**   3
Nathan C. Sheets and Andrew Z. Wang

Chapter 2   **Use of Radiation and Isotopes in Insects**   23
Thiago Mastrangelo and Julio Walder

Chapter 3   **Radioisotopes in Drug Research and Development:**
Focus on Positron Emission Tomography   49
Sosuke Miyoshi, Keisuke Mitsuoka,
Shintaro Nishimura and Stephan A. Veltkamp

Chapter 4   **Medical Radioisotopes Production: A Comprehensive**
Cross-Section Study for the Production of Mo and Tc
Radioisotopes Via Proton Induced Nuclear
Reactions on ^{nat}Mo   71
A. A. Alharbi, A. Azzam, M. McCleskey,
B. Roeder, A. Spiridon,E. Simmons, V.Z. Goldberg,
A. Banu, L. Trache and R. E. Tribble

Chapter 5   **Use of Radioactive Precursors for Biochemical**
Characterization the Biosynthesis of Isoprenoids in
Intraerythrocytic Stages of *Plasmodium falciparum*   95
Emilia A. Kimura, Gerhard Wunderlich,
Fabiana M. Jordão, Renata Tonhosolo,
Heloisa B. Gabriel, Rodrigo A. C. Sussmann,
Alexandre Y. Saito and Alejandro M. Katzin

Chapter 6   **Application of Radioisotopes in Biochemical Analyses:**
Metal Binding Proteins and Metal Transporters   115
Miki Kawachi, Nahoko Nagasaki-Takeuchi,
Mariko Kato and Masayoshi Maeshima

Chapter 7   **The Use of Radioisotopes to Characterise
the Abnormal Permeability of Red Blood
Cells from Sickle Cell Patients**   127
Anke Hannemann, Urszula Cytlak, Robert J. Wilkins,
J. Clive Ellory, David C. Rees and John S. Gibson

Chapter 8   **Undesirable Radioisotopes Induced by Therapeutic
Beams from Medical Linear Accelerators**   149
Adam Konefał

Chapter 9   **Boron Studies in Interdisciplinary Fields
Employing Nuclear Track Detectors (NTDs)**   173
László Sajo-Bohus, Eduardo D. Greaves and
József K. Pálfalvi

Part 2   **Radioisotopes and Radiology in Medical Science**   197

Chapter 10   **Radiolabelled Nanoparticles for Diagnosis
and Treatment of Cancer**   199
Dimple Chopra

Chapter 11   **Production and Selection of Metal PET
Radioisotopes for Molecular Imaging**   223
Suzanne V. Smith, Marian Jones and Vanessa Holmes

Chapter 12   **Nuclear Medicine in the Imaging and
Management of Breast Cancer**   249
Luciano Izzo, Sara Savelli, Andrea Stagnitti and
Mario Marini

Chapter 13   **3-Dimensional CT Lymphography in Identifying
the Sentinel Node in Breast Cancer**   263
Junko Honda, Chieko Hirose, Masako Takahashi,
Sonoka Hisaoka, Miyuki Kanematsu, Yoshimi Bando
and Mitsunori Sasa

Chapter 14   **Axillary Reverse Mapping in Breast Cancer**   273
Masakuni Noguchi, Miki Yokoi, Yasuharu Nakano,
Yukako Ohno and Takeo Kosaka

Chapter 15   **Lymphedema: Clinical Picture,
Diagnosis and Management**   289
Tanja Planinšek Ručigaj and Vesna Tlaker Žunter

Chapter 16   **Targeting the Causes of Intractable Chronic Constipation
in Children: The Nuclear Transit Study (NTS)**   305
Yee Ian Yik, David J. Cook, Duncan M. Veysey,
Stephen J. Rutkowski, Coral F. Tudball, Brooke S. King,
Timothy M. Cain, Bridget R. Southwell and John M. Hutson

Permissions

List of Contributors

Preface

I am honored to present to you this unique book which encompasses the most up-to-date data in the field. I was extremely pleased to get this opportunity of editing the work of experts from across the globe. I have also written papers in this field and researched the various aspects revolving around the progress of the discipline. I have tried to unify my knowledge along with that of stalwarts from every corner of the world, to produce a text which not only benefits the readers but also facilitates the growth of the field.

The book concentrates primarily on radioisotopes and its applications in bioscience and in medical sciences. This book elucidates creation of radioisotopes for medicine, the usage of radioisotopes in various fields, the utilization of radiation in drug research, among others. Topics like 3D and computed tomography (CT) scan, SS nuclear medicine in imaging, detection and management of cancer, and various other topics are also discussed. This book will benefit clinicians and researchers dealing with the various applications of radioisotopes.

Finally, I would like to thank all the contributing authors for their valuable time and contributions. This book would not have been possible without their efforts. I would also like to thank my friends and family for their constant support.

Editor

Part 1

Radioisotopes and Radiations in Bioscience

Radioisotopes and Nanomedicine

Nathan C. Sheets and Andrew Z. Wang
University of North Carolina – Chapel Hill
United States of America

1. Introduction

Nanomedicine, the medical application of nanotechnology, is poised to make a major impact on the diagnosis and treatment of many diseases. It is a relatively new branch of science that involves harnessing the unique properties of particles that are nanometers in scale (nanoparticles) for both diagnostic and therapeutic applications. Nanoparticles (NPs) can be engineered with precise sizes, shapes, compositions and surface chemistries for specific applications. In particular, nanoparticles are uniquely suited for the treatment of cancers and cardiovascular diseases. NPs can be engineered with surface targeting ligands against cancer markers or atherosclerosis markers, which in turn allow the specific detection of as well as specific drug delivery to these diseases. In the case of cancer, nanoparticles also preferentially accumulate in tumors due to the enhanced permeability and retention (EPR) effect (Noguchi et al. 1998). Although nanomedicine is a relatively new field, it has been quickly translated into clinical medicine and has led to a number of clinically-approved agents. Today, there are more than twenty nanoparticle therapeutics in clinical use. Furthermore, a 2006 global survey conducted by the European Science and Technology Observatory (Duconge et al.) revealed that more than 150 companies are developing nanoscale therapeutics, highlighting the high enthusiasm for nanoparticle-based therapeutics (Wagner et al. 2006).

While nanomedicine has many facets, radioisotopes have been an integral component. Radioisotopes have been extensively utilized in the development of nanoparticle-based therapeutics. Radiolabeling has been the predominant method for the accurate assessment of biodistribution, circulation half-life, and pharmacokinetics of nanoparticles. In addition, there has been great interest in developing nanoparticles with dual diagnostic and therapeutic functionalities (theranostics) and radioisotopes have been a natural fit for the diagnostic component. Lastly, therapeutic radioisotopes can also be delivered by nanoparticles, especially for the treatment of cancer. In this chapter, we will review the various applications of radioisotopes in nanomedicine in detail. We also plan to discuss the methods of incorporating radioisotopes into nanoparticles as well as preclinical and clinical data on radioisotope containing nanoparticle diagnostics and therapeutics.

2. Nanoparticle platforms

Nanomedicine is a fairly new and rapidly developing branch of science. First described in 1965, liposomes are one of the first structures manufactured on a nanometer scale to be applied in medicine (Bangham 1993). Over the past 15 years, there has been a rapid growth of interest in NPs probably due in large part to the inherent flexibility in the construction and design of the particles.

Liposomal agents were the first NP platform to make it into the clinic and have since been joined by a number of other nanoparticle classes. These various platforms include polymer-therapeutic conjugates, polymeric micelles, dendrimers, nanoshells, nucleic acid-based nanoparticles and carbon nanotubes (Wang et al. 2008). Liposomes are spherical lipid containers enclosing an aqueous space. Depending on design, they can range in size from tens of nanometers up to micrometers in size (Torchilin 2005). The outer sphere is composed of a phospholipid bilayer which acts a hydrophobic shell around the aqueous core. This allows for entrapment of hydrophilic compounds within the liposomes. Conversely, hydrophobic drugs can be incorporated into the lipid bilayer itself. The outside surface of the bilayer is exposed and can therefore act as a point of interaction between a liposome and a biosystem. The liposomal surface can be functionalized with targeting agents such as antibodies, protective polymers such as polyethylene glycol to prevent opsonization by the immune system, or diagnostic labels such as radioisotopes (Torchilin 2005).

Most of the next generation nanoparticle therapeutics are biodegradable polymer nanoparticles, which have been extensively investigated as therapeutic carriers (Kim et al. 2005). Polymeric nanoparticles are generally formulated by the self-assembly of block-copolymers consisting of two or more polymer chains with different hydrophobicity. These copolymers spontaneously assemble into a core-shell structure in an aqueous environment through the hydrophobic hydrophilic interactions. Specifically, the hydrophobic blocks form the core to minimize their exposure to aqueous surroundings while the hydrophilic blocks form the shell to stabilize the core (Farokhzad et al. 2004). This results in a structure ideally suited for drug delivery. The hydrophobic core is capable of carrying therapeutics with high loading capacity (5-25 % weight). The hydrophilic shell not only provides a steric protection for the micelle, but also provides functional groups for further particle surface modifications. Polymeric nanoparticles have been formulated to encapsulate either hydrophilic or hydrophobic small drug molecules, and macromolecules such as proteins and nucleic acids (Gu et al. 2008).

Fullerene carriers are another well-studied nanoparticle platform which includes both spherical carbon "buckyballs" and carbon nanotubes (CNTs). Both have potential as carriers for radioisotopes, in particular because of their ability to shield the reactive radioisotopes from the biosystem while at the same time allowing for modification of the outside of the nanotube for targeted therapy (Mackeyev et al. 2005). CNTs have cylindrical architecture with an extremely high length-to-diameter ratio. Until recently this system had not been successfully tested *in vivo*. Researchers at Oxford carried out the first experiments in a mouse model with I-125-filled, single-walled carbon nanotubes. Though iodine is rapidly taken up by thyroid tissue, researchers were able to completely redirect the CNTs to lung. This suggests that the iodine was entirely shielded from the circulation. This particular platform also set a record for radiation ionizing dose per gram of carrier at 800% (70% represents a reasonable prior *in vivo* benchmark). By surface functionalizing this platform, researchers hope to show versatility of targeting which may allow for extremely high doses of radiation to be delivered very selectively (Hong et al. 2010; Strano 2010).

Other platforms of nanoparticle delivery have been extensively studied including dendrimers, metal nanoparticles, nanoemulsions and nucleic acid based nanoparticles (Wang et al. 2008). The interest in these other nanoparticle platforms is growing and several dendrimer-based antiviral therapies are currently in clinical trials. Investigators have also developed metallic nanoshells used for controlled release of chemotherapy. Each of these platforms has unique properties and holds promise for almost every branch of medicine. Much of this work continues in the preclinical stages with a few drugs now making it to clinical trials. By improving the stealth properties and optimizing the payload capacities of these various nanomolecules, there is great hope and high expectations that this branch of science can continue to make wide-ranging contributions to medicine.

3. Common isotopes for nanomedicine and incorporation strategies

Selection of an isotope for imaging or therapy depends on a number of parameters. For imaging, radioisotopes should have decay properties conducive to detection. In positron emission tomography (PET) imaging, the energy is invariably 511 keV due to positron decay; however, for single-photon emission computed tomography (SPECT) imaging ideal energy is 100 keV to 200 keV(Phillips et al. 2009). The photons must be of sufficient energy to penetrate the body while still allowing for adequate collimation for spatial localization. The half-life of the isotope is also important to consider, particularly in relation to the duration of expected nanoparticle circulation. For instance, fluorine-18 as a component of fluorodeoxyglucose-positron emission tomography (FDG-PET) is most commonly used in clinical practice. With a half-life of 110 minutes, this allows for adequate time to allow uptake of the FDG while it retains enough activity to be clinically detected after getting a patient to the scanner, but not so much as to require long periods of detection time or as to retain substantial *in vivo* activity for more than a day. The same principles guide selection of laboratory isotope half-life, particularly when selecting PET-agents which have a wider range of half-life. Commercially available PET agents range in half-life from 2 minutes up to 4 days (Table 1) (Delbeke & Martin 2001; 2004). Naturally, nanoparticles with longer expected circulation time will require PET agents with longer half-life while those with relatively short circulation times should be tracked using isotopes with shorter half-life to improve photon yield over time. This selection process is complicated by variation in stability amongst different isotopes-nanoparticle combinations (Phillips et al. 2009). It is important to consider both *in vitro* and *in vivo* stability because unstable labeling can lead to incorrect measurements of biodistribution via isotope loss. Many methods have been described for labeling nanoparticles, most commonly liposomal platforms. The method of labeling can also greatly affect stability of the conjugate (Phillips et al. 2009).

To be useful in the laboratory setting, radiolabeled nanoparticles ideally need to meet certain criteria. Important factors include ease and efficiency of preparation as well as stability of isotope-nanoparticle conjugate without *in vitro* or *in vivo* release of the radioisotope. Steady advancements in this area have led to increasing stability of isotope labeled liposomes, some of which are being translated into the clinic for diagnostic molecular imaging (Hahn et al. 2011).

Due to the relatively short half-life of the commonly used radioisotopes, nanoparticles should be labeled shortly before being used. This is referred to as "after loading" or "remote-labeling." Though not ideal, a very simple strategy for radiolabeling liposomes is to incubate the premanufactured liposomes with a lipophilic isotope (Phillips et al. 2009). This leads to emulsion of the isotope into the lipid bilayer; however, while this method is fairly simplistic, it results in a very unstable conjugation. Several more effective "after loading" techniques have been described that yield more stable isotope-liposome conjugations (Harrington et al. 2001).

Selected radioisotopes used for nanoparticle detection and imaging			
Radionuclide	Detection Method	T½	Photon Energy (Bult et al.)
Gallium-67	SPECT	3.3 d	93, 185, 300
Indium-111	SPECT	2.8 d	171, 245
Iodine-123	SPECT	13.2 h	159
Rhenium-186	SPECT	3.72 d	137, 132
Technetium-99m	SPECT	6.01 h	141
Iodine-125	SPECT	59 d	36
Carbon-11	PET	20 min	511
Copper-64	PET	12.7 h	511
Copper-62	PET	3.5 h	511
Fluorine-18	PET	109 min	511
Gallium-68	PET	68 min	511
Iodine-124	PET	4.2 d	511
Oxygen-15	PET	2 min	511
Zirconium-89	PET	78 h	511
Hydrogen-3	Scintillation	12.3 y	(β=18.5 keV)
Carbon-14	Scintillation	5730 y	(β=156 keV)

Table 1. Common radioisotopes for nanoparticle imaging. All are generated by cyclotron with the exception of Tc-99m which comes from Molybdenum-99/Tc-99m generators, resulting in a substantially lower relative cost because it is easier to obtain.

Radiolabeling via surface chelation was first described 30 years ago and is still used today. Hnatowich et al. first described chelation of both Ga-67 and Tc-99m via diethylenetriamine pentacetic acid (DTPA), a metal chelator (Hnatowich et al. 1981). More recently, Torchilin et al. published a method of radiolabeling liposomes with In-111 using phosphatidylethanolamine (PE) conjugated to DTPA (DTPA-PE). For Tc-99m a separate method has been developed by Laverman et al. (Laverman et al. 1999). A technetium chelator, N-hydroxysuccinimidyl hydrazine nicotinate hydrochloride (HYNIC), is conjugated to a free amino group of the liposome and incorporated via surface chelation to premanufactured liposomes. The liposomes are then incubated with Tc-99m prior to the start of the study (Laverman et al. 1999). External conjugation suffers from the potential for *in vivo* interactions between the labels and other biomolecules as well as surface crowding that could affect interactions with other functional groups on the liposome.

To help overcome this problem, other methods of "after loading" have been developed, specifically internal chelation. This approach utilizes liposomes that encapsulate a reactive species. The radiolabel is conjugated to a lipophilic chelator that can pass through the lipid bilayer (Harrington et al. 2001). Upon reaching the inside of the liposome, the complex interacts with the internal molecule, effectively entrapping it within the liposome. Numerous investigators have developed methods of radiolabeling based on this principle of internal chelation. Examples include In-111-oxine incubated with liposomes encapsulating DTPA leading to anchoring of the isotope (Harrington et al. 2001); Tc-99m-hexamethylpropyleneamine oxime incubated with liposomes encapsulating glutathione causing conversion to a hydrophilic form of the Tc-99m conjugate(Phillips et al. 1992);

Rhenium-186-N,N-bis(2-mercaptoethyl)-N',N'-diethylethylenediamine reacting with encapsulated cysteine (Bao et al. 2003); and halogenated I-123 combining with intra-liposomal arginine (Mougin-Degraef et al. 2006). Each of the reactions secures the label within the core of the nanoparticle which in general results in greater stability of the radiolabeled liposome.

Because of the wide range of methods used for incorporation of therapeutic radionuclides, discussion of these strategies is reserved for the specific examples cited later in the chapter (Table 2).

Selected radioisotopes under study for nanoparticle radiotherapy				
Radionuclide	Nanoparticle Platform	Notable Decays	$T_{1/2}$	Reference
Iodine-131	Synthetic polymer	γ, β	8 d	(Klutz et al. 2009)
Holmium-166	Acetylacetone	γ, β	27 h	(Bult et al. 2010)
Lutetium-177	Liposome, fullerene cage	β	161 h	(Helbok et al. 2010)
Rhenium-186	Liposome	γ, β	89 h	(French et al. 2010)
Rhenium-188	Liposome, iron oxide	γ, β	17 h	(Liang et al. 2007)
Yttirium-90	Lipid-polymer hybrid	β	64 h	(Wang et al. 2010)
Indium-111	Biotinylated streptavidin	Auger, γ	67 h	(Natarajan et al. 2008b)
Iodine-125	Biotinylated streptavidin	Auger, γ	60 d	(Liu et al. 2010)
Actinium-225	Mineral lattice	a	10 d	(Woodward et al. 2011)
Gold-198	Dendrimer	β	2.7 d	(Khan et al. 2008)

Table 2. Radioisotopes that have been tested in preclinical models. Radioisotopes that decay by both electron and photon emission have potential as theranostic agents.

4. Radioisotope applications in nanomedicine

Radionanoparticles have three core applications: drug development, imaging and therapy. We will discuss these applications and how they are being combined to provide unique avenues for the treatment of disease.

4.1 Radioisotopes for nanoparticle biodistribution and drug development

Nowhere have radioisotopes been more influential in nanomedicine than in their use for preclinical drug development. By successfully incorporating radioisotopes into the structure of a nanoparticle in such a way as to minimally influence external structure, nanoparticles can be tracked by radioactive decay (Ting et al. 2010). A variety of methods using an array of different isotopes have been developed for this purpose (Phillips et al. 2009). Gamma photon imaging, in the form of positron emission tomography and single photon emission computed tomography (Racker & Spector), are the predominant methods of tracking *in vivo* particle biodistribution. Particle distribution is tracked via detection of emitted photons from isotope decay. In the case of PET, annihilation photons of 511 keV (generated when a positron and shell electron collide) emitted at 180 degrees from one another are detected and used to localize the particle (Chen & Conti 2010). SPECT uses emission of a single photon combined with collimation to allow photons of predictable trajectories to be recorded; thus, the particle decay can be localized within the body. In clinical practice, both methods have issues with spatial resolution; however, the level of localization is adequate for determining *in vivo* biodistribution.

The choice of radioisotope depends on a number of characteristics including the type of NP platform, ease of preparation and stability of the isotope-NP conjugate as well as availability, expense and decay characteristics of the radioisotope (Phillips et al. 2009). For instance, Tc-99m is readily available in most hospitals, and accounts for about 80% of the nuclear medicine tests worldwide (IAEA 2011); thus, it is commonly used for nanoparticle labeling. Carbon-14 was one of the first elements used to track the fate of nanoparticles by measuring urinary excretion of the isotope (Bazile et al. 1992). Tritium also has a long history of being used for tracking the fate of nanoparticles (Kukowska-Latallo et al. 2005; Takahama et al. 2009). These two radioisotopes are popular due to their long half-lives and ease of incorporation. Both radioisotopes are beta emitters and thus require dissection for *in vivo* quantification (Shokeen et al. 2008). Over the past two decades, SPECT and PET detection have emerged as facile ways of tracking internal biodistribution without the need for dissection to quantify uptake. Copper-64 is a commonly used agent for PET imaging that can be incorporated into a variety of nanoparticles for *in vivo* tracking (Fukukawa et al. 2008; Pressly et al. 2007; Rossin et al. 2005; Schipper et al. 2007; Xie et al. 2010a; Xie et al. 2010b). It's conjugation to nanoparticles is much easier than that of the more common clinical radioisotope fluorine-18 and is the primary reason it is used over this more clinically useful isotope. Indium-111 and Technitium-99m both have long track records for use in drug development as SPECT imaging agents (DeNardo et al. 2007; Elbayoumi et al. 2007; Kobayashi et al. 1999; Lin et al. 2009; Natarajan et al. 2008a; Natarajan et al. 2008b; Phillips et al. 2009). Similarly, several iodine radioisotopes (I-123, I-124, I-125) have been used for SPECT (Chrastina & Schnitzer 2010; Geze et al. 2007; Gratton et al. 2007; Kennel et al. 2008; Kumar et al. 2010; Malik et al. 2000; Medina et al. 2011). Other PET agents such as Gallium-68 are in use as well (Helbok et al. 2010; Phillips et al. 2009). Many of these radioisotopes are being used in molecular imaging agents.

4.2 Molecular imaging with nanoparticles

Nanoparticles have received a great deal of attention in the past decade in large part due to their potential for use in molecular imaging. As targeting specificity improves, methods of detection should simultaneously improve, leading to earlier detection of conditions such as heart disease and cancer. While radiolabeled liposomes have been studied for some time, only recently have researchers developed multifaceted imaging agents with the potential for human use. Active work in this area involves PET, SPECT, MRI and quantum dots. Molecular imaging has evolved a great deal over the past decade, and one of the most intriguing aspects of nanoparticle molecular imaging is the potential for designing combination particles for use as multifaceted imaging agents. To date, most combinations of the various imaging modalities have been combined at some level with varying levels of success.

Single modality carriers

There is good rationale for functionalizing nanoparticles with PET or SPECT agents for imaging. While glucose analog agents such as FDG show tumor uptake through the Warburg effect (Gatenby & Gillies 2004), specificity may be increased via the enhanced permeability and retention (EPR) effect (Noguchi et al. 1998). Additionally, nanocarriers may be functionalized with targeting ligands to further improve specificity for their target (Wang et al. 2008).

Despite its widespread use in the clinic for imaging, fluorine-18 is difficult to efficiently functionalize to targeted nanoparticles for this purpose. Due to its long half-life relative to other isotopes (carbon-11, nitrogen-13, oxygen-15), F-19 is a F-19 is a better candidate candidate for PET nanoparticles as there is sufficient time to functionalize the NP with radioisotope prior to administration. In 2008, Matson et al. were the first to demonstrate efficient synthesis of F-19 labeled micelle nanoparticles (Matson & Grubbs 2008). This may pave the way for better imaging agents using NPs.

Elements such as copper-64 are much easier to functionalize to NP because of their longer half-life (12.7 h), and thus to this point it has been used more frequently than fluorine. Chakrabarti et al. designed a peptide nucleic acid probe radiolabeled with copper-64 to target mutated *KRAS* mRNA. *KRAS* is mutated at a very high level in pancreatic cancer and mutated *KRAS* mRNA is expressed at a high level in these tumors suggesting it may be a good target for early detection or targeted therapy. To achieve cellular uptake, peptide nanoparticles were combined with an IGF1-analog peptide to allow for internalization by IGF1 receptor-mediated endocytosis. IGF1 receptor is commonly upregulated in human cancers, including pancreatic cancer. Authors designed a series of radiohybridization probes linked with copper-64 to target various single nucleotide polymorphisms in *KRAS* mRNA. The probe had good selectivity for *KRAS*-mutated in mice engrafted with AsPC1, a pancreatic cancer cell line (Chakrabarti et al. 2007). Similarly, Petersen et al. designed copper-64 labeled liposomes for PET imaging in a mouse model (Petersen et al. 2011).

Authors have described the targeting of tumor neovasculature with both SPECT and PET nanoparticles. Hu et al. designed an indium-111 labeled perfluorocarbon nanoparticle targeted to $\alpha_v\beta_3$, an integrin characteristically overexpressed in tumor neovasculature. In a rabbit model, the authors demonstrated a high sensitivity, low-resolution signal for tumor vasculature (Hu et al. 2007). Perfluorocarbons NP may also be used for MRI (Schmieder et al. 2005) allowing for a dual modality probe in the future that would increase this resolution while maintaining sensitivity. Ruggiero et al. imaged tumor vasculature using zirconium-89 labeled carbon nanotubes to target the monomeric vascular-endothelial cadherin epitobe that is frequently overexpressed in tumor vasculature (Ruggiero et al. 2010). The versatile carbon nanotube may prove to be both an efficient imaging and therapeutic nanoparticle platform. These dual modality (theranostic) particles will be discussed in detail later. More recently, Sun et al. published imaging data on the clinically available nanoparticle albumin-bound paclitaxel (Abraxane, Celegene). To judge early response to therapy, researchers attached either the PET agent 18F-FPPRGD2, a novel radiopharmaceutical shown to target $\alpha_v\beta_3$(Mittra et al. 2010), or standard 18F-FDG to Abraxane. The authors concluded that the novel agent was more sensitive than 18F-FDG for early detection of tumor response, likely due to down-regulation of $\alpha_v\beta_3$ in tumor endothelial cells when responding to Abraxane therapy (Sun et al. 2011).

The utility of nanoparticle SPECT versus PET agents was directly compared by Cheng et al. Tc-99m and F-18 were each conjugated individually to an anti-HER-2 nanoparticle, and the agents were tested in a mouse model with HER-2-positive xenografts. Despite the better resolution of Tc-99m, F-18 was 15 times more sensitive for *in vivo* tumor detection. Unfortunately, covalent attachment of F-18 is a much more difficult process than chelation of Tc-99m to the nanoparticle platform (Cheng et al. 2010); however, due to its much higher sensitivity, at this time, PET appears to be the choice modality for clinical applications.

Dual modality carriers

One of the exciting features of nanoparticle platforms is their versatility. Researchers are actively investigating almost every permutation of the various imaging modalities in combination with nanoparticle carriers. The list of potential modalities includes PET, SPECT, MRI and quantum dots (QD). While a full description of investigations into nanoparticle MRI and QD is beyond the scope of this chapter, each will be described as they pertain to multifaceted imaging nanoparticles involving PET and SPECT.

QD are nano-sized semiconductors with unique optical properties specific to the size and possibly the shape of the semiconductor crystal. In medical imaging, quantum dots are being studied as fluorophores, particles that absorb and emit energy of characteristic wave lengths. QD have been used *in vitro* for applications such as cell labeling and fluorescence *in situ* hybridization (Jiang et al. 2007; Wu et al. 2003). To date, *in vivo* use has been mostly limited to non-specific applications such as qualitative imaging of lymph nodes and vasculature in animal models (Hikage et al. 2010). Progress for medical applications has been hindered by technical challenges such as toxicity and characterization of the *in vivo* fluorescence signal (Aillon et al. 2009; Zhang & Badea 2011).

Despite these challenges, several researchers have described multifaceted nanoparticles for imaging with either PET or SPECT in combination with QD. Cai et al. designed cadmium QD functionalized with RGD peptide, a $\alpha_v\beta_3$ antagonist, and labeled with copper-64. This allowed for simultaneous near-infrared fluorescence (NIRF) and PET imaging. The probe was tested in a murine glioblastoma model and correlation between the *in vivo* PET signal and *ex vivo* QD fluorescence was excellent. On histologic examination, the nanoparticle exhibited specificity for tumor neovasculature. This study allowed for accurate determination of the probe biodistribution. A major advantage of this probe may be the gain in sensitivity with PET. This allows for the use of lower levels of cadmium QD to achieve accurate imaging. In the future, this will be of key importance for quantum dots to reduce the potential for toxicity from metals used in the semiconductors (Cai et al. 2007).

Similarly, Duconge et al. reported the first F-18-labeled QD using a PEG-phospholipid QD micelles. Pegylation allowed for prolonged circulation ($T_{1/2}$ of 2 h) and reduced RES uptake. Researchers were able to study the biodistribution both by whole body PET and by QD fluorescence using *in vivo* fibered confocal fluorescence microscopy. The authors conclude that this system effectively utilizes two molecular imaging modalities simultaneously: PET for whole body imaging and fluorescence imaging for subcellular localization of tumor cells (Duconge et al. 2008). SPECT-QD combinations have also been described (Liang et al. 2010). While one can envision clinical applications of these probes for invasive procedures, they still suffer from the inability to assess the fluorescence signal in deep tissue without direct surgical intervention.

This is not the case for dual MRI/PET probes which both allow for non-invasive imaging *in vivo*. While several researchers have looked at MRI/SPECT (Lijowski et al. 2009; Park et al. 2011; Torres Martin de Rosales et al. 2011; Zielhuis et al. 2006), the much improved sensitivity of positron emitting probes makes PET a much more enticing option for dual modality imaging. MRI allows for excellent spatial resolution that PET is lacking while PET improves upon the suboptimal sensitivity of MRI for functional imaging. The combination is appealing because of this complimentary nature (Cheon & Lee 2008).

Several investigators have designed iron oxide nanoparticles with copper-64 for dual modality imaging (Glaus et al. 2010; Jarrett et al. 2008; Patel et al. 2011). Jarrett et al. had previously demonstrated the use of an MRI compatible PET scanner to obtain simultaneous images using separate imaging probes (Catana et al. 2006; Judenhofer et al. 2007; Wu et al. 2009). Preclinical results are awaited using these untargeted nanoparticles in a model system for dual modality imaging. Several authors have developed and tested a similar iron oxide nanoparticles labeled with copper-64 and functionalized the carriers with integrin–binding $\alpha_v\beta_3$ RGD peptide (Lee et al. 2008; Yang et al. 2011). Researchers have tested the particles *in vivo* in a mouse model; however, the MRI characterization was performed without copper-64 due to limitations of the MR scanner (Lee et al. 2008). Despite this, imaging results between the two modalities were comparable.

Trimodality imaging agents and beyond

With clinical application of multifunctional imaging nanoparticles on the horizon, researchers have continued to push the limits of what is achievable with a single nanoparticle agent. Several authors have described trimodality imaging nanoparticles used for PET, MR and fluorescence imaging (Devaraj et al. 2009; Ko et al. 2011; Nahrendorf et al. 2008; Zhou et al. 2011), and at least one group has published on the synthesis of a tetramodality particle that allows for bioluminescence and fluorescence imaging as well as PET and MRI using a cobalt-ferrite nanoparticle (Hwang do et al. 2009).

4.3 Therapeutic radionanoparticles

While much of the early work on therapeutic nanoparticles focused on drug delivery, there is increasing interest in targeting diseases, specifically cancer, via therapeutic doses of internal radiation. Improvements in functionality and streamlining of manufacturing are leading to multidimensional particles capable of delivering multiple therapies via a single carrier.

Internal radiation therapy already has a track record as a useful modality in the clinic. Iodine-131 therapy has long been a standard ablative therapy for both non-malignant thyroid diseases as well as thyroid cancer. In the past 5 years several radioimmunotherapies (Davies et al.) using monoclonal antibodies have been approved for clinical use. Radiolabeled anti-CD-20-antibody therapies Y-90-ibritumomab tiuxetan (Zevalin, Spectrum Pharmaceuticals) and I-131-tositumomab (Bexxar, GlaxoSmithKline) have shown promising results in clinical trials for the treatment of relapsed/refractory non-Hodgkin's lymphoma (Davies et al. 2004; Vose et al. 2005; Vose et al. 2000) and are now being tested as part of initial treatment in clinical trials (Fisher 2005; NA 2006; Press et al. 2006). A limitation of RIT is that only one radioisotope can be attached per antibody (Jestin et al. 2007). Yttrium-90-microspheres, an agent in clinical use for ablation of hepatic malignancies, overcome this first problem but due to their size must be injected directly into the hepatic arterial supply to result in meaningful tumor ablation without substantial toxicity (Cosimelli et al. 2010; Hendlisz et al. 2010; Kennedy & Salem 2010). Other transarterial embolic agents tagged with Lu-177, Rh-188 and Ho-166 have been developed (Chakraborty et al. 2008; Hamoudeh et al. 2007; Kim et al. 2006), but all range in size from 300 nm up to 60 μm. Nanoparticles may be able to overcome both the size and payload problems and provide a systemic route to selectively deliver a larger radioisotope and/or drug payload.

Radioisotopes with various decay properties are being utilized for therapeutic nanoparticles. With penetration ranges of 1-10 mm and energies 0.1-2.2 MeV, beta-emitting isotopes such as those in clinical use for RIT may be ideal candidates for nanoparticle radiation. The range of beta particles allows the so-called "cross-fire" effect as it is greater than several cell diameters. However, the decay property should match the goal of therapy (Sofou 2008). For instance, particles that decay through alpha emission may be attractive alternatives for targeting intravascular micrometastases due to high linear energy transfer. These particles are limited by a penetration range of 50-100 μm making them less well suited for bulky tumor deposits. Auger electrons, with a path length of less than 10 nm such as those produced by Indium-111, have been suggested as potential therapeutic agents for small volume disease if carriers can be internalized in high enough numbers to result in meaning radiation dose deposition to the tumor cell DNA. One theoretical appeal of auger-mediated therapy is that the short path length also suggests a low potential for toxicity (Tavares & Tavares 2010).

Many of the studies involving therapeutic radionanoparticles have been theoretical and qualitative in nature. Such studies have looked at modeling of radio-loaded nanoparticles with various decay properties with mixed theoretical results. Bouchat et al. modeled Y-90 labeled antibody biodistribution by Monte Carlo code and showed that RIT would not result in tumor doses sufficient to treat solid tumors such as lung cancer (biological effective dose < 60 Gy). Using a similar modeling system, they showed that radioactive nanoparticles with ~100 Y-90 atoms per nanoparticle can substantially increase the biologic effective dose deposited to a solid tumor (Bouchat et al. 2010). Additionally, radioisotopes that are usually insoluble in body fluids can be encapsulated within nanoparticles, allowing them to be sequestered until delivery at the target (Khan et al. 2008). Hrycushko et al. have modeled liposomes tagged with both beta emitters Re-186 and Re-188 for the post-surgical treatment of breast cancer, showing this to be a potential method of delivering focal radiation to the locations at highest risk for harboring residual cancer cells while sparing normal tissue (Hrycushko et al. 2011a; Hrycushko et al. 2011b).

Preclinical development of therapeutic radionanoparticles has been primarily directed at the treatment of cancer often involving concurrent beta decay (for therapy) and gamma decay (for imaging). Rapid advances in the ability to functionalize nanoparticle platforms for diagnostic and therapeutic purposes have led investigators to develop dual purpose particles, a field now known as theranostics. Many of the therapeutic nanoparticles fall into this category owing to multiple decay pathways of the incorporated radioisotopes, most often involving concurrent both beta decay for therapy and gamma decay for imaging. While some authors consider dual imaging nanoparticles to be theranostic agents (Ma et al. 2006), this section will deal only with those agents that have been considered for therapeutic use. To date, the study of therapeutic radionanoparticles has involved a wide range of nanoparticle platforms combined with an even vaster array of radioisotopes. The section is organized by type of radiation decay (beta, alpha, and auger) as each has its own unique applications and shortcomings when used for internal radiation therapy (Sofou 2008).

Beta decay radionanoparticles

Radioisotopes decaying through beta emission are perhaps the best studied agents for nanoparticle radiotherapy. With a path length of 1-10 mm, beta particles traverse multiple cells while interacting, leading to a "cross fire" effect. Therefore, tagged-NPs do not require cellular internalization to provide effective cell kill. The particle range may also allow for potential cell kill in areas of tumor lacking good vascular access.

Significant research has gone into gold nanoparticles for imaging and enhancement of external beam irradiation, but until recently, little successful work had been published using beta-emitting gold nanoparticles for internal radiation therapy. Though first described for this purpose several decades ago, nanoparticle technology was not as robust at that time, and efforts to deliver internal radiation with gold were discontinued (Hainfeld et al. 1990). Recently, Khan et al. demonstrated the antitumor effect of dendrimer nanoparticles carrying therapeutic loads of Au-198. Researchers generated the radioactive composite nanodevice (CND) by simultaneous neutron activation and radiation polymerization of dendrimers tagged with Au-197. Neutron capture formed Au-198 which could then decay by beta emission. In a melanoma model, tumors of mice injected with the Au-198 CND decreased in size by 45% compared to the untagged CND with no observed toxicity (Khan et al. 2008). Chanda et al. described a similar Au-198 NP with good therapeutic efficacy and low toxicity in a murine prostate cancer model (Chanda et al. 2010).

Several other platforms have been developed for delivery of internal beta irradiation. Schultz et al. described a fullerene cage nanoparticle tagged with Lu-177 for targeting Il-13, an interleukin characteristically overexpressed on glioblastoma cells (Shultz et al. 2010). Diener et al. devised a similar fullerene system incorporating Pb-212, a beta emitter with an alpha-emitting daughter radionuclide, Bi-212 (Diener et al. 2007). Reactive ionic radioisotopes such as Lu and Pb must be shielded from the microenvironment to prevent interaction. Fullerene cages are attractive nanoparticles for this purpose because the ions are completely enclosed within the platform. Ho-166 is another appealing candidate radioisotope due to its concurrent gamma and beta decays as well as its high attenuation coefficient and paramagnetic properties. This combination suggests it to have wide-ranging potential as a true theranostic agent. With this in mind, Bult et al. successfully developed an acetylacetone nanoparticle labeled with Ho-166 (Bult et al. 2010); though, the particle has yet to be tested as a therapeutic agent in an animal model.

Alpha decay radionanoparticles

While beta-emitters have received the bulk of the attention for radionanoparticles, alpha emitters possess unique qualities that make them viable therapy candidates. One such radioisotope is Ac-225, which has received attention as an *in vivo* radiation generator. Ac-225 decays through two chains of short-lived radioisotopes, yielding 4 alpha particles and ending in stable Bi-209. A challenge of these generators is to prevent leakage of the daughter radionuclides during circulation, and nanoparticles are being studied as a way to retain decay chain products for delivery at the target site. Two groups have developed nanoparticles capable of daughter particle retention to a limited extent. Sofou et al. designed a multivesicular liposomal carrier of Ac-225 to enhance retention of daughter radionuclides. Though retention was not as high as hoped, when the particles were targeted to an ovarian cell line *in vitro* using immunolabeling with anti-HER2/neu, they were internalized at an impressive rate (Sofou et al. 2007). More recently, Woodward et al.developed a lanthum phosphate nanoparticle (LaPO$_4$ NP) for Ac-225 encapsulation. The particles were conjugated with antibodies to lung endothelium. This system showed better retention of the daughter nuclides in the LaPO$_4$ NP and excellent dose deposition in mouse lung tissue (Woodward et al. 2011). Effective targeting and daughter retention suggest this model may have therapeutic potential in humans.

Auger decay radionanoparticles

Due to their short range, auger electrons rely on high selective cellular uptake and incorporation in cellular DNA to result in meaning tumoricidal effect. To date, In-111, I-123 and I-125 have been studied as potential agents for this purpose (Chow et al. 2008; Liu et al. 2010; Liu et al. 2009; Mamede et al. 2003; Reske et al. 2007).

Combined modality particles for cancer treatment

Radiation is a major treatment modality for cancer along with surgery and chemotherapy. Over the past 20 years, a large body of evidence has shown that chemotherapy can synergistically enhance the tumoricidal effects of radiation. Many cancers are now treated with concomitant chemotherapy and radiation with absolute benefits in overall survival on the order of 5-10%, but at the cost of increased toxicity of treatment (Auperin et al. 2010; Green et al. 2001; Pignon et al. 2009). Advancements in disease targeting and release properties make nanoparticles an intriguing choice for combining these two therapies *in vivo* with the hope of reducing toxicity and enhancing efficacy, similar to what is seen with nanoparticle chemotherapy agents already on the market such as Doxil and Abraxane.

Wang et al. described such a combined modality nanoparticle, ChemoRad NP (Wang et al. 2010). Using a lipid-polymer hybrid nanoparticle platform, researchers synthesized a particle containing docetaxel chemotherapy within the core and used metal chelation with DMPE-DPTA to attach Yttrium-90, a beta-emitter that has been effectively used in the clinic when conjugated to anti-CD-20 monoclonal antibodies for treatment of Hodgkin's lymphoma. Both agents were attached in therapeutic quantities with 5% docetaxel by weight in the NP and 100 µCi Y-90/mg NP. This activity of Yttrium is enough to deliver 6 Gy of radiation to a tumor target based on pharmacokinetic studies of the NP. The surface of the NP was conjugated via DSPE-PEG with the aptamer A10, a ligand that targets prostate specific membrane antigen (PSMA). In a preliminary study of the molecule, *in vitro* profiling showed the NP to have excellent stability and controlled release properties while successfully targeting tumor cells using the aptamer ligand. Comparison of the combined modality NP to single-agent NP and non-targeted NP as controls showed that the ChemoRad NP has higher therapeutic efficacy than the control options. ChemoRad NP is believed to be the first targeted nanoparticle for delivery of chemoradiation therapy.

Chow et al. report on a second combined modality particle combining Indium-111 and vinorelbine in a pegylated liposome carrier (Chow et al. 2008). This nanoparticle platform did not have a targeting ligand, but showed tumor selectively by the EPR effect on mice with colorectal carcinoma xenografts. The agent showed antitumor efficacy both when chemotherapy and radiation were included and when the single agents alone were used. Because Indium-111 produces both a gamma photon and auger electron, this suggests the auger electron may contribute meaningfully to tumor kill. Because of this decay feature, In-111 has potential as a dual diagnostic and therapeutic (theranostic) agent.

5. Summary

Advances in nanotechnology have inspired the development of nanoparticles for medical applications. Radioisotopes have grown to be an integral part of this field both for purposes of drug development and now diagnostic and therapeutic applications. The flexibility of NPs enables the utilization of radioisotopes in novel applications. The wide range of ongoing nanoparticle research and the explosion of interest in the field are both testaments to the versatility and potential of this expanding branch of science in which radioisotopes promise to continue playing an important role.

6. References

Aillon, K. L., Y. Xie, et al. (2009). "Effects of nanomaterial physicochemical properties on in vivo toxicity." Advanced drug delivery reviews 61(6): 457-466.

Auperin, A., C. Le Pechoux, et al. (2010). "Meta-analysis of concomitant versus sequential radiochemotherapy in locally advanced non-small-cell lung cancer." Journal of clinical oncology : official journal of the American Society of Clinical Oncology 28(13): 2181-2190.

Bangham, A. D. (1993). "Liposomes: the Babraham connection." Chem Phys Lipids 64(1-3): 275-285.

Bao, A., B. Goins, et al. (2003). "186Re-liposome labeling using 186Re-SNS/S complexes: in vitro stability, imaging, and biodistribution in rats." Journal of nuclear medicine : official publication, Society of Nuclear Medicine 44(12): 1992-1999.

Bazile, D. V., C. Ropert, et al. (1992). "Body distribution of fully biodegradable [14C]-poly(lactic acid) nanoparticles coated with albumin after parenteral administration to rats." Biomaterials 13(15): 1093-1102.

Bouchat, V., V. E. Nuttens, et al. (2010). "Radioimmunotherapy with radioactive nanoparticles: biological doses and treatment efficiency for vascularized tumors with or without a central hypoxic area." Med Phys 37(4): 1826-1839.

Bult, W., R. Varkevisser, et al. (2010). "Holmium nanoparticles: preparation and in vitro characterization of a new device for radioablation of solid malignancies." Pharm Res 27(10): 2205-2212.

Cai, W., K. Chen, et al. (2007). "Dual-function probe for PET and near-infrared fluorescence imaging of tumor vasculature." J Nucl Med 48(11): 1862-1870.

Catana, C., Y. Wu, et al. (2006). "Simultaneous acquisition of multislice PET and MR images: initial results with a MR-compatible PET scanner." Journal of nuclear medicine : official publication, Society of Nuclear Medicine 47(12): 1968-1976.

Chakrabarti, A., K. Zhang, et al. (2007). "Radiohybridization PET imaging of KRAS G12D mRNA expression in human pancreas cancer xenografts with [(64)Cu]DO3A-peptide nucleic acid-peptide nanoparticles." Cancer Biol Ther 6(6): 948-956.

Chakraborty, S., T. Das, et al. (2008). "Preparation and preliminary studies on 177Lu-labeled hydroxyapatite particles for possible use in the therapy of liver cancer " Nucl Med Biol 35(5): 589-597.

Chanda, N., P. Kan, et al. (2010). "Radioactive gold nanoparticles in cancer therapy: therapeutic efficacy studies of GA-198AuNP nanoconstruct in prostate tumor-bearing mice." Nanomedicine 6(2): 201-209.

Chen, K. & P. S. Conti (2010). "Target-specific delivery of peptide-based probes for PET imaging." Advanced drug delivery reviews 62(11): 1005-1022.

Cheng, D., Y. Wang, et al. (2010). "Comparison of 18F PET and 99mTc SPECT imaging in phantoms and in tumored mice." Bioconjugate chemistry 21(8): 1565-1570.

Cheon, J. & J. H. Lee (2008). "Synergistically integrated nanoparticles as multimodal probes for nanobiotechnology." Accounts of chemical research 41(12): 1630-1640.

Chow, T. H., Y. Y. Lin, et al. (2008). "Diagnostic and therapeutic evaluation of 111In-vinorelbine-liposomes in a human colorectal carcinoma HT-29/luc-bearing animal model." Nucl Med Biol 35(5): 623-634.

Chrastina, A. & J. E. Schnitzer (2010). "Iodine-125 radiolabeling of silver nanoparticles for in vivo SPECT imaging." Int J Nanomedicine 5: 653-659.

Cosimelli, M., R. Golfieri, et al. (2010). "Multi-centre phase II clinical trial of yttrium-90 resin microspheres alone in unresectable, chemotherapy refractory colorectal liver metastases." British journal of cancer 103(3): 324-331.

Davies, A. J., A. Z. Rohatiner, et al. (2004). "Tositumomab and iodine I 131 tositumomab for recurrent indolent and transformed B-cell non-Hodgkin's lymphoma." Journal of clinical oncology : official journal of the American Society of Clinical Oncology 22(8): 1469-1479.

Delbeke, D. & W. H. Martin (2001). "Positron emission tomography imaging in oncology." Radiologic clinics of North America 39(5): 883-917.

Delbeke, D. & W. H. Martin (2004). "Metabolic imaging with FDG: a primer." Cancer journal 10(4): 201-213.

DeNardo, S. J., G. L. DeNardo, et al. (2007). "Thermal dosimetry predictive of efficacy of 111In-ChL6 nanoparticle AMF--induced thermoablative therapy for human breast cancer in mice." J Nucl Med 48(3): 437-444.

Devaraj, N. K., E. J. Keliher, et al. (2009). "18F labeled nanoparticles for in vivo PET-CT imaging." Bioconjug Chem 20(2): 397-401.

Diener, M. D., J. M. Alford, et al. (2007). "(212)Pb@C(60) and its water-soluble derivatives: synthesis, stability, and suitability for radioimmunotherapy." Journal of the American Chemical Society 129(16): 5131-5138.

Duconge, F., T. Pons, et al. (2008). "Fluorine-18-labeled phospholipid quantum dot micelles for in vivo multimodal imaging from whole body to cellular scales." Bioconjug Chem 19(9): 1921-1926.

Elbayoumi, T. A., S. Pabba, et al. (2007). "Antinucleosome antibody-modified liposomes and lipid-core micelles for tumor-targeted delivery of therapeutic and diagnostic agents." J Liposome Res 17(1): 1-14.

Farokhzad, O. C., S. Jon, et al. (2004). "Nanoparticle-aptamer bioconjugates: a new approach for targeting prostate cancer cells." Cancer Res 64(21): 7668-7672.

Fisher, R. I. (2005). "Overview of Southwest Oncology Group Clinical Trials in non-Hodgkin Lymphoma. S0016. A phase III trial of CHOP vs CHOP + rituximab vs CHOP + iodine131-labeled monoclonal anti-B1 antibody (tositumomab) for treatment of newly diagnosed follicular NHL." Clinical advances in hematology & oncology : H&O 3(7): 544-546.

French, J. T., B. Goins, et al. (2010). "Interventional therapy of head and neck cancer with lipid nanoparticle-carried rhenium 186 radionuclide." J Vasc Interv Radiol 21(8): 1271-1279.

Fukukawa, K., R. Rossin, et al. (2008). "Synthesis and characterization of core-shell star copolymers for in vivo PET imaging applications." Biomacromolecules 9(4): 1329-1339.

Gatenby, R. A. & R. J. Gillies (2004). "Why do cancers have high aerobic glycolysis?" Nature reviews. Cancer 4(11): 891-899.

Geze, A., L. T. Chau, et al. (2007). "Biodistribution of intravenously administered amphiphilic beta-cyclodextrin nanospheres." Int J Pharm 344(1-2): 135-142.

Glaus, C., R. Rossin, et al. (2010). "In vivo evaluation of (64)Cu-labeled magnetic nanoparticles as a dual-modality PET/MR imaging agent." Bioconjug Chem 21(4): 715-722.

Gratton, S. E., P. D. Pohlhaus, et al. (2007). "Nanofabricated particles for engineered drug therapies: a preliminary biodistribution study of PRINT nanoparticles." J Control Release 121(1-2): 10-18.

Green, J. A., J. M. Kirwan, et al. (2001). "Survival and recurrence after concomitant chemotherapy and radiotherapy for cancer of the uterine cervix: a systematic review and meta-analysis." Lancet 358(9284): 781-786.

Gu, F., L. Zhang, et al. (2008). "Precise engineering of targeted nanoparticles by using self-assembled biointegrated block copolymers." Proc Natl Acad Sci U S A 105(7): 2586-2591.

Hahn, M. A., A. K. Singh, et al. (2011). "Nanoparticles as contrast agents for in-vivo bioimaging: current status and future perspectives." Analytical and bioanalytical chemistry 399(1): 3-27.

Hainfeld, J. F., C. J. Foley, et al. (1990). "Radioactive gold cluster immunoconjugates: potential agents for cancer therapy." Int J Rad Appl Instrum B 17(3): 287-294.

Hamoudeh, M., H. Salim, et al. (2007). "Preparation and characterization of radioactive dirhenium decacarbonyl-loaded PLLA nanoparticles for radionuclide intra-tumoral therapy." Eur J Pharm Biopharm 67(3): 597-611.

Harrington, K. J., S. Mohammadtaghi, et al. (2001). "Effective targeting of solid tumors in patients with locally advanced cancers by radiolabeled pegylated liposomes." Clinical cancer research : an official journal of the American Association for Cancer Research 7(2): 243-254.

Helbok, A., C. Decristoforo, et al. (2010). "Radiolabeling of lipid-based nanoparticles for diagnostics and therapeutic applications: a comparison using different radiometals." J Liposome Res 20(3): 219-227.

Hendlisz, A., M. Van den Eynde, et al. (2010). "Phase III trial comparing protracted intravenous fluorouracil infusion alone or with yttrium-90 resin microspheres radioembolization for liver-limited metastatic colorectal cancer refractory to standard chemotherapy." Journal of clinical oncology : official journal of the American Society of Clinical Oncology 28(23): 3687-3694.

Hikage, M., K. Gonda, et al. (2010). "Nano-imaging of the lymph network structure with quantum dots." Nanotechnology 21(18): 185103.

Hnatowich, D. J., B. Friedman, et al. (1981). "Labeling of preformed liposomes with Ga-67 and Tc-99m by chelation." J Nucl Med 22(9): 810-814.

Hong, S. Y., G. Tobias, et al. (2010). "Filled and glycosylated carbon nanotubes for in vivo radioemitter localization and imaging." Nat Mater 9(6): 485-490.

Hrycushko, B. A., A. N. Gutierrez, et al. (2011a). "Radiobiological characterization of post-lumpectomy focal brachytherapy with lipid nanoparticle-carried radionuclides." Phys Med Biol 56(3): 703-719.

Hrycushko, B. A., S. Li, et al. (2011b). "Postlumpectomy focal brachytherapy for simultaneous treatment of surgical cavity and draining lymph nodes." Int J Radiat Oncol Biol Phys 79(3): 948-955.

Hu, G., M. Lijowski, et al. (2007). "Imaging of Vx-2 rabbit tumors with alpha(nu)beta3-integrin-targeted 111In nanoparticles." Int J Cancer 120(9): 1951-1957.

Hwang do, W., H. Y. Ko, et al. (2009). "Development of a quadruple imaging modality by using nanoparticles." Chemistry 15(37): 9387-9393.

IAEA (2011). "Radiopharmaceuticals: Production and Availability."

Jarrett, B. R., B. Gustafsson, et al. (2008). "Synthesis of 64Cu-labeled magnetic nanoparticles for multimodal imaging." Bioconjug Chem 19(7): 1496-1504.

Jestin, E., M. Mougin-Degraef, et al. (2007). "Radiolabeling and targeting of lipidic nanocapsules for applications in radioimmunotherapy." Q J Nucl Med Mol Imaging 51(1): 51-60.

Jiang, Z., R. Li, et al. (2007). "Detecting genomic aberrations by fluorescence in situ hybridization with quantum dots-labeled probes." Journal of nanoscience and nanotechnology 7(12): 4254-4259.

Judenhofer, M. S., C. Catana, et al. (2007). "PET/MR images acquired with a compact MR-compatible PET detector in a 7-T magnet." Radiology 244(3): 807-814.

Kennedy, A. S. & R. Salem (2010). "Radioembolization (yttrium-90 microspheres) for primary and metastatic hepatic malignancies." Cancer journal 16(2): 163-175.

Kennel, S. J., J. D. Woodward, et al. (2008). "The fate of MAb-targeted Cd(125m)Te/ZnS nanoparticles in vivo." Nucl Med Biol 35(4): 501-514.

Khan, M. K., L. D. Minc, et al. (2008). "Fabrication of {198Au0} radioactive composite nanodevices and their use for nanobrachytherapy." Nanomedicine 4(1): 57-69.

Kim, J. K., K. H. Han, et al. (2006). "Long-term clinical outcome of phase IIb clinical trial of percutaneous injection with holmium-166/chitosan complex (Milican) for the treatment of small hepatocellular carcinoma." Clinical cancer research : an official journal of the American Association for Cancer Research 12(2): 543-548.

Kim, S. H., J. H. Jeong, et al. (2005). "Target-specific cellular uptake of PLGA nanoparticles coated with poly(L-lysine)-poly(ethylene glycol)-folate conjugate." Langmuir 21(19): 8852-8857.

Klutz, K., V. Russ, et al. (2009). "Targeted radioiodine therapy of neuroblastoma tumors following systemic nonviral delivery of the sodium iodide symporter gene." Clin Cancer Res 15(19): 6079-6086.

Ko, H. Y., K. J. Choi, et al. (2011). "A multimodal nanoparticle-based cancer imaging probe simultaneously targeting nucleolin, integrin alphavbeta3 and tenascin-C proteins." Biomaterials 32(4): 1130-1138.

Kobayashi, H., C. Wu, et al. (1999). "Evaluation of the in vivo biodistribution of indium-111 and yttrium-88 labeled dendrimer-1B4M-DTPA and its conjugation with anti-Tac monoclonal antibody." Bioconjug Chem 10(1): 103-111.

Kukowska-Latallo, J. F., K. A. Candido, et al. (2005). "Nanoparticle targeting of anticancer drug improves therapeutic response in animal model of human epithelial cancer." Cancer Res 65(12): 5317-5324.

Kumar, R., I. Roy, et al. (2010). "In vivo biodistribution and clearance studies using multimodal organically modified silica nanoparticles." ACS Nano 4(2): 699-708.

Laverman, P., E. T. Dams, et al. (1999). "A novel method to label liposomes with 99mTc by the hydrazino nicotinyl derivative." J Nucl Med 40(1): 192-197.

Lee, H. Y., Z. Li, et al. (2008). "PET/MRI dual-modality tumor imaging using arginine-glycine-aspartic (RGD)-conjugated radiolabeled iron oxide nanoparticles." J Nucl Med 49(8): 1371-1379.

Liang, M., X. Liu, et al. (2010). "Multimodality nuclear and fluorescence tumor imaging in mice using a streptavidin nanoparticle." Bioconjugate chemistry 21(7): 1385-1388.

Liang, S., Y. Wang, et al. (2007). "Surface modified superparamagnetic iron oxide nanoparticles: as a new carrier for bio-magnetically targeted therapy." J Mater Sci Mater Med 18(12): 2297-2302.

Lijowski, M., S. Caruthers, et al. (2009). "High sensitivity: high-resolution SPECT-CT/MR molecular imaging of angiogenesis in the Vx2 model." Investigative radiology 44(1): 15-22.

Lin, Y. Y., J. J. Li, et al. (2009). "Evaluation of pharmacokinetics of 111In-labeled VNB-PEGylated liposomes after intraperitoneal and intravenous administration in a tumor/ascites mouse model." Cancer Biother Radiopharm 24(4): 453-460.

Liu, X., K. Nakamura, et al. (2010). "Auger-mediated cytotoxicity of cancer cells in culture by an 125I-antisense oligomer delivered as a three-component streptavidin nanoparticle." J Biomed Nanotechnol 6(2): 153-157.

Liu, X., Y. Wang, et al. (2009). "Auger radiation-induced, antisense-mediated cytotoxicity of tumor cells using a 3-component streptavidin-delivery nanoparticle with 111In." J Nucl Med 50(4): 582-590.

Ma, D., Z. J. Jakubek, et al. (2006). "A new approach towards controlled synthesis of multifunctional core-shell nano-architectures: luminescent and superparamagnetic." Journal of nanoscience and nanotechnology 6(12): 3677-3684.

Mackeyev, Y. A., J. W. Marks, et al. (2005). "Stable containment of radionuclides on the nanoscale by cut single-wall carbon nanotubes." The journal of physical chemistry. B 109(12): 5482-5484.

Malik, N., R. Wiwattanapatapee, et al. (2000). "Dendrimers: relationship between structure and biocompatibility in vitro, and preliminary studies on the biodistribution of 125I-labelled polyamidoamine dendrimers in vivo." J Control Release 65(1-2): 133-148.

Mamede, M., T. Saga, et al. (2003). "Radiolabeling of avidin with very high specific activity for internal radiation therapy of intraperitoneally disseminated tumors." Clin Cancer Res 9(10 Pt 1): 3756-3762.

Matson, J. B. & R. H. Grubbs (2008). "Synthesis of fluorine-18 functionalized nanoparticles for use as in vivo molecular imaging agents." J Am Chem Soc 130(21): 6731-6733.

Medina, O. P., N. Pillarsetty, et al. (2011). "Optimizing tumor targeting of the lipophilic EGFR-binding radiotracer SKI 243 using a liposomal nanoparticle delivery system." Journal of controlled release : official journal of the Controlled Release Society 149(3): 292-298.

Mittra, E., M. Goris, et al. (2010). "First in man studies of [18F]FPPRGD2: A novel PET radiopharmaceutical for imaging {alpha}v{beta}3 integrin levels." J NUCL MED MEETING ABSTRACTS 51(2_MeetingAbstracts): 1433-.

Mougin-Degraef, M., E. Jestin, et al. (2006). "High-activity radio-iodine labeling of conventional and stealth liposomes." Journal of liposome research 16(1): 91-102.

NA (2006). "A phase III trial of CHOP vs CHOP + rituximab vs CHOP + 1-131 tositumomab for newly diagnosed follicular non-Hodgkin lymphoma." Clinical advances in hematology & oncology : H&O 4(7): 536-538.

Nahrendorf, M., H. Zhang, et al. (2008). "Nanoparticle PET-CT imaging of macrophages in inflammatory atherosclerosis." Circulation 117(3): 379-387.

Natarajan, A., C. Gruettner, et al. (2008a). "NanoFerrite particle based radioimmunonanoparticles: binding affinity and in vivo pharmacokinetics." Bioconjug Chem 19(6): 1211-1218.

Natarajan, A., C. Y. Xiong, et al. (2008b). "Development of multivalent radioimmunonanoparticles for cancer imaging and therapy." Cancer Biother Radiopharm 23(1): 82-91.

Noguchi, Y., J. Wu, et al. (1998). "Early phase tumor accumulation of macromolecules: a great difference in clearance rate between tumor and normal tissues." Japanese journal of cancer research : Gann 89(3): 307-314.

Park, S. I., B. J. Kwon, et al. (2011). "Synthesis and characterization of 3-[131I]iodo-L-tyrosine grafted Fe3O4@SiO2 nanocomposite for single photon emission computed tomography (SPECT) and magnetic resonance imaging (MRI)." Journal of nanoscience and nanotechnology 11(2): 1818-1821.

Patel, D., A. Kell, et al. (2011). "The cell labeling efficacy, cytotoxicity and relaxivity of copper-activated MRI/PET imaging contrast agents." Biomaterials 32(4): 1167-1176.

Petersen, A. L., T. Binderup, et al. (2011). "64Cu loaded liposomes as positron emission tomography imaging agents." Biomaterials 32(9): 2334-2341.

Phillips, W. T., B. A. Goins, et al. (2009). "Radioactive liposomes." Wiley Interdiscip Rev Nanomed Nanobiotechnol 1(1): 69-83.

Phillips, W. T., A. S. Rudolph, et al. (1992). "A simple method for producing a technetium-99m-labeled liposome which is stable in vivo." International journal of radiation applications and instrumentation. Part B, Nuclear medicine and biology 19(5): 539-547.

Pignon, J. P., A. le Maitre, et al. (2009). "Meta-analysis of chemotherapy in head and neck cancer (MACH-NC): an update on 93 randomised trials and 17,346 patients." Radiotherapy and oncology : journal of the European Society for Therapeutic Radiology and Oncology 92(1): 4-14.

Press, O. W., J. M. Unger, et al. (2006). "Phase II trial of CHOP chemotherapy followed by tositumomab/iodine I-131 tositumomab for previously untreated follicular non-Hodgkin's lymphoma: five-year follow-up of Southwest Oncology Group Protocol S9911." Journal of clinical oncology : official journal of the American Society of Clinical Oncology 24(25): 4143-4149.

Pressly, E. D., R. Rossin, et al. (2007). "Structural effects on the biodistribution and positron emission tomography (PET) imaging of well-defined (64)Cu-labeled nanoparticles comprised of amphiphilic block graft copolymers." Biomacromolecules 8(10): 3126-3134.

Racker, E. & M. Spector (1981). "Warburg effect revisited: merger of biochemistry and molecular biology." Science 213(4505): 303-307.

Reske, S. N., S. Deisenhofer, et al. (2007). "123I-ITdU-mediated nanoirradiation of DNA efficiently induces cell kill in HL60 leukemia cells and in doxorubicin-, beta-, or gamma-radiation-resistant cell lines." J Nucl Med 48(6): 1000-1007.

Rossin, R., D. Pan, et al. (2005). "64Cu-labeled folate-conjugated shell cross-linked nanoparticles for tumor imaging and radiotherapy: synthesis, radiolabeling, and biologic evaluation." J Nucl Med 46(7): 1210-1218.

Ruggiero, A., C. H. Villa, et al. (2010). "Imaging and treating tumor vasculature with targeted radiolabeled carbon nanotubes." International journal of nanomedicine 5: 783-802.

Schipper, M. L., Z. Cheng, et al. (2007). "microPET-based biodistribution of quantum dots in living mice." J Nucl Med 48(9): 1511-1518.

Schmieder, A. H., P. M. Winter, et al. (2005). "Molecular MR imaging of melanoma angiogenesis with alphanubeta3-targeted paramagnetic nanoparticles." Magnetic resonance in medicine : official journal of the Society of Magnetic Resonance in Medicine / Society of Magnetic Resonance in Medicine 53(3): 621-627.

Shokeen, M., N. M. Fettig, et al. (2008). "Synthesis, in vitro and in vivo evaluation of radiolabeled nanoparticles." The quarterly journal of nuclear medicine and molecular imaging : official publication of the Italian Association of Nuclear Medicine 52(3): 267-277.

Shultz, M. D., J. C. Duchamp, et al. (2010). "Encapsulation of a radiolabeled cluster inside a fullerene cage, (177)Lu(x)Lu((3-x))N@C(80): an interleukin-13-conjugated radiolabeled metallofullerene platform." J Am Chem Soc 132(14): 4980-4981.

Sofou, S. (2008). "Radionuclide carriers for targeting of cancer." Int J Nanomedicine 3(2): 181-199.

Sofou, S., B. J. Kappel, et al. (2007). "Enhanced retention of the alpha-particle-emitting daughters of Actinium-225 by liposome carriers." Bioconjugate chemistry 18(6): 2061-2067.

Strano, M. S. (2010). "Biomedical materials: Nanoscale radiosurgery." Nat Mater 9(6): 467-468.

Sun, X., Y. Yan, et al. (2011). "18F-FPPRGD2 and 18F-FDG PET of response to Abraxane therapy." Journal of nuclear medicine : official publication, Society of Nuclear Medicine 52(1): 140-146.

Takahama, H., T. Minamino, et al. (2009). "Prolonged targeting of ischemic/reperfused myocardium by liposomal adenosine augments cardioprotection in rats." J Am Coll Cardiol 53(8): 709-717.

Tavares, A. A. & J. M. Tavares (2010). "Evaluating 99mTc Auger electrons for targeted tumor radiotherapy by computational methods." Med Phys 37(7): 3551-3559.

Ting, G., C. H. Chang, et al. (2010). "Nanotargeted radionuclides for cancer nuclear imaging and internal radiotherapy." J Biomed Biotechnol 2010.

Torchilin, V. P. (2005). "Recent advances with liposomes as pharmaceutical carriers." Nature reviews. Drug discovery 4(2): 145-160.

Torres Martin de Rosales, R., R. Tavare, et al. (2011). "(99m)Tc-Bisphosphonate-Iron Oxide Nanoparticle Conjugates for Dual-Modality Biomedical Imaging." Bioconjugate chemistry 22(3): 455-465.

Vose, J. M., P. J. Bierman, et al. (2005). "Phase I trial of iodine-131 tositumomab with high-dose chemotherapy and autologous stem-cell transplantation for relapsed non-Hodgkin's lymphoma." Journal of clinical oncology : official journal of the American Society of Clinical Oncology 23(3): 461-467.

Vose, J. M., R. L. Wahl, et al. (2000). "Multicenter phase II study of iodine-131 tositumomab for chemotherapy-relapsed/refractory low-grade and transformed low-grade B-cell non-Hodgkin's lymphomas." Journal of clinical oncology : official journal of the American Society of Clinical Oncology 18(6): 1316-1323.

Wagner, V., A. Dullaart, et al. (2006). "The emerging nanomedicine landscape." Nature biotechnology 24(10): 1211-1217.

Wang, A. Z., F. Gu, et al. (2008). "Biofunctionalized targeted nanoparticles for therapeutic applications." Expert Opin Biol Ther 8(8): 1063-1070.

Wang, A. Z., K. Yuet, et al. (2010). "ChemoRad nanoparticles: a novel multifunctional nanoparticle platform for targeted delivery of concurrent chemoradiation." Nanomedicine (Lond) 5(3): 361-368.

Woodward, J., S. J. Kennel, et al. (2011). "LaPO(4) Nanoparticles Doped with Actinium-225 that Partially Sequester Daughter Radionuclides." Bioconjug Chem.

Wu, X., H. Liu, et al. (2003). "Immunofluorescent labeling of cancer marker Her2 and other cellular targets with semiconductor quantum dots." Nature biotechnology 21(1): 41-46.

Wu, Y., C. Catana, et al. (2009). "PET Performance Evaluation of an MR-Compatible PET Insert." IEEE transactions on nuclear science 56(3): 574-580.

Xie, H., Z. J. Wang, et al. (2010a). "In vivo PET imaging and biodistribution of radiolabeled gold nanoshells in rats with tumor xenografts." Int J Pharm 395(1-2): 324-330.

Xie, J., K. Chen, et al. (2010b). "PET/NIRF/MRI triple functional iron oxide nanoparticles." Biomaterials 31(11): 3016-3022.

Yang, X., H. Hong, et al. (2011). "cRGD-functionalized, DOX-conjugated, and (64)Cu-labeled superparamagnetic iron oxide nanoparticles for targeted anticancer drug delivery and PET/MR imaging." Biomaterials 32(17): 4151-4160.

Zhang, X. & C. T. Badea (2011). "Highly efficient detection in fluorescence tomography of quantum dots using time-gated acquisition and ultrafast pulsed laser." Proceedings - Society of Photo-Optical Instrumentation Engineers 7896.

Zhou, J., M. Yu, et al. (2011). "Fluorine-18-labeled Gd3+/Yb3+/Er3+ co-doped NaYF4 nanophosphors for multimodality PET/MR/UCL imaging." Biomaterials 32(4): 1148-1156.

Zielhuis, S. W., J. H. Seppenwoolde, et al. (2006). "Lanthanide-loaded liposomes for multimodality imaging and therapy." Cancer biotherapy & radiopharmaceuticals 21(5): 520-527.

2

Use of Radiation and Isotopes in Insects

Thiago Mastrangelo and Julio Walder
Centre for Nuclear Energy in Agriculture (CENA/USP)
Brazil

1. Introduction

One of the first successful experiments to verify the effects of an ionizing radiation over an insect (*Lasioderma serricorne* F., the cigarette beetle) was performed by Runner in 1916. Soon afterwards, Muller demonstrated in 1927 that rays from radioactive substances could induce genetic damage and a larger number of dominant lethal mutations in *Drosophila melanogaster* Meigen, which were expressed through a reduction in the hatchability of the eggs laid by treated females or fathered by treated males. However, the economic entomologists became really aware that sterility in insects could be quite easily achieved through ionizing radiations only after 1950, when Muller made a great effort to publicize the biological effects of radiation. This was the beginning of a new branch of Science, the Radioentomology.

More than 2800 references in literature were published along the past six decades. Over 300 species of arthropods, mostly of economic importance, have already been subjected to irradiation studies for basic research, pest control applications (*e.g.*, the autocidal control known as the Sterile Insect Technique-SIT, and in support of biological control programs), and for disinfestation of commodities (for quarantine and phytosanitary purposes) (Bakri et al., 2005). Besides that, insects may be labelled with stable or radioactive isotopes for ecology or nutrition studies.

2. Mode of action of radiation in insects

In general, the mode of action of ionizing radiations in living cells consists basically in a chain of oxidative reactions along the radiation path and the formation of free damaging peroxy-radicals, which alter irreversibly the organic molecules. At the cytological level, sterilization is the result of the germ cell chromosome fragmentation (dominant lethal mutations, translocations, and other chromosomal aberrations), leading to the production of imbalanced gametes and, subsequently, inhibition of mitosis and the death of fertilized eggs. Beside the reproductive sterility induced by direct lesion of the genetic material by radiation, LaChance et al. (1967) reported that there are other causes of reproductive sterility that might have a cytological or physiological basis.

According to the law of Bergonie and Tribondeau, cells are most sensitive to radiation when they are dividing. The most radiosensitive cells are, therefore, those with a high mitotic rate, with a long mitotic future and that are of the stem or germ cell type (Casarett, 1968).

Apparently, neither DNA content, chromosome number, nor chromosome arm number can be responsible for the differences in radiosensitivity of cells (Jacquet & Leonard, 1983). Even thou, there is a relationship between the interphase nuclear volume and cell sensitivity to radiation, that is used in vertebrate animals and plants to predict their sensitivity to chronic irradiation, which express that the larger the nuclear volume, the greater the sensitivity (Casarett, 1968; Sparrow et al., 1963).

The radiosensitivity of the mitotically active reproductive cells has different sterilization and killing susceptibility regarding the developmental stage and the division phase. According to Proverbs (1969), spermatocytes and spermatogonia are more radiosensitive than the spermatids and spermatozoa. Dey & Manna (1983) found that the chromosomes of the bug *Physopelta schlanbuschi* were more sensitive to X rays at the spermatogonial metaphase and anaphase I. The sensitivity of the mitotically active reproductive cells in female insects can be increased by the presence of nurse cells, which possesses polytene chromosomes with huge nucleus of unraveled chromatin when undergoing endomitosis (LaChance & Leverich, 1962; LaChance & Bruns, 1963). Thus, female insects are, in general, more radiosensitive than males (Bakri et al., 2005; Hooper, 1989).

Somatic cells are less sensitive to radiation than stem or gonial cells, as they are generally differentiated cells that have lost their ability to divide at the adult stage, what explains why lethal doses are usually higher than the sterilizing doses for insects (Bakri et al., 2005). The effect of radiation on somatic cells can be expressed by the development of abnormalities, reduction of adult lifespan, flight ability, mating propensity, nutrition, and, ultimately, death of the insect (FAO/IAEA/USDA, 2003).

3. Radioresistance in insect orders

Insects are more radioresistant than higher vertebrates, but less resistant than bacteria, protozoa and viruses (Harrison & Anderson, 1996; Rice & Baptist, 1974; Ravera, 1967; Whicker & Schultz, 1982). One of the reasons for this is the Dyar's Rule, *i.e.* arthropods have a discontinuous growth and most cell divisions happen only during the moulting process (Behera et al., 1999). The sensitivity to radiation varies widely among the insect orders, *e.g.* some species from Orthoptera are sterilized at doses below 5 Gy, whilst some Lepidoptera requires more than 300 Gy (IDIDAS, 2010).

According to Bakri et al. (2005), more than 200 species of insects of economic importance from at least eight taxonomic orders have been irradiated for radiobiological studies, integrated pest management (IPM) programs or phytosanitary purposes. Almost 80% of these species belongs to only three orders, *i.e.* Diptera, Lepidoptera and Coleoptera. The radiation dose values for sterilization summarized in the Table 1 were chosen to serve as standard criteria in analysing the differences in radiosensitivity among the insect orders.

ORDER	Sterilization Doses (Gy)
Coleoptera	13 - 200
Dictyoptera	5 – 140
Diptera	10 – 200
Hemiptera	10 - 200
Hymenoptera	80 - 100
Lepidoptera	40 – 400
Orthoptera	4 - 30
Thysanoptera	100 -200

Table 1. Estimated sterilization doses reported for nymphs or pupae of insects from different taxonomic orders (IDIDAS, 2010).

In Diptera, doses ranging from 20 to 160 Gy are usually required for sterilization. Tachinids are the most sensitive, whereas some species from the Drosophilidae family are quite radioresistant. Most fly species are often irradiated at the late pupal stage due to the easiness in handling and even shipping. In the Culicidae family, most of the studies were conducted treating the adults instead of pupae because of their higher resistance in handling and to dehydration. Most of the Area-Wide IPM programs integrating the SIT targeted fruit flies from the Tephritidae family, which are generally sterilized at 90-150 Gy (Bakri et al., 2005; IDIDAS, 2010).

Lepidopterans are relatively more resistant to radiation, with sterilization doses ranging from 40 to 400 Gy. Major differences between lepidopteran and the other insect orders are that their chromosomes usually show diffuse centromere (holokinetic) and also do not present the classical breakage-fusion-bridge cycle which is characteristic of the dominant lethal radioinduced in Diptera, for instance. These two characteristics increase the tolerance of lepidopteran chromosomes to telomere loss without the drastic effects that this has on chromosomes of other orders (Robinson, 2005). As fully sterilizing doses reduce significantly the competitiveness of sterile moths, Proverbs (1962) proposed the use of substerilizing doses which revealed to lead to a F_1 progeny that is more sterile than the parents and whose sex ratio is biased toward males. Such F_1 sterility phenomenon has been exploited in some Area-Wide IPM programs using the SIT against the pink bollworm (*Pectinophora gossypiella* Saunders, pupae sterilized at 100-150 Gy), the codling moth (*Cydia pomonella* L., newly emerged males are partially sterilized at 100-250 Gy), the cactus moth (*Cactoblastis cactorum* Berg, adults partially sterilized at 200 Gy) and the diamondback moth (*Plutella xylostella* L., pupae partially sterilized at 100 Gy) (IDIDAS, 2010).

The sterilization doses for Coleoptera range from 13 to 200 Gy. In Coccinellidae, the same family of the important natural enemy *Cycloneda sanguine* L., a sterilization dose of 35 Gy was reported for pupae of the Mexican bean beetle, *Epilachna varivestis* Mulsant (Heneberry et al., 1964). Since the boll weevil *Anthonomus grandis grandis* Boheman was first recorded in the USA in 1892, it was estimated to have caused billion of dollars in economic damages and it was responsible for almost one-third of the insecticides applied in US agriculture. Therefore, it was the species of this group most studied, aiming the application of the SIT. The males could be sterilized at 80 Gy, but the competitiveness and longevity of the weevils were severely reduced. Fecundity of females was reduced at 50 Gy, but they kept laying fertile eggs up to 200 Gy. Nevertheless, these problems were solved if the insects were sterilized after being treated with the anti-leukemia drug busulfan or under low doses of fractioned radiation, and egg hatch was avoided by using chitin synthesis inhibitor (Haynes & Smith, 1992; Klassen & Earle, 1970; McKibben et al., 2001). Other detrimental impacts of radiation, such as the collapse of the epithelial tissue of the midgut, were also reported for the West Indian sweetpotato weevil, *Euscepes postfasciatus* Fairmare, target of an Area-Wide-IPM program in Japan (Kumano et al., 2010).

The sterilizing dose values reported in Table 1 were difficult to combine due basically to the inexistence of neither standard dosimetry nor even experimental procedures. Furthermore, several factors can also affect insect radiosensitivity, especially the biological factors (developmental stage, age, sex, size, weight, nutritional stage, diapauses and genetic differences) and physical conditions (atmosphere, temperature and irradiation dose rate) (Bakri et al., 2005).

4. Radioisotopes in the sterile insect technique and biological control

4.1 The sterile insect technique

The use of sterile insects (Sterile Insect Technique – SIT) to control or eradicate pest populations was a revolutionary initiative in Entomology conceived at the beginning of the twenty century.

During the 1930's, the New World Screwworm fly (NWS), *Cochliomyia hominivorax* Coquerel, was a serious pest of warm-blooded animals, including humans, in North America. The larval stages of the NWS are obligate parasites, feeding on the living flesh of the host (myiasis), what caused serious livestock production losses. In 1935, an infestation with 230,000 myiasis cases happened in USA (James, 1947) and Parish (1942) reported 6,148 myiasis cases during the screwworm seasons from 1936 to 1940 on Texas. The economic losses at the southeast and southwest regions of USA were estimated in US$ 20 million/year and US$ 50-100 million/year respectively (Baumhover, 1966). In view of this, scientists of the US Department of Agriculture (USDA) started summing efforts to control this plague.

Melvin & Bushland (1936) developed techniques to colonize and an artificial diet to rear NWS, what made available large number of insects for studies. E.F. Knipling then observed that females of NWS usually mated only once and realized that if sexual sterility could be induced in males and if large numbers of them could be continuously released in the field for several successive generations, the wild NWS population could decrease up to the suppression level (Knipling, 1955). Furthermore, if the wild population was isolated, the ratio number of sterile flies: number of fertile flies would become so high that probably not even a single fertile mating would occur and that wild population would finally be eradicated.

In 1927, H.J. Muller demonstrated that ionizing radiation could induce in *Drosophila* dominant lethal mutations that could be verified by the reduction in viability of the eggs laid by irradiated females or normal females mated with irradiated males. But only in 1946, A.W. Lindquist called Knipling's attention to the fact that Muller had reported a means of sterilizing insects. Bushland & Hopkins (1951) then conducted the first NWS irradiations at the X-ray Therapy Section of Brooke Army Hospital and found that, when 6 day old pupae were exposed to 50 Gy, the adult flies that emerged were sterile and could compete almost equally with non-irradiated insects.

The first field evaluation pilot test was performed during 1951-1953 at the Sanibel Island (47 km^2), 4 km from the coast of Florida, using ^{32}P-labelled flies for a release-recapture experiment. The percentage of radioactive egg masses was also assessed. The results corroborated the laboratory studies and just after 8 weeks of releases (ca. 39 sterile males/km^2/week), 100% of the egg masses collected from wounded sentinel goats were sterile. Nevertheless, eradication was not achieved because wild fertile flies continued reinfestating the island from the mainland (Baumhover et al., 1955).

To prove the SIT reliability once for all, an eradication trial was initiated on the Curaçao Island (435 km^2), 6.5 km from Venezuela coast, on 1954. The NWS flies were reared in Florida and irradiated pupae were packaged in paper bags, air shipped to Curaçao and released by air twice per week. The sterile males started being released (ca. 155 sterile males/km^2/week) on August 1954 and eradication had been accomplished within just 14 weeks (Baumhover et al., 1955).

The success of the NWS eradication experiment on the Curaçao Island led to the implementation of eradication programs in the Southeastern (1957-1959) and Southwestern (1962-1966) United States. In 1966, the entire USA was declared free of the NWS, but the country remained vulnerable to the influx of fertile flies from Mexico. So, in 1972, the *Comisión México-Americana para la Erradicación del Gusano Barrenador del Ganado* (COMEXA) was created. In 1977, a mass-rearing facility with capacity to produce 500 million sterile flies/week, built at Tuxtla Gutiérrez, Mexico, reached full production. By 1984, the NWS had been eradicated from USA to the Isthmus of Tehuantepec, Mexico. The eradication operations continued through Central America in the following years with new countries been declared free: Mexico (1991), Belize and Guatemala (1994), El Salvador (1995), Honduras (1996), Nicaragua (1999) and Costa Rica (2000). Panama was declared free from the NWS in 2001 and a buffer zone of 30,000 km^2 was set at the Darien Gap by the release of 40-50 million sterile males (Klassen &Curtis, 2005; Wyss, 2000). The estimated annual producer benefits in the USA, Mexico and Central America were US$ 796 million, US$ 292 million and US$ 77.9 million respectively (Wyss, 2000).

According to the International Plant Protection Convention (IPPC), the SIT can be defined currently as "a method of pest control using area-wide inundative releases of sterile insects to reduce fertility of a field population of the same species" (FAO, 2005).

Sterility in insects can be induced by chemosterilants when added to rearing diets, sprayed over the insect or even in attractant baits in the field. However, most of the chemosterilants are mutagenic or teratogenic, what leads to human health and environmental issues, especially the integrity of ecological food chains (Hayes, 1968), besides the fact that the insects can develop resistance (Fitt, 2008). Thus, for the past five decades, exposure to ionizing radiation has been the main method of inducing sterility in mass-reared insects for Area-Wide IPM programs that integrate the SIT.

Once the absorbed dose is achieved, the irradiation process reaches high accuracy. Other advantages of the irradiation are: insects can be irradiated inside packaging materials, the sterile insects may be released immediately after the irradiation, radiation does not leave residues that could be harmful to humans or the environment, temperature rise in the irradiation process is usually insignificant (Bakri et al., 2005). The induction of radioactivity in irradiated material for SIT operations is avoided by ensuring that the energy applied is bellow 5 million electron volts (MeV) for photons (gamma or X rays) or 10 MeV for electrons (FAO/IAEA/WHO, 1999; IAEA, 2002).

In the 1950's, the scientists used more X rays to investigate the effects of ionizing radiation in insects. Bushland & Hopkins (1953) reviewed initial literature about the effects of X rays on arthropods. However, the penetration and dose rate achieved by most X ray machines at that time were much lower than that of isotopic sources, limiting drastically the number of insects that could be sterilized per batch (Lindquist, 1955).

Most of the facilities dedicated to produce sterile insects to Area-Wide IPM programs have used irradiators with gamma radiation from the radioisotopes ^{60}Co or ^{137}Cs. The ^{60}Co is produced when natural cobalt (100% ^{59}Co) absorbs neutrons in nuclear reactors, while the ^{137}Cs can be obtained by chemical separation from spent nuclear fuel such as plutonium or uranium. These radioisotopes can be then encapsulated in stainless steel to become source pencils, for example. The half-lifes of the ^{60}Co and ^{137}Cs are 5.27 and 30.07 years respectively. The ^{60}Co emits photons with two energies (1.17 and 1.33 MeV), while ^{137}Cs emits a monoenergetic photon of 0.66 MeV. Therefore, cesium sources require almost four times more activity than cobalt sources to provide the same throughput (Bakri et al., 2005).

The commercially available irradiators for SIT programs are usually of two types, self-contained or large scale panoramic type, both having as irradiation source several source pencils arranged in many different ways, but typically in a circular array or in a plane or a single rod. So far, the self-contained irradiators were the most commonly used in SIT facilities. Generally, the radiation source is kept inside a protective shielded chamber, which receives the material to be irradiated by a mechanism that rotates or lowers the canister of insects from the load position to the irradiation position. The canister must be moved (as in a turntable) or hold in a way that the dose delivered gets relatively uniform. The operator controls the dose delivered to the insects by positioning them correctly in the canister and calculating the time of exposure, since the dose rate from ^{60}Co or ^{137}Cs pencils changes with time and is determined by the current activity of the source. For example, cobalt sources have their activity reduced by 12% annually, so the operator has to compensate this loss by increasing the exposure time (Bakri et al., 2005). When the time of exposure becomes too long, the irradiation source must be replaced and reload with new pencils of high activity.

After the September 11th attacks, the fear of terrorism have provoked an increase in delays and denials of transboundary shipments of radioisotopes, what is constraining the reloading of existing sources and the acquisition of new ones. Between September 2007 and March 2008, for instance, almost 70 reports of delays and denials of shipments of radioactive materials were forwarded to the IAEA, being 13 related to ^{60}Co (Mastrangelo et al., 2010). In addition, the production of the isotopic irradiator most commonly used at SIT facilities, the self-contained Gamma Cell 220 ^{60}Co irradiator (MDS Nordion International Inc., Ottawa, ON, Canada) has been discontinued. Therefore, due to the growing complexities of the transboundary shipment of radioisotopes and the fear of "dirty bombs", there are serious doubts about the future availability of small scale irradiators (Mastrangelo et al., 2010).

Two alternatives to gamma radiation are high-energy electrons (with energy < 10 MeV) and X rays (generated from electron beams with energies bellow 7.5 MeV) (US FDA, 2004). The high energy electrons are generated by electron accelerators and X rays can come from breaking radiation (*i.e.*, rapid deceleration of a beam of electrons before striking a material with a high atomic number, or "bremsstrahlung"), in which way that any radioactive materials are involved. Since they have relative biological effectiveness (RBE) normalized to gamma rays close to one for most insect life stages and doses, many studies have demonstrated that they produce similar effect on insects (Adem et al., 1978; Bushland & Hopkins 1953; Dohino et al., 1994; Lindquist, 1955; Mastrangelo et al., 2010). Actually, almost a hundred low-energy self-contained X ray irradiators are already operating successfully at medical institutes in USA and at two SIT facilities. Such machines have also other advantages as no radiation is produced when switched off, no generation of radioactive waste, better public acceptance and simpler regulatory requirements. It seems that these new technologies may address the demands of national Area-Wide IPM programs that are in expansion around the world.

Currently, about 36 facilities are producing millions of sterile insects per week for national Area-Wide Integrated Pest Control programs or making research on SIT against screwworms, fruit flies, moths and tsetse flies (IDIDAS, 2010). One of the largest biofactories of the world (part of the MOSCAMED program) is located at El Piño, Guatemala, where is produced almost 2 billion sterile males per week of the Mediterranean fruit fly, *Ceratitis capitata* Wiedemann. At Metapa de Domínguez, Mexico, a mass-rearing facility produces the fruit flies *Anastrepha ludens* Loew (300 million/week), *Anastrepha obliqua* Macquart (30 million/week) and the parasitoid *Diachasmimorpha longicaudata* Ashmead (50 million/week) (Rull et al., 1996). These programs generate substantial direct and indirect benefits to the horticulture industry, health sector and the overall society.

As a result of the containment barrier at the Mexico-Guatemala border by the MOSCAMED program, Mexico's gross revenue from horticultural products have tripled since 1994 to more than US$ 3.5 billion/year, with a economic return over these years of 167 dollars for each dollar invested in the program (Enkerlin, 2005). The pink bollworm, *Pectinophora gossypiella* Saunders, has been excluded of the San Joaquin Valley, USA, by an ongoing SIT containment program since 1968, which cost approximately US$ 12.5/ha/season for each cotton grower, but it was estimated that growers pest control costs would increase by US$ 200/ha if the program was not in place, besides an additional 2.2 million kg of pesticide that would have to be used every year (Bloem et al., 2005). In 1994-1997, the population of a tsetse fly, *Glossina austeni* Newstead, vector of trypanosomosis ("sleeping sickness"), was eradicated by the SIT from Unguja Island of Zanzibar, what allowed a local increase in the amount of small farmers raising indigenous cattle (from 31% in 1985 to 94% in 2002) and in the proportion of domestic versus imported cattle slaughtered for meat (29% to 66% on 1986-1995) (Feldman et al., 2005).

4.2 Biological control

The biological control can be defined as the action of predators, parasites or pathogens in maintaining a pest population density at a lower average than would occur in their absence, eventually reaching a level below the crop's economic damage threshold (De Bach, 1964). This control method differs from the other forms of pest control by acting in a density-dependent manner with the pest population at the point that may become, in many agroecosystems, a self-sustained strategy, restoring the functional biodiversity (Altieri, 1994). As such, the biological control is one of the most environmentally and sustainable control tactics for insect pests and can be applied as part of IPM programs (Altieri, 1994). Basically, the applied biological control can be done through conservation management and classical or inundative techniques.

The conservation approach consists on practices to attract, protect or enhance the natural enemies' populations, as mixing plant cultivars or providing flowering borders which increase the diversity of habitats and serve as alternative food resource (Rabb et al., 1976). In classical biological control, exotic biological agents (*e.g.*, predators, parasitoids or entomopathogens) are introduced into the target area usually in inoculative releases, keeping as main concern the host specificity of the exotic agent (Louda et al., 2003). The third type of biological control, the inundative tactic, involves the mass-rearing and release of the biological agent, exotic or not, in very large numbers, often several times each season as they generally do not establish permanently in the environment. Constraints of the later tactic include the high cost of rearing, adequate quality control and political regulations which complicate shipping and trade (van Lenteren, 2003). Nuclear techniques present several potential applications that can increase the trade, safety, efficiency and cost effectiveness of biological control tactics (Hendrichs et al., 2009).

Reproductively inactivated host insects can be placed in the field at strategic locations as sentinels to monitor wild populations of indigenous biological agents or to explore new exotic agents (Jordao-Paranhos et al., 2003). Sterile F_1 larvae from irradiated *Lymantria dispar* L. had been used to monitor the density and type of parasitoids and entomophatogens in forests (Novotny & Zubrik, 2003). Sterilized *Musca domestica* L. pupae can be used in traps to monitor wild populations of pteromalid parasitoids under conditions of livestock production (Zapater et al., 2009).

Radiation can also be useful in screening classical biological agents under field conditions, when host specificity doubts still remain (Hendrichs et al., 2009) after pre-release studies under quarantine conditions. In other words, the exotic biological agent could be radiosterilized and then released in the field without the risk of establishing any permanent breeding population, allowing assessment of host associations under actual field conditions (Carpenter et al., 2001). As example under evaluation, the exotic herbivore *Episimus unguiculus* Clarke was in quarantine in Florida for the eventual biological control of the Brazilian pepper tree *Schinus terebinthifolius* Raddi (Moeri et al., 2009).

Radiosterilized preys or hosts could also be released in the field prior to the pest outbreak to serve as supplemental food to increase the native population of natural enemies or inoculatively released biological agents. Sterile eggs from irradiated cotton bollworm *Helicoverpa armigera* Hubner and diamondback moth *Plutella xylostella* L. served as hosts for wild egg parasitoids (Wang et al., 2009). In sugarcane fields, the provision of irradiated host eggs to *Trichogramma chilonis* Ishii early in the season allowed building-up its populations and such approach is providing now the control of many sugarcane borers species in a 40,000 ha sugarcane area in Pakistan (Fatima et al., 2009).

Radiation can increase rearing efficiency and parasitoid quality or make non-habitual hosts (*i.e.*, cheaper to mass rear) by suppressing host immune system. Some physiological processes in the host (*e.g.*, defence mechanisms and hormone metabolism) can be selectively modified by radiation (Vey & Causse, 1979). The immune response (haemolymph melanisation and haemocytic encapsulation) of haemolymph from *Galleria mellonella* L. was severely reduced when irradiated, and *G. mellonella* larvae irradiated at 65 Gy were found to serve as potential hosts for the rearing of *Venturia canescens* Gravenhorst (Genchev et al., 2007). Irradiated *Sitotroga cerealella* Olivier eggs, when used as prey substitute, increased larvae viability, fecundity and the sex ratio of the predator *Chrysoperla carnea* Stephens (Hamed et al., 2009). Irradiation of the wasp *Glyptapanteles liparidis* Bouché was used to study the action of its polydnavirus and venom, which were injected along with the sterilized eggs, in *L. dispar* larvae (Hoch et al., 2009). Radiation hormesis was reported in the parasitoids *Habracon hebetor* Say, *T. chilonis* and *V. canescens* after exposure to very low doses of radiation (Genchev et al., 2008; Wang et al., 2009).

The limited shelf life of prey and hosts may limit their use during mass production of biological agents, but, in some cases, radiation can delay normal insect development, extending the time window for feeding or host parasitisation. The parasitisation period of third instar *Anastrepha* spp. larvae irradiated at 45 Gy was extended, as the efficiency of the parasitism by *Diachasmimorpha longicaudata* Ashmead increased (Cancino et al., 2009b). Irradiation of the carambola fruit fly *Bactrocera carambolae* Drew & Hancock eggs extended the larval period suitable for parasitisation by *Psyttalia incise* Sylvestri and *Fopius vandenboschi* Fullaway (Kuswadi et al., 2003). The irradiated parasitoid *Cotesia flavipes* Cameron could be stored as pupae up to 2 months at 10 °C without apparent loss of quality (Fatima et al., 2009).

Insect mass-rearing facilities usually produce significant amounts of by-products (*e.g.*, large batches of sub-standard insects and specimens from the quality control tests). Instead of being discarded, these by-products can be processed or irradiated to rear biological agents (Hendrichs et al., 2009; Nakashima et al., 1996). In fruit fly mass-rearing facilities, discarded larvae or pupae could be used to rear some kinds of parasitoids (Cancino et al., 2009a).

Irradiation can overcome problem with trade barriers related to shipment of biological agents, since accidental inclusion of hitchhikers or fertile pest specimens in the shipments are possible, by eliminating the need to sort parasitoids from non-parasitized hosts and avoiding the emergence of adult pests from non-parasitized immature stages (Hendrichs et al., 2009). Irradiation can be used to eliminate the risk of introducing fertile spider mites *Tetranychus urticae* Koch, which are provided along with several shipped species of predatory mites (Baptiste et al., 2003). Fruit fly larvae of *A. obliqua*, *A. serpentina* Wiedemann and *A. ludens* are routinely irradiated in the mass-rearing of tens of millions of parasitoids (Cancino et al., 2009a), ensuring shipments clean of the adult stage of these pests.

The release of sterile or half sterile insects together with biological agents has been known to have synergistic effects for population suppression when applied simultaneously, because the sterile insects impact on the adult stage, while the biological agents target mostly the immature stages (Knipling, 1992). Saour (2009) demonstrated the synergistic effects of combining F_1 sterility with egg parasitoids, by releasing *Trichogramma principium* Sugonyaev & Sorokina together with moths irradiated at 250 Gy, what reduced significantly *Phthorimaea operculella* Zeller progeny. Field-cage evaluations in citrus orchards in South Africa revealed that releases of irradiated moths combined with releases of *Trichogrammatoidea cryptophlebiae* Nagaraja provided synergistic suppression of false codling moth populations (Carpenter et al., 2004).

5. Radiation as quarantine treatment against insect pests

In international agricultural markets, the use of radiation as a method for the prevention of quarantine insects represents an important alternative post-harvest pest control, reducing the need for chemical fumigants and other similar toxic products. The US Food and Drug Administration (FDA) has approved radiation up to 1 kGy to control insects in foods and to extend the shelf life of fresh fruits and vegetables (US FDA, 2004).

The advantages of radiation include the no resistance development by pest insects, the absence of residual radioactivity and few significant changes in the physicochemical properties or the nutritive value of the treated products (Lapidot et al., 1991). A major disadvantage is that it is the only commercially applied quarantine treatment that does not result in significant acute mortality. This issue is very important because when inspectors find live quarantine pests from the major phytosanitary treatments, which are based on heat, cold or methyl bromide fumigation, the entire consignment is rejected or retreated regardless of certification of treatment. In this case, the inspectors assume that the treatment was not properly done, the shipment was contaminated with infested commodity or that the cargo was reinfested after treatment. In addition, live adults found in survey traps could trigger restrictive and costly regulatory responses in importing countries (Hallman et al., 2010).

Nevertheless, Hallman (2004a) stated that the objective of irradiation is not acute mortality but prevention of development or reproduction, as most commodities do not tolerate the usual dose ranges required to reach it (usually ≥ 1 kGy). Therefore, the inhibition of further development must be considered as a measure of efficacy of phytosanitary irradiation (FAO, 2003). The US Animal and Plant Health Inspection Service (APHIS) has not objected to live adults because of a comprehensive process of validation and certification of irradiation treatment facilities with monitoring of dosimetry and dose application during preclearance programs (Hallman et al., 2010).

The APHIS and the International Plant Protection Convention (IPPC) have approved phytosanitary irradiation treatments for more than 20 insect pest species (FAO, 2009). Tephritid fruit flies are one of the most invasive quarantine pests, attacking 21 of the 24 fresh commodities exported to the United States and to sterilize or disrupt normal development of early stages, doses ranging from 70 to 100 Gy are sufficient for *Anastrepha* species, whilst *Bactrocera* spp. and other species may require doses in the range of 100 to 150 Gy (Follett, 2009). After fruit flies, tortricid moths are the most important pests of quarantine concern for fruit and vegetables. Several studies have shown that a dose of 200 Gy could be sufficient to control codling moth (*Cydia pomonella* L.), *Ecdytolopha aurantiana* Lima and oriental fruit moth (*Grapholita molesta* Busck) (Arthur, 2004; Hallman, 2004b; IDIDAS, 2010; Mansour, 2003). Curculionid weevils are another important group of pests and available studies suggest that a dose ≤ 150 Gy may be sufficient to control cowpea weevil (*Callosobruchu chinensis* L.), *Euscepes postfasciatus* Fairmaire and boll weevil (*Anthonomus grandis* Boheman) (Davich & Lindquist, 1962; Follett, 2006; Gao et al., 2004).

Hallman et al. (2010) analyzed several factors that could affect phytosanitary irradiation efficacy, such as hypoxia, insect life stage, host, dose rate, temperature, diapause and genotypes. After dose itself, hypoxia can be considered the most important factor that abates the effects of radiation on living organisms because lesser radioinduced radicals are produced (Hallman & Hellmich, 2010; von Sonntag, 1987). Cryptically-feeding Tephritidae and Curculionidae that occur as immature inside host plants (practically hypoxic conditions) may present increased radiotolerance (Hallman & Loaharanu, 2002). As radiotolerance also increases as insects develop, a phytosanitary irradiation treatment must be effective against the most tolerant stage that could be present on the commodity (Hallman, 2000). About the differences in hosts, dose rates and temperatures, their characteristics are not so relevant as long as the required minimum dose is absorbed (APHIS, 2005). Jessup et al. (1992) reported the desinfestation of six different fruits from *Bactrocera tryoni* Froggatt third instars at 75 Gy. Temperature did not affect radiation efficacy for a tephritid and a crambid within the cold storage range for fresh commodities (Hallman 2004b; Hallman & Hellmich, 2009). Insects in diapause may be more susceptible to radiation (Hallman, 2000). According to Hallman et al. (2010), most studies that made direct comparisons among populations of the same species did not show significant differences in response to radiation. Cornwell (1966) analyzed 35 irradiated strains of *Sitophilus granaries* L. and found no differences in sterility. The adult emergence of laboratory and wild strains of three tephritid species irradiated as third instars did not differ significantly (Follett & Armstrong, 2004).

Quarantine entomologists are constantly looking for a generic radiation quarantine treatment, which could be able to control a broad group of pests without adversely affecting the quality of a wide range of commodities (Follett & Neven, 2006). Such doses would necessarily be set at the minimum absorbed dose required for the most tolerant organism within that group (Hallman & Phillips, 2008). In 2006, the APHIS accepted as phytosanitary treatments the generic doses of 150 Gy for all Tephritidae and 400 Gy for all insects other than Lepidoptera for commodities entering the United States (USDA-APHIS, 2006). The generic dose of 150 Gy is currently being used for mangoes and citrus fruit exported from Mexico to the United States, and the dose of 400 Gy is applied for Mexican guavas, Indian mangoes and dragon fruit (*Hylocereus undatus* Britton & Rose) from Vietnam, all exported to the United States (Hallman et al., 2010). Australia is also using a generic treatment (250 Gy for insects) to send mangoes and litchi to New Zealand (MAF, 2009).

Hallman & Phillips (2008) suggested that a generic dose of 600 Gy for all insects in ambient atmospheres would be efficacious to attend quarantine purposes, owing to the high radiotolerance of the Angoumois grain moth (*Sitotroga cerealella* Olivier).

6. Isotopes as markers for insects

By the late 1940s, isotopic releases from nuclear operations had demonstrated the utility of radiotracers for studying the dynamics of biological systems, and by the early 1950s ecologists were using radioisotopes to new areas of experimental research. Labeling insects with radiotracers in order to study dispersal, population densities, behavior and food intake became a very popular insect-marking method from the 1950s to the 1970s (Table 2).

	Type of Study	Radioisotope	References
Pest Management	Synthetic diets for mass-rearing		Radeleff et al. (1952); Strong & Landes (1965)
	Efficiency of Sterile Males	^{32}P	Baumhover et al. (1955); Haisch (1970)
	Efficiency of Predators		Smith (1965); Van Dinther & Mensink (1971); Edney et al. (1974); Moore et al. (1974)
Ecology	Energy Flow in Communities	^{134}Cs, ^{45}Ca	Crossley (1963); Odum (1963); Reichle (1967); Williams & Reichle (1968)
	Heterotrophic Productivity	^{22}Na, ^{86}Rb	Van Hook et al. (1970); Kowal & Crossley (1971)
Nutrition	Growth	^{32}P	Radeleff et al. (1952); Gordon (1972); Klein & Kogan (1974); Rapport & Turner (1975); Baily (1976)
	Food Utilization	^{14}C, ^{32}P	Evans (1939); Day & Irzykiekwiez (1953); Oertel et al. (1953); Kasting & McGinnis (1965); Waldbauer (1968); Dietz & Lambremont (1970); Devine (1978)

Table 2. Radiotracers most commonly used in entomological studies.

Labeling procedures involved rearing the insects in labeled larval or adult diet, spraying, submersion, attaching radioactive objects to the body, and even indirect methods, in which a plant or animal was rendered radioactive (Barnes, 1959; Radeleff et al., 1952). The detection methods included trapping, killing, and examining specimens at bait stations with a Geiger counter or radioautography (Dissanaike et al., 1957; Jensen & Fay, 1951).

Screwworm *Cochliomyia hominivorax* larvae had been successfully reared on a ground meat medium labeled with 0.5 μc of $^{32}P/g$ (Radeleff et al., 1952). Long & Lilly (1958) attached pieces of radioactive wire made of ^{60}Co and gold plated to the outside of the wireworm

Melanotus communis Gyllenhal with a plastic cement and, after their release into soil, their location was determined by an end-window Geiger tube and rate meter. Strips of ^{182}Ta, glued to the prothorax of the insect, had been used to label coccinellid larvae (Banks, 1955). Barnes (1959) sprayed foliage with a solution of corn protein hydrolysate with 50 µc of ^{32}P/mL to label the walnut husk fly *Rhagoletis completa,* and 15% of the flies captured in the orchard were labeled, with flies averaging 8022 counts/minute.

When screwworm larvae were reared in wounds of goats which had received intravenously injection of ^{32}P at 0.1 µc/g body weight, the adult flies showed 2500 counts/minute (Radeleff et al., 1952). Plants that had grown in water containing 100 µc of ^{35}S/L were used to mark Lepidoptera larvae (Kettlewell, 1952).

Schoof & Siverly (1954) labeled the American cockroach, *Periplaneta americana* L., in order to assess its moviments within a sewage system. The authors captured about 6500 insects and sprayed them with a solution containing ^{32}P and casein as sticker. Almost 14% of the insects were recovered and, curiously, they noticed that only a single specimen was found elsewhere in the sewage system, leading to the conclusion that the involvement of *P. americana* in disease transmission out of the sewage system was practically negligible.

The advantages of using radioisotopes rather than conventional markers were the relative permanence (dyes often rubbed off and molting usually eliminates external labels), the rapidly of checking, and the possibility of tracing the labeled insects that were out of sight (*e.g.,* underground). Several tracers were studied, such as ^{60}Co, ^{89}Sr, ^{65}Zn, ^{144}Ce, ^{131}I and ^{45}Ca, but ^{32}P was far the most applied radioisotope for tagging due to its short half-life, safety, activity and easy of detection (O'Brien & Wolfe, 1964).

One of the earliest examples of using inorganic ^{32}P as a label was reported by Hasset & Jenkins (1949), who reared *Aedes aegypti* mosquitoes in water containing up to 1 µc/mL and obtained adults with up to 10,000 counts/minute each. Fredeen et al. (1953) applied 0.2 µc of ^{32}P/mL in vats used to rear blackflies, and 800,000 larvae were labeled and released, with counts up to 50,000 counts/minute each. Odum (1963) used ^{32}P to isolate individual food chains of several natural communities.

The short half-life of 14 days of ^{32}P, however, was a disadvantage in studies where prolonged observations were necessary. Cerium-144 was used as persistent label for fleas, mosquitoes, cockroaches, ticks and other insects, as it has a half-life of 282 days and its daughter, ^{144}Pr, has a half-life of 18 minutes, but emits an energetic beta particle of 2.97 MeV, which could be easily detected (Quan et al., 1957). Babers et al. (1954) needed to make observation on *A. grandis* for over a 5 month period and found that dipped weevils from a solution of $Co^{60}Cl$ (5 µc/mL) with detergent averaged 710 counts/minute each. Although the radioisotope ^{35}S has a convenient half-life of 87 days, it is a weak beta emitter (0.17 MeV) and, therefore, a poor choice for labeling (Kettlewell, 1952).

A requirement for the use of such tracers was that the behavior of the labeled insect should not be affected, and some studies investigated the conditions that could affect the tagged insects, such as age, stage, radiotracer concentrations and so on. Hasset & Jenkins (1951) performed a detailed study of the conditions affecting mosquitoes labelled with ^{32}P and compared stages, ^{32}P concentrations and age. Younger larvae were too radiosensitive and pupae absorbed less phosphorus. The maximum concentration in water was 0.1 µc/mL and females took up three times as much as males, concentrating the ^{32}P about 75 times over the concentration in the medium. Toxicity to the insect was also a serious problem to be considered in many studies (Quarterman et al., 1955). The radioisotopes ^{45}Ca and ^{131}I were very toxic when fed to adult house-flies at 1 µc/mL of milk, whereas ^{32}P was satisfactory (Quarterman et al., 1954).

Jensen & Fay (1951) compared the effectiveness of feeding ^{32}P to adults and to larvae of houseflies and secondary screwworms (*Cochliomyia macellaria* F.) and observed that larvae fed on a medium with 0.1 µc of ^{32}P/g gave adults averaging 100 counts/minute, whilst adults fed on milk with 1 µc/mL gave counts of 1100 for males and 2000 for females.

Much work was done on the measurement of the feeding rate of insects, as in the transfer of food within colonies, water intake and transfer of plant juices between plants.

One of the methods used for feeding rate determination, which was proved very fruitful for studying especially primary consumers or predators, was based on the radiotracer conservation through the system "food-insect body-removed products". If insects were fed with uniformly labelled food, the food uptake could be calculated from the amount of tracer measured in the body and that measured in the products removed by secretion (like honey dew), excretion, respiration of CO_2, (whether ^{18}O or ^{14}C was used), water transpiration (whether ^{18}O or ^3H was used), egg production and others (Buscarlet, 1983; Kasting & McGinnis, 1965). In grain beetles fed on tapioca labelled with inulin-^{14}C, which was not assimilated by the insect, the ingestion rate could be estimated from the ^{14}C turnover rate constant and from the expendable solids (Devine, 1978).

Oertel et al. (1953) studied food transmission in bee colonies. Drone bees were exposed in a cage with unlabeled sucrose syrup available and separated by a screen from worker bees which had fed on sucrose-^{14}C syrup. The drone bees became radioactive as a result of being fed sucrose by the workers. Alibert (1959) studied the termite *Calotermes flavicollis* F. and verified a quick transmission of food between the workers: in 35 hours, all insects were labeled.

McEnroe (1961) used ^{32}P to evaluate the water intake of individual mites. *Tetranychus telarius* L. was fed on water containing inorganic ^{32}P and the author verified that it took in from 1.3 to 4.6 µmL in one hour (about 25% of the mite's body weight).

Many researchers had been interested in following the feeding behavior of phytophagous insects, in particular disease vectors as aphids. Miss Hamilton (1935) described the use of the α-emitter, polonium, as a tracer in *Myzus persicae* Sulzer transmission, measuring its activity with a gold-leaf electroscope. Day & Irzykiekwiez (1953) fed aphids either on leaves from cabbages cultivated in a ^{32}P solution or through a plastic membrane on a sucrose solution labeled with ^{32}P and verified that *Myzus* took up to 69 µg of plant material in only one hour (35% of its weight; data computed from the average radioactivity in leaf tissues and the radioactivity taken up by the insects) and up to 7% of the total uptake was excreted in an hour. When imbibing sucrose, *Myzus* took up only about 3% of its plant value, and, finally, the authors shown that the aphids did reinject imbibed material (but only up to 0.5% of the imbibed dose was reinjected in a day).

Lawson et al. (1954) assessed the transmission of tobacco plant juices by *Myzus* sp. by feeding the aphids on tobacco plants grown in soil treated with ^{32}P phosphoric acid and then placing them on unlabeled plants. Several spots of radioactivity were shown after 6 days, not caused by external honeydew or by absorption of honeydew, and translocation of radioactivity to other leaves of the plant was also found.

Labeled insects were also used as devices to study parasites. Larvae of the mosquito *Armigerea obturbans* Walker were reared in water containing 1 µc of ^{32}P/mL, and the subsequent adults, averaging 7×10^5 counts/min., were let to feed on cows or men infected with microfilaria. The filarial larvae, *Setaria digitata* Linstow, gave 174 counts/minute, which was sufficient to allow tracking its passage through host tissues either by counting or radioautographically (Dissanaike et al., 1957).

Recent stricter environmental protection laws coupled with the development of simpler, less expensive and reliable methods have reduced the usefulness of the radioactive isotopes as insect markers.

A substitute for many radionuclide methods is the stable isotope methods, as they pose no health or environmental risks. Isotopic signatures are natural differences in stable isotope composition of organisms caused by discrimination against the heavier isotopes during some biological processes. Over the past twenty years, much progress was made in isotope ratio mass spectrometry in terms of detection, accuracy and automation. Stable isotopes do not decay and occur naturally in the environment. Other advantages are the analysis costs (depending on the isotope and the matrix, the cost per sample may range from US$ 5-100.00), shipping stable isotope samples is simple, safe and inexpensive (IAEA, 2009).

Natural isotopic signatures are already used in wide range of research areas and, therefore, were standardized to an internationally accepted scale. Because of the large differences in the abundances of the carbon and nitrogen isotopes ($^{12}C \approx 1.1\%$, $^{13}C \approx 98.9\%$, $^{15}N \approx 0.3663\%$ and $^{14}N \approx 99.63\%$), $^{13}C/^{12}C$ and $^{15}N/^{14}N$ ratios are generally expressed in the delta notation in parts per thousand (per mil ‰) relative to the international standard Vienna PeeDee Belemnite (VPDB) and atmospheric nitrogen, respectively (IAEA, 2009). These stable isotope markers meet the usual criteria for use in insect studies: retention, no effect on behavior, durability, easily applied, clearly identifiable, and not expensive (Hagler & Jackson, 2001).

The isotopic signature of an organism is mainly dependent on what it eats and as natural processes also lead to distinctive isotopic signals, the formation of the so called isotopic landscapes (i.e., isoscapes) is even useful in tracing insect movement, mating patterns, and in studies about the use of natural resources (Hershey et al., 1993; Helinski et al., 2007; Peterson, 1987). For example, some insects that are reared on C4 sugar based diets will be isotopically different in ^{13}C signatures from the wild populations that naturally feed on C3 plants (Hood-Nowotny et al., 2006).

While most of the fruit fly species feed on C3 plants in the wild, which have a carbon isotope signature of around -28‰ versus VPDB, almost all mass-rearing facilities use cane sugar in the larval and adult diet, which is a C4 sugar source (with a signal of around -11‰ versus VPDB). Hood-Nowotny et al. (2009) demonstrated that this difference in isotopic signatures between wild and released factory-reared flies could be a reliable and intrinsic secondary marker to complement existing marking methods from Area-Wide IPM programs.

Insects can also be marked by adding an enriched compound to the diet, such as ^{15}N labelled glycine (Caquet, 2006; Fisher et al., 2003; Markow et al., 2000; McNeill et al., 1998; Nienstedt et al., 2000; Nienstedt et al., 2004). The natural enemies, *Cotesia plutellae* Kurdjumov and *Hippodamia convergens* Guérin-Meneville, that foraged at the flowers of ^{15}N-marked plants showed detectable quantities of the marker (Steffan et al., 2000). Plant material enriched with ^{15}N was added to the diet of navel orangeworms *Amyelois transitella* Walker and both the orangeworms and wasps *Goniozus legneri* Gordh that parasitized them gave detectable levels of ^{15}N in their systems (Steffan et al., 2000).

Among the disadvantages of the stable isotopes methodology are the cost of isotope ratio mass spectrometers (some cost more than US$ 100,000), the controlled environment required by the equipment and skilled personnel to use the mass spectrometer (IAEA, 2009).

7. Conclusion

Radioisotopes allowed the rise of an entire new branch of the study of insects, the Radioentomology. Even today, the Sterile Insect Technique and phytosanitary irradiation treatments protect horticultural markets and livestock of many countries. Furthermore, radiotracers had a very important role in revealing the characteristics and dynamics of several biological systems.

8. Acknowledgment

We are thankful to the Centre for Nuclear Energy in Agriculture (CENA/USP) and Fundação de Amparo à Pesquisa do Estado de São Paulo (FAPESP) for funding authors' projects and this work.

9. References

Adem, F.L.; Watters; Uribe-Rendon, R. & Piedad, A.D. (1978). Comparison of ^{60}Co gamma irradiation and accelerated electrons for suppressing emergence of *Sitophilus* spp. in stored maize. *Journal of Stored Products Research*, vol.14, pp. 135–142, ISSN 0022-474X.

Alibert, J. (1959). Les Echanges Trophallactiques Chez le Termite a Cou Jaune (*Calotermes flavicollis* Fabr.) Études a l'Aide du Phosphore Radioactif. *Comptes Rendus de l'Académie des Sciences*, v.248, pp.1040-1042, ISSN 0764-4442.

Altieri, M. A. (1994). *Biodiversity and pest management in agroecosystems*. Haworth Press, ISBN 1560220376, New York, USA.

[APHIS] Animal and Plant Health Inspection Service. 2005. Treatments for fruits and vegetables. Fed. Reg. 70: 33857-33873.

Arthur, V. (2004). Use of gamma irradiation to control three lepidopteran pests in Brazil, In: *Irradiation as a phytosanitary treatment of food and agricultural commodities*. IAEA-TECDOC 1427, pp.45-50, International Atomic Energy Agency, Vienna, Austria.

Babers, F.H.; Roan, C.C. & Walker, R.L. (1954). Tagging Boll Weevils with Radioactive Cobalt. *Journal of Economic Entomology*, vol. 47, pp.928-929, ISSN 1938-291X

Bailey C. G. (1976). A quantitative study of consumption and utilization of various diets in the Bertha armyworm, *Mamestra configurata* (Lepidoptera: Noctuidae). *Canadian Entomologist* 108, 1319-1326, ISSN 0008-347X

Bakri, A.; Mehta, K. & Lance, D.R. (2005). Sterilizing insects with ionizing radiation. In: *Sterile insect technique: principles and practice in area-wide integrated pest management*. V.A. Dyck.; J. Hendrichs & A.S. Robinson, (Eds.), 233-269, Springer, ISBN 1-4020-4050-4, The Netherlands.

Banks, C.J. (1955). The use of radioactive Tantalum in studies of the behavior of small crawling insects on plants. *British Journal of Animal Behavior*, vol.3, pp.158-159, ISSN 0950-5601

Baptiste, S.J., Bloem, K., Reitz, S., and Mizell III, R. (2003). Use of Radiation to Sterilize Two-spotted Spider Mite (Acari: Tetranychidae) Eggs Used as a Food Source for Predatory Mites. *Florida Entomologist*, vol.86, pp.389-394, ISSN 0015-4040.

Barnes, M.M. (1959). Radiotracer labeling of a natural tephritid population and flight range of the walnut husk fly. *Annals of the Entomological Society of America*, v.52, pp.90-92, ISSN 0013-8746

Baumhover, A. H. (1966). Eradication of the Screwworm Fly, an Agent of Myiasis. *Journal of the American Medical Association*, vol.196, no. 3, pp.240-248, ISSN 00987484

Baumhover A.H.; Graham, A.J.; Bitter, B.A.; Hopkins, D.E.; New, W.D.; Dudley, F.H.; Bushland, R.C. (1955). Screw-worm control through release of sterilized flies. *Journal of Economic Entomology*, vol.48, pp.462-466, ISSN 1938-291X

Behera, M. K.; Behera, R. & Patro, B. (1999). Application of Dyar's rule to the development of *Macrosiphoniella sanborni* (Gill.) (Aphididae: Homoptera). *Agricultural Science Digest*, vol.19, pp.179-182, ISSN 0253-150X

Bloem, K.A.; Bloem, S. & Carpenter, J.E. (2005). Impact of moth suppression/eradication programmes using the sterile insect technique or inherited sterility, In: *Sterile Insect Technique. Principles and Practice in Area-wide Integrated Pest Management*, V.A. Dyck; J. Hendrichs & A.S. Robinson, (Eds.), 677-700, Springer, ISBN 1-4020-4050-4, Dordrecht, The Netherlands.

Buscarlett, L.A. (1983). The use of radioactive tracers for insect feeding rate determination. *The International Journal of Applied Radiation and Isotopes*, v.34, n.5, pp.855-859, ISSN 0020-708X

Bushland, R.C. & Hopkins, D.E. (1951). Experience with screwworm flies sterilized by X-rays. *Journal of Economic Entomology*, v. 44, pp. 725-731, ISSN 1938-291X

Bushland, R.C. & Hopkins, D.E. (1953). Sterilization of screwworm flies with X-rays and gamma rays. *Journal of Economic Entomology*, v. 46, pp. 648-656, ISSN 1938-291X

Cancino, J.; Ruíz, L.; López, P. & Sivinski, J. (2009a), The Suitability of *Anastrepha* spp. and *Ceratitis capitata* (Diptera: Tephritidae) Larvae as Hosts of *Diachasmimorpha longicaudata* and *Diachasmimorpha tryoni* (Hymenoptera: Braconidae): Effects of Host Age and Radiation Dose and Implications for Quality Control in Mass Rearing. *Biocontrol Science and Technology*, vol.19, S1, pp.81-94, ISSN 1360-0478.

Cancino, J.; Ruíz, L.; Sivinski, J.; Gálvez, F.O. & Aluja, M.(2009b). Rearing of Five Hymenopterous Larval-prepupal (Braconidae, Figitidae) and Three Pupal (Diapriidae, Chalcidoidea, Eurytomidae) Native Parasitoids of the Genus *Anastrepha* (Diptera: Tephritidae) on Irradiated *A. ludens* Larvae and Pupae. *Biocontrol Science and Technology*, vol.19, S1, pp.193-209, ISSN 1360-0478

Caquet, T. (2006). Use of carbon and nitrogen stable isotope ratios to assess the effects of environmental contaminants on aquatic food webs. *Environmental Pollution*, v.141, pp.54-59, ISSN 0269-7491

Carpenter, J.E.; Bloem, S. & Bloem, K.A. (2001). Inherited Sterility in *Cactoblastis cactorum* (Lepidoptera: Pyralidae). *Florida Entomologist*, vol.84, pp.537-542, ISSN 0015-4040

Carpenter, J.E.; Bloem, S. & Hofmeyr, S. (2004). Acceptability and Suitability of Eggs of False Codling Moth (Lepidoptera: Tortricidae) from Irradiated Parents to Parasitism by *Trichogrammatoidea cryptophlebiae* (Hymenoptera: Trichogrammatidae). *Biological Control*, vol.30, pp. 351-359, ISSN 1090-2112

Casarett, A. P. (1968). *Radiation biology*. Prentice-Hall, ISBN 0137503563, Englewood Cliffs, NJ.

Cornwell, P. B. (1966). Susceptibility of laboratory and wild strains of the grain weevil *Sitophilus granarius* (L.) to gamma radiation, In: *The entomology of radiation disinfestation of grain*, P.B. Cornwell, (Ed.), 19-26, Pergamon, ISBN 0080104193, Oxford, United Kingdom.

Crossley D. A. (1963). Movement and accumulation of radiostrontium and radiocaesium in insects. In: *Radioecology*, V. Schultz & A.W. Klement, (Eds.), 103-105, Reinhold, New York.

Davich, T. B. & Lindquist, D.A. (1962). Exploratory studies on gamma radiation for the sterilization of the boll weevil. *Journal of Economic Entomology*, vol.55, pp.164-167, ISSN 1938-291X

Day, M.F. & Irzykiekwiez, H. (1953). Feeding behavior of the aphids *Myzus persicae* (Sulz.) on radioactive plants. *Annals of Applied Biology*,v.40, pp.537-545, ISSN 1744-7348.

DeBach, P. (1964). *Biological control of insect pests and weeds*. Reihold, ISBN 0470204567, New York.

Devine, T. L. (1978). The turnover of the gut contents (traced with inulin-carboxyl-^{14}C), tritiated water and ^{22}Na in three stored product insects. *Journal of Stored Products Research, vol.* 14, pp.189-211, ISSN 0022-474X

Dey, S. K. & Manna, G. K. (1983). Differential stage sensitivity to x-rays in a bug *Physopelta schlanbuschi. National Academy Science Letters*, vol.16, pp.101-103, ISSN 0250-541X

Dissanaike, A.S.; Dissanaike, G.A.; Niles, W.J.& Surendranathan, R.(1957). Further Studies on Radioactive Mosquitoes and Filarial Larvae using Autoradiographic Technique. *Experimental Parasito*logy, v.6, pp.261-270, ISSN 1090-2449

Dietz, A. & Lambremont, E. (1970). A Method of Studying Food Consumption of Live Honey Bee Larvae by Liquid Scintillation Counting. *Annals of the Entomological Society of America*, vol.63, pp.1340-1342, ISSN 0013-8746

Dohino, T.; Tanabe, K. & Hayashi, T. (1994). Comparison of lethal effects of electron beams and gamma rays on eggs of two spotted spider mite, *Tetranychus urticae* Koch (Acari: Tetranychidae). *Research Bulletin of the Plant Protection Service Japan*, vol.30, pp.69–73, ISSN 0387-0707

Edney, E. B.; Allen, W. & McFarlane, J. (1974). Predation by terrestrial isopods. *Ecology*, vol.55, pp. 428-433, ISSN 0012-9658

Enkerlin, W.R. (2005). Impact of fruit fly control programmes using the sterile insect technique. In: *Sterile Insect Technique. Principles and Practice in Area-wide Integrated Pest Management*, V.A. Dyck, J. Hendrichs & A.S. Robinson, (Eds.), 651-676, Springer, ISBN 1402040504, Dordrecht, The Netherlands.

Evans, A. C. (1939). The utilization of food by certain lepidopterous larvae. *Transactions of the Royal Entomological Society of London, vol.* 89, pp.13-22, ISSN 1365-2311

[FAO] Food and Agricultural Organization. (2003). Guidelines for the use of irradiation as a phytosanitary measure. ISPM #18. Food and Agricultural Organization, Rome, Italy.

[FAO] Food and Agricultural Organization. (2005). Glossary of phytosanitary terms. Provisional additions. Rome: FAO/IPPC.

[FAO] Food and Agricultural Organization. (2009). Phytosanitary treatments for regulated pests. ISPM 28. Food and Agricultural Organization, Rome, Italy. DOI: 10.1111/j.1365-2338.2009.02224.x

[FAO/IAEA/WHO] Food and Agriculture Organization/ International Atomic Energy Agency/World Health Organization. (1999). *High-dose irradiation: wholesomeness of food irradiated with doses above 10 kGy*. Joint FAO/IAEA/WHO Study Group, Technical Report Series 890. World Health Organization, Geneva, Switzerland.

[FAO/IAEA/USDA] Food and Agriculture Organization/ International Atomic Energy Agency/US Department of Agriculture (2003). *Product quality control and shipping procedures for sterile mass-reared tephritid fruit flies*. Version 5.0., IAEA, Vienna, Austria.

Fatima, B.; Ahmad, N.; Memon, R.M.; Bux, M. & Ahmad, Q. (2009). Enhancing Biological Control of Sugarcane Shoot Borer, *Chilo infuscatellus* (Lepidoptera: Pyralidae), through Use of Radiation to Improve Laboratory Rearing and Field Augmentation of Egg and Larval Parasitoids. *Biocontrol Science and Technology*, vol.19, S1, pp.277 – 290, ISSN 1360-0478

Fischer, R.C.; Wanek, W.; Richter, A. & Mayer, V. (2003). Do ants feed plants? A ^{15}N labeling study of nitrogen fluxes from ants to plants in the mutualism of *Pheidole* and *Piper*. *Journal of Ecology*, v.91, pp.126–134, ISSN 1365-2745

Feldmann, U.; Dyck, V.A.; Mattioli, R.C. & Jannin, J. (2005). Potential impact of tsetse fly control involving the sterile insect technique. In: *Sterile insect technique. Principles and practice in area-wide integrated pest management*, V.A. Dyck; J. Hendrichs & A. S. Robinson, (Eds.), 701-726, Springer, ISBN 1-4020-4050-4, Dordrecht, The Netherlands.

Fitt, G.; Andow, D.; Huan, N.H.; Capaldo, D.M.F.; Omoto, C.; Tho, N.; Son, N.H.; Tuyen, B. C. (2008). Resistance Risk Assessment and Management for Bt Cotton in Vietnam. In: *Environmental Risk Assessment of Genetically Modified Organisms: Challenges and Opportunities with Bt Cotton in Vietnam*, D.A. Andow; A. Hilbeck; N.V. Tuat., (Eds.), 296-329, CAB International, ISBN 1845933907, Oxfordshire, UK.

Follett, P.A. (2006). Irradiation as a methyl bromide alternative for postharvest control of *Omphisa anastomosalis* (Lepidoptera: Pyralidae) and *Euscepes postfasciatus* and *Cylas formicarius elegantulus* (Coleoptera: Curculionidae) in sweet potatoes. *Journal of Economic Entomology*, vol.99, pp.32-37, ISSN 1938-291X

Follett, P.A. (2009). Generic radiation quarantine treatments: the next steps. *Journal of Economic Entomology*, vol.102, pp.1399-1406, ISSN 1938-291X

Follett, P. A. & Armstrong, J.W. (2004). Revised irradiation doses to control melon fly, Mediterranean fruit fly, and oriental fruit fly (Diptera: Tephritidae) and a generic dose for tephritid fruit flies. *Journal of Economic Entomology*, vol.97, pp.1254-1262, ISSN 1938-291X

Follett, P. A. & Neven, L.G. (2006). Current trends in quarantine entomology. *Annual Review of Entomology*, vol.51, pp.359-385, ISSN 0066-4170

Fredeen, F.J.H.; Spinks, J.W.T.; Anderson, J.R.; Arnason, A.P. & Rempel, J.G. (1953). Mass tagging of black flies (Diptera: Simuliidae) with radiophosphorus. *Canadian Journal of Zoology*, vol.31, pp.1-15, ISSN 1480-3283.

Gao, M.; Wang, C.; Li, S. & Zhang, S. (2004). Irradiation as a phytosanitary treatment for *Trogoderma granarium* Everts and *Callosobruchus chinensis* L. in food and agricultural products. In: *Irradiation as a phytosanitary treatment of food and agricultural commodities*. IAEA-TECDOC-1427, pp.75-85, International Atomic Energy Agency, ISBN 92-0-113804-0, Vienna, Austria.

Genchev, N.P.; Milcheva-Dimitrova, R.Y. & Kozhuharova, M.V. (2007). Use of Gamma Radiation for Suppression of the Hemocytic Immune Response in Larvae of *Galleria mellonella* (Lepidoptera) against *Venturia canescens* (Hymenoptera). *Journal of Balkan Ecology*, vol.10, pp.411-419, ISSN 1311-0527

Genchev, N.P.; Balevski, N.; Obretenchev, D.A & Obretencheva, A.D. (2008). Stimulation Effects of Low Doses of Gamma Radiation on Adults *Habrobracon hebetor* Say (Hymenoptera: Braconidae). *Journal of Balkan Ecology*, vol.11, pp.99-102, ISSN 1311-0527

Gordon, H. T. (1972). Interpretation of insect quantitative nutrition. In: *Insect and Mite Nutrition*, Rodriguez, J.G., (Ed.), 73-105, North Holland, ISBN 0444104372, Amsterdam.

Hagler, J. R. & Jackson, C. G. (2001). Methods of marking insects: current techniques and future prospects. *Annual Review of Entomology*, vol.46, pp.511-543, ISSN 0066-4170

Haish A. (1970). Some observations on decreased vitality of irradiated Mediterranean fruit fly. In: *Sterile-Male-Technique for Control of Fruit Flies*, IAEA, (Ed.), 71-76, STI/PUB/276, IAEA, Vienna, Austria.

Hallman, G. J. (2000). Expanding radiation quarantine treatments beyond fruit flies. *Agricultural and Forest Entomology*, vol.2, pp.85-95, ISSN 1461-9563

Hallman, G.J. & Loaharanu, P. (2002). Generic radiation quarantine treatments against fruit flies (Diptera: Tephritidae) proposed. *Journal of Economic Entomology*, vol.95, pp.893-901, ISSN 1938-291X

Hallman, G.J. (2004a). Ionizing irradiation quarantine treatment against oriental fruit moth (Lepidoptera: Tortricidae) in ambient and hypoxic atmospheres. *Journal of Economic Entomology*, v.97, pp.824–827, ISSN 1938-291X

Hallman, G. J. (2004b). Irradiation disinfestation of apple maggot (Diptera: Tephritidae) in hypoxic and low temperature storage. *Journal of Economic Entomology*, vol.97, pp.1245-1248, ISSN 1938-291X

Hallman, G.J. & Phillips, T.W. (2008). Ionizing irradiation of adults of Angoumois grain moth (Lepidoptera: Gelechiidae) and Indianmeal moth (Lepidoptera: Pyralidae) to prevent reproduction and implications for a generic irradiation treatment for insects. *Journal of Economic Entomology*, v.101, pp.1051–1056, ISSN 1938-291X

Hallman, G.J. & Hellmich, R.L. (2009). Ionizing radiation as a phytosanitary treatment against European corn borer (Lepidoptera: Crambidae) in ambient, low oxygen, and cold conditions. *Journal of Economic Entomology*, vol.102, pp.64-68, ISSN 1938-291X

Hallman, G.J. & Hellmich, R.L. (2010). Modified atmosphere storage may reduce efficacy of irradiation phytosanitary treatments. *Acta Horticulturae*, vol.857, pp.159-162, ISSN 0567-7572

Hallman, G.J.; Levang-Brilz, N.M.; Zettler, J.L. & Winborne, I.C. (2010). Factors Affecting Ionizing Radiation Phytosanitary Treatments, and Implications for Research and Generic Treatments. . *Journal of Economic Entomology*, v.103, pp.1950–1963, ISSN 1938-291X

Hamed, M.; Nadeem, S. & Riaz, A. (2009). Use of Gamma Radiation for Improving the Mass Production of *Trichogramma chilonis* and *Chrysoperla carnea*. *Biocontrol Science and Technology*, v.19, S1, pp.43-48, ISSN 1360-0478

Hamilton, M.A. (1935). Further experiments on the artificial feeding of *Myzus persicae* (Sulz.). *Annals of Applied Biology*, vol.22, pp.243-258, ISSN 1744-7348.

Harrison, F.L. & Anderson, S.L. (1996). Taxonomic and development aspects of radiosensitivity, *Proceedings of the Symposium: Ionizing Radiation*, B. Amiro; R. Avadhanula; G. Johansson; C. M. Larsson & M. Luning, (Eds.), pp. 65-88, the Swedish Radiation Protection Institute (SSI) and The Atomic Energy Control Board (AECB) of Canada, Stockholm, Sweden, 20-24 May, 1996.

Hasset, C.C. & Jenkins, D.W. (1949). Production of radioactive mosquitoes. *Science*, vol.110, pp.109-110, ISSN 1095-9203.

Hasset, C.C. & Jenkins, D.W. (1951). The Uptake and Effect of Radiophosphorus in Mosquitoes. *Physiological Zoology*, v.24, pp.257, ISSN 0031-935X

Hayes, W.J. (1968). Toxicological aspects of chemosterilants. In: *Principles of Insect Chemosterilisation*, G.C. LaBrecque & C.N. Smith, (Eds.), 315-347, Appleton Century Crofts, ISBN , New York, USA.

Haynes, J. W. & Smith, J.W. (1992). Competitiveness of boll weevils (Coleoptera: Curculionidae) sterilized with low doses of fractionated irradiation. *Journal of Entomological Science*, vol.27, pp.421-426, ISSN 0749-8004

Hendrichs, J.; Bloem K.; Hoch, G.; Carpenter, J.E.; Greany, P. & Robinson, A. (2009). Improving the cost-effectiveness, trade and safety of biological control for agricultural insect pests using nuclear techniques. *Biocontrol Science and Technology*, v. 19, S1, pp.3-22, ISSN 1360-0478

Helinski, M.E.H.; Hood-Nowotny, R.; Mayr, L. & Knols, B.G.J. (2007). Stable isotope-mass spectrometric determination of semen transfer in malaria mosquitoes. *The Journal of Experimental Biology*, v.210, pp.1266–1274, ISSN 0022-0949

Henneberry, T. J.; Smith, F. F. & McGovern, W. L. (1964). Some Effects of Gamma Radiation and a Chemosterilant on the Mexican Bean Beetle. *Journal of Economic Entomology*, vol.57, n°.6, pp.813-815, ISSN 1938-291X

Hershey, A.E.; Pastor, J.; Peterson, B.J. & Klang, G.W. (1993). Stable isotopes resolve the drift paradox for *Baetis* mayflies in an Arctic river. *Ecology*, v.74, pp.2315–2325, ISSN 0012-9658

Hoch, G.; Marktl, R.C. & Schopf, A.(2009). Gamma Radiation-induced Pseudoparasitization as a Tool to Study Interactions between Host Insects and Parasitoids in the System *Lymantria dispar* (Lep., Lymantriidae)-*Glyptapanteles liparidis* (Hym., Braconidae). *Biocontrol Science and Technology*, vol.19, S1, pp.23-34, ISSN 1360-0478

Hood-Nowotny, R.; Mayr, L. & Knols, B.G.J. (2006). Use of carbon-13 as a population marker for *Anopheles arabiensis* in a sterile insect technique (SIT) context. *Malaria Journal*, v.5, pp.6-13, ISSN 1475-2875

Hood-Nowotny, R.; Mayr, L.; Islam, A.; Robinson, A. & Caceres, C. (2009). Routine Isotope Marking for the Mediterranean Fruit Fly (Diptera: Tephritidae). *Journal of Economic Entomology*, v.102, n.3, pp.941-947, ISSN 1938-291X

Hooper, G.H.S. (1989). The effect of ionizing radiation on reproduction. In: *Fruit Flies, their biology, natural enemies and control*, A.S. Robinson & G. Hooper, (Eds.), 153–164, World Crop Pests, v.3A, Elsevier, ISBN 0-444-42750-3, Amsterdam, The Netherlands.

[IAEA] International Atomic Energy Agency. (2002). *Natural and induced radioactivity in food*. IAEA, ISSN 1011–4289, Vienna, Austria.

[IAEA] International Atomic Energy Agency (2009). *Manual for the Use of Stable Isotopes in Entomology*. International Atomic Energy Agency, ISBN 978-92-0-102209-7, Vienna, Austria.

IDIDAS. (2010). *International Database on Insect Disinfestation and Sterilization*. International Atomic Energy Agency, Vienna, Austria. Available from: http://www-infocris.iaea.org/IDIDAS/start.htm).

Jacquet, P. & A. Leonard (1983). Studies on the differential radiosensitivity of the male and female pronuclei in the mouse zygote, C4-08. In: *Proceedings, Congress: Radiation Research, Somatic and Genetic effects. The 7th International Congress of Radiation*

Research, J.J. Broerse; G.W. Barendsen & A.J. Van der Kogel (Eds.), Amsterdam, The Netherlands, July 3-8, 1983.

James, M.T. (1947). The flies that cause myiasis in man. *U.S. Department of Agriculture Miscellaneous Publication,* n° 631, pp.1-175.

Jensen, J.A. & Fay, R.W. (1951). Tagging of Adult House Flies and Flesh Flies with Radioactive Phosphorus. *American Journal of Tropical Medicine and Hygiene,* v.31, pp.523-530, ISSN 0002-9637.

Jessup, A.J.; Rigney, C.J.; Millar, A.; Sloggett, R.F. & Quinn, N.M. (1992). Gamma irradiation as a commodity treatment against the Queensland fruit fly in fresh fruit. In: *Use of irradiation as a quarantine treatment of food and agricultural commodities.* International Atomic Energy Agency, pp. 13-42, Vienna, Austria.

Jordao-Paranhos, B.A.; Walder, J.M.M. & Papadopoulos, N.T. (2003). A Simple Method to Study Parasitism and Field Biology of the Parasitoid *Diachasmimorpha longicaudata* (Hymenoptera: Braconidae) on *Ceratitis capitata* (Diptera: Tephritidae). *Biocontrol Science and Technology,* vol.13, pp.631-639, ISSN 1360-0478

Kasting, R. & McGinnis, J. (1965). Measuring consumption of food by an insect with carbon-14 labeled compounds. *Journal of Insect Physiology,* vol. 11, pp.1253-1260, ISSN 0022-1910

Klassen, W. (2005). Area-wide integrated pest management and the sterile insect technique, In: *Sterile Insect Technique. Principles and Practice in Area-wide Integrated Pest Management,* V.A. Dyck, J. Hendrichs & A.S. Robinson (Eds.), 39-68, Springer, ISBN 1-4020-4050-4, Dordrecht, The Netherlands.

Klassen, W. & Earle, N.W. (1970). Permanent sterility induced in boll weevils with busulfan without reducing production of pheromone. *Journal of Economic Entomology,* vol.63, pp.1195–1198, ISSN 1938-291X

Klassen, W. & Curtis, C.F. (2005). History of the sterile insect technique. In: *Sterile insect technique – principles and practice in area-wide integrated pest management,* V.A. Dyck, J. Hendrichsc& A.S. Robinson (Eds.), 3–38, Springer, ISBN1-4020-4050-4, Dordrecht, The Netherlands.

Klein, I. & Kogan, M. (1974). Analysis of food intake, utilization and growth in phytophagous insects a computer program. *Annals of the Entomological Society of America,* vol.67, pp.295-297, ISSN 0013-8746

Kettlewell, H.B.D. (1952). Use of radioactive tracer in the study of insect populations (Lepidoptera). *Nature,* vol.170, pp.584-585, ISSN 0028-0836

Knipling E.F. 1955. Possibilities of insect control or eradication through the use of sexually sterile males. *Journal of Economic Entomology,* vol.48, n°4, pp.459-462, ISSN 1938-291X

Knipling, E.F. (1992). *Principles of Insect Parasitism Analyzed from New Perspectives: Practical Implications for Regulating Insect Populations by Biological Means,* USDA Agriculture Handbook No. 693, ISBN 9780160378140.

Kowal, N. E. & Crossley, D. A. (1971). The ingestion rates of microarthropods in pine more, estimated with radioactive calcium. *Ecology,* vol. 52, pp. 444-452, ISSN 0012-9658.

Kumano, N.; Kuriwada, T.; Shiromoto, K.; Haraguchi, D. & Kohama, T. (2010). Evaluation of partial sterility in mating performance and reproduction of the West Indian sweetpotato weevil, *Euscepes postfasciatus. Entomologia Experimentalis et Applicata,* vol.136, n°1, pp.45–52, ISSN 1570-7458

Kuswadi, A.N.; Himawan, T.; Indarwatmi, M. & Nasution, I.A. (2003). The Use of Gamma Irradiation to Support the Colonization and Production of Natural Enemies of *Bactrocera carambolae* (Drew & Hancock). Report of 3rd Research Coordination Meeting, IAEA, Vienna, Austria.

LaChance, L. E. & Leverich, A.P. (1962). Radiosensitivity of developing reproductive cells in female *Cochliomyia hominivorax*. *Genetics*, vol.47, pp.721-735, ISSN 1943-2361

LaChance, L.E. & Bruns, S.B. (1963). Oogenesis and radiosensitivity in *Cochliomyia hominivorax* (Diptera: Calliphoridae). *Biology Bull*etin, vol.124, pp.65-83, ISSN 1608-3059

LaChance, L. E.; Schmidt, C.H. & Bushland, R.C. (1967). Radiation-induced sterilization, In: *Pest control: biological, physical, and selected chemical methods,* W.W. Kilgore & R. L. Doutt (Eds.), 147-196, Academic, New York, USA.

Lapidot, M.; Saveanu, S.; Padova, R. & Ross, I. (1991). Insect disinfestation by irradiation, In: *Insect Disinfestation of Food and Agricultural Products by Irradiation,* IAEA (Ed.), 93–103, ISBN 9789201111913, Vienna, Austria.

Lawson, F.R.; Lucas, G.B. & Hall, N.S. (1954). Translocation of radioactive phosphorus injected by the green peach aphid into tobacco plants. *Journal of Economic Entomology*, vol.47, pp.749-752, ISSN 1938-291X

Lindquist, A.W. (1955). The use of gamma radiation for control or eradication of the screwworm. *Journal of Economic Entomology*, vol.48, pp.467-469, ISSN 1938-291X

Long, W.H. & Lilly, J.H. (1958). Wireworm behavior in response to chemical seed treatment. *Journal of Economic Entomology*, vol.51, pp.291-295, ISSN 1938-291X

Louda, S.M.; Pemberton, R.W.; Johnson, M.T. & Follett, P.A. (2003). Non-target Effects: the Achilles' Heel of Biological Control?. *Annual Review of Entomology*, vol.48, pp.365-396, ISSN 0066-4170

Loosjes, M. (2000). The sterile insect technique for commercial control of the onion fly. In: *Area-wide Control of Fruit Flies and Other Insect Pests. Proc. Int. Conf. on Area-wide Control of Insect Pests and the Fifth Int. Symp. on Fruit Flies of Economic Importance,* K.H. Tan (Ed.), pp. 181-184 , Penerbit Universiti Sains, ISBN 9838611956, Pulau Pinang, Malaysia.

[MAF] Ministry of Agriculture and Fisheries, New Zealand. (2009). Import health standard commodity sub-class: fresh fruit/vegetables mango, *Mangifera indica* from Australia. (http://www.biosecurity.govt.nz/Ρles/ihs/mango-au.pdf).

Mansour, M. (2003). Gamma irradiation as a quarantine treatment for apples infested by codling moth (Lep., Tortricidae). *Journal of Applied Entomology*, vol.127, pp.137-141, ISSN 1439-0418

Markow, T.A.; Anwar, S. & Pfeiler, E. (2000). Stable isotope ratios of carbon and nitrogen in natural populations of *Drosophila* species and their hosts. *Functional Ecology*, v.14, pp.261–266, ISSN 1365-2435

Mastrangelo, T.; Parker, A.G ; Jessup, A. ; Pereira, R.; Orozco-Dávilla, D.; Islam, A.; Dammalage, T. & Walder, J. M. (2010). A new generation of X ray irradiators for insect sterilization. *Journal of Economic Entomology*, v. 103, pp. 85-94, ISSN 1938-291X

McEnroe, W.D. (1961). The control of water loss by the two-spotted spider mite (*Tetranychus telarius*). *Annals of the Entomological Society of America*, v.54, pp.883-886, ISSN 1938-2901

McKibben, G. H.; Villavaso, E.J.; McGovern, W.L. & Grefenstette, W.J. (2001). United States Department of Agriculture - research support, methods development and program implementation. In: *Boll weevil eradication in the United States through 1999*, W.A. Dickerson; A.L. Brashear; J.T. Brumley; F.L. Carter; W.J. Grefenstette & F.A. Harris, (Eds.), 101-136, The Cotton Foundation, ISBN 9780939809066, Memphis, TN, USA.

McNeill, A.M.; Zhu, C. & Fillery, I.R.P. (1998). A new approach to quantifying the N benefit from pasture legumes to succeeding wheat. *Australian Journal of Agricultural Research*, v.49, pp.427–436, ISSN 0004-9409

Melvin, R. & Bushland, R.C. (1936). A method of rearing *Cochliomyia americana* C. and *P.* on artificial media. *Bureau of Entomology and Plant Quarantine (Report)* ET-88. Washington, D.C., USDA.

Moeri, O.; Cuda, J.P.; Overholt, W.A.; Bloem, S. & Carpenter, E. (2009). F_1 Sterile Insect Technique: A Novel Approach for Risk Assessment of *Episimus unguiculus* (Lepidoptera: Tortricidae), a Candidate Biological Control Agent of *Schinus terebinthifolius* in the continental USA. *Biocontrol Science and Technology*, vol.19, S1, pp. 303-315, ISSN 1360-0478

Moore, S. T.; Schuster; M. F. & Harris; F. A. (1974). Radioisotope Technique for Estimating Lady Beetle Consumption of Tobacco Budworm Eggs and Larvae. *Journal of Economic Entomology*, vol. 67, pp. 703-709, ISSN 1938-291X

Muller, H.J. (1927). Artificial transmutation of the gene. *Science*, v. 66, pp. 84-87, ISSN 1095-9203.

Nakashima, Y.; Hirose, Y. & Kinjo, K. (1996). Rearing *Orius sauteri* (Poppius) on Diet of Freeze-dried Larval Powder of Melon Fly, *Bactrocera cucurbitae* Coquillett. *Japanese Journal of Applied Entomology and Zoology*, vol.40, pp.80-82, ISSN 1347-6068

Nienstedt, K.M. & Poehling, H.M. (2000). [15]Nmarked aphids for predation studies under field conditions. *Entomologia Experimentalis et Applicata*, v.94, pp.319–323, ISSN 1570-7458

Nienstedt, K.M. & Poehling, H.M. (2004). Prey to predator transfer of enriched [15]Ncontents: basic laboratory data for predation studies using ^{15}N as marker. *Entomologia Experimentalis et Applicata*, v.112, pp. 183–190, ISSN 1570-7458

Novotny, J. & Zubrik, M. (2003). Sterile Insect Technique as a Tool for Increasing the Efficacy of Gypsy Moth Biocontrol, *Proceedings: Ecology, Survey and Management of Forest Insects*, M.L. McManus & A.M. Liebhold, (Eds.), pp. 80-86, Krakow, Poland, , USDA Forest Service, Northeastern Research Station, General Technical Report NE-311, USA, September 1-5 2002.

O'Brien, R.D. & Wolfe, L.S. (1964). *Radiation, radioactivity and insects*. Academic Press, ISBN 9780125239509, New York, USA.

Odum, E. P. (1963). Experimental Isolation of Food Chains in an Old-Field Ecosystem with the Uses of Phosphorus-32, *Proceedings of the First International Symposium on Radioecology*, V. Schultz & A. Klement Jr. (Eds.), pp.113-120, Reinhold Publishing, NY.

Oertel, E.; Emerson, R.B. & Wheeler, H.E. (1953). Transfer of radioactivity from worker to drone honey bees after ingestion of radioactive sucrose. *Annals of the Entomological Society of America*, v.46, pp.596-598, ISSN 1938-2901.

Parish, H. E. (1942). Factors Predisposing Animals to Screwworm Infestation in Texas. *Journal of Economic Entomology*, vol.35, No. 6, pp. 899-903, ISSN 1938-291X

Proverbs, M. D. (1962). Progress on the use of induced sexual sterility for the control of the codling moth *Carpocapsa pomonella* (L.) (Lepidoptera: Olethreutidae). *Proceedings of the Entomological Society of Ontario*, vol.92, pp. 5-11, ISSN 0071-0768

Proverbs, M. D. (1969). Induced sterilization and control of insects. *Annual Review of Entomology*, vol.14, pp.81-102, ISSN 0066-4170

Quan, S.F.; Hartwell, W.V.; Scott, K.G. & Peng, C.T. (1957). Cerium[44] as a tag for arthropods of medical importance. *Transactions of the Royal Society of Tropical Medicine and Hygiene*, vol.51, pp.87-88, ISSN 00359203.

Quarterman, K.D.; Mathis, W. & Kilpatrick, J.W. (1954). Urban fly dispersal in the area of Savannah, Georgia. *Journal of Economic Entomology*, vol.47, pp.405-412, ISSN 1938-291X

Quarterman, K.D.; Jensen, J.A.; Mathis, W. & Smith, W.W. (1955). Flight dispersal of rice field mosquitoes in Arkansas. *Journal of Economic Entomology*, vol.48, pp.30-32, ISSN 1938-291X

Rabb, R.L.; Stinner, R.E. & van den Bosch, R. (1976). Conservation and augmentation of natural enemies. In: *Theory and Practice of Biological Control*, C.B. Huffaker & P.S. Messenger, (Eds.), 233–254, Academic Press, ISBN 0-12-360350-1, New York, NY, USA.

Radeleff, R.D.; Bushland, R.C. & Hopkins, D.E. (1952). Phosphorus-32 labelling of the screwworm fly. *Journal of Economic Entomology*, v.45, pp.509-514, ISSN 1938-291X

Rapport D. J. & Turner J.E. (1975). Feeding rates and population growth. *Ecology 56*, 942-949, ISSN 0012-9658

Ravera, O. (1967). The effect of x-ray on the demographic characteristics of *Physa acuta* (Gasteropoda: basommatophora). *Malacologia*, v.5, pp.95-109, ISSN 0076-2997

Reichle, D.E. (1967). Radioisotope Turnover and Energy Flow in Terrestial Isopod Populations. *Ecology*, vol. 48, pp.349-366, ISSN 0012-9658.

Rice, T.R. & Baptist, J.P.(1974). Ecologic effects of radioactive emissions from nuclear power plants. In: *Human and ecological effects of nuclear power plants*, L. A. Sagan (Ed.), 373-439, Charles C. Thomas, ISBN 0398029296, Springfield, IL, USA.

Robinson, A.S. (2005). Genetic basis of the Sterile Insect Technique, In: *Sterile insect technique: principles and practice in area-wide integrated pest management*, V.A. Dyck; J. Hendrichs & A.S. Robinson, (Eds.), 95-114, Springer, ISBN 9781402040504, Berlin, Germany.

Rull, J.; Reyes, J. & Enkerlin, W. (1996). The Mexican fruit fly eradication campaign: Largest fruit fly industrial complex in the world, In: *Fruit fly pests: a world assessment of their biology and management*, B.A. McPheron & G. Steck, (Eds.), 561-563, St. Lucie Press, ISBN 978-1574440140, Delray Beach, Florida.

Runner, G.A. (1916). Effect of Roentgen rays on the tobacco, or cigarette, beetle, and the results of experiments with a new form of Roentgen tube. *Journal of Agricultural Research*, vol. 6, No. 2, pp. 383-388, ISSN 0095-9758

Saour, G. (2009). Effect of Early Oviposition Experience on Host Acceptance in *Trichogramma* (Hymenoptera: Trichogrammatidae) and Application of F_1 Sterility and *T. principium* to Suppress the Potato Tuber Moth (Lepidoptera: Gelechiidae). *Biocontrol Science and Technology*, v. 19, S1, pp. 225-234, ISSN 1360-0478

Schoof, H.F. & Siverly, R.E. (1954). The occurrence of movement of *Periplaneta americana* (L.) within an urban sewage system. *American Journal of Tropical Medicine and Hygiene*, vol.3, pp.367-371, ISSN 0002-9637

Smith, B. C. (1965). Effects of Food on the Longevity, Fecundity, and Development of Adult Coccinellids (Coleoptera: Coccinellidae). *Canadian Entomologist 97*, 910-919, ISSN 0008-347X

Sparrow, A. H.; Schairer, L.A. & Sparrow, R.C. (1963). Relationships between nuclear volumes, chromosome numbers and relative radiosensitivities. *Science*, vol.141, pp.163- 166, ISSN 1095-9203

Steffan, S.A.; Daane, K.M. & Mahr, D.L. (2001). ^{15}N-enrichment of plant tissue to mark phytophagous insects, associated parasitoids, and flower visiting entomophaga. *Entomologia Experimentalis et Applicata*, vol.98, pp. 173–180, ISSN 1570-7458

Strong, F. E. & Landes, D. A. (1965). Feeding and Nutrition of *Lygus Hesperus* (Hemiptera: Miridae) II. An estimation of normal feeding rates. *Annals of the Entomological Society of America*, vol. 58, pp. 309-314, ISSN 0013-8746

[USDA-APHIS] U.S. Department of Agriculture-Animal and Plant Health Inspection Service. (2006). *Treatments for fruits and vegetables.* USDA-APHIS, Federal Register 71, pp.4451–4464.

[US FDA] United States Food and Drug Administration. (2004). *Irradiation in the production, processing and handling of food: final rule.* US FDA, Fed. Reg. 69, pp.76844–76847.

Van Dinther, J.B.M. & Mensink, F.T. (1971). *Meded. Fac. Landbouwwet. Rijksuniv. Gent.* 36, 283-293, ISSN 0368-9697

Van Hook, R. I.; Reichle, D. E. & Auerbach, S.I. (1970). ORNL-4509, *UC-48-Biology and Medicine* (Oak Ridge National Laboratory).

Van Lenteren, J.C. (2003). *Quality Control and Production of Biological Control Agents: Theory and Testing Procedures*, CABI Publishing, ISBN 0851996884, Wallingford, UK.

Vargas-Terán, M.; Hofmann, H.C. & Tweddle, N.E.(2005). Impact of screwworm eradication programmes using the sterile insect technique. In: *Sterile Insect Technique, Principles and Practice in Area-Wide Integrated Pest Management*, V. A. Dick, J. Hendrichs & A. S. Robinson (Eds.), 629–650, Springer, ISBN 978-1402040504, The Netherlands.

Vey, A. & Causse, R. (1979). Effect de l'Exposition aux Rayons Gamma sur la Re´action Haemocytaire Multicellulare des Larves de *Mamestra brassica* (Lep.: Noctuidae). *Entomophaga*, vol.24, pp.41-47, ISSN 0013-8959

Von Sonntag, C. (1987). *The chemical basis of radiation biology.* Taylor & Francis, ISBN 9780850663754, London, United Kingdom.

Waldbauer G. P. (1968). The consumption and utilization of food by insects. *Advances in Insect Physiololy*, vol.5, pp.229-288, ISSN 0065-2806

Wang, E.; Lu, D.; Liu, X. & Li., Y. (2009). Evaluating the Use of Nuclear Techniques for Colonization and Production of *Trichogramma chilonis* in Combination with Releasing Irradiated Moths for Control of Cotton Bollworm, *Helicoverpa armigera*. *Biocontrol Science and Technology*, vol. 19, S1, pp. 235-242, ISSN 1360-0478

Williams, E. C. & Reichle, D.E. (1968). Radioactive tracers in the study of energy turnover by a grazing insect (*Chrysochus auratus* Fab.; Coleoptera. Chrysomelidae). *Oikos*, vol. 19, pp. 10-18, ISSN 0030-1299

Whicker, F. W. & Schultz, V. (1982). *Radioecology: nuclear energy and the environment.* CRC, ISBN 978-0849353543, Boca Raton, FL, USA.

Wyss, J. H. (2000). Screw-worm eradication in the Americas— overview, In: *Are-awide Control of Fruit Flies and Other Insect Pests*, K.H. Tan, (Ed.), 79-86, Penerbit Universiti Sains Malaysia, ISBN 9838611956, Pulau Pinang, Malaysia.

Zapater, M.C.; Andiarena, C.E; Pérez-Camargo, G. & Bartoloni, N. (2009). Use of Irradiated
    Musca domestica Pupae to Optimize Mass Rearing and Commercial Shipment of the
    Parasitoid *Spalangia endius* (Hymenoptera: Pteromalidae). *Biocontrol Science and
    Technology*, vol.19, S1, pp. 261-270, ISSN 1360-0478

3

Radioisotopes in Drug Research and Development: Focus on Positron Emission Tomography

Sosuke Miyoshi[1], Keisuke Mitsuoka[1],
Shintaro Nishimura[1] and Stephan A. Veltkamp[2]
[1]Bioimaging Research Laboratories, Drug Discovery Research,
Astellas Pharma Inc., Tsukuba,
[2]Global Clinical Pharmacology & Exploratory Development,
Astellas Pharma Europe BV, Leiderdorp,
[1]Japan
[2]Netherlands

1. Introduction

The use of radioisotopes is important in pharmaceutical research and development (R&D). They are frequently used in non-clinical and clinical studies for the development of compounds for different therapeutic areas, such as central nervous system (CNS) diseases (e.g. Dementia, Alzheimer's disease (AD), and Parkinson's disease), oncology, and metabolic diseases (e.g. diabetes mellitus).

Pharmaceutical companies invest a lot of time and money in research on new treatment strategies for diseases with a high medical need, such as oncology and metabolic diseases. A large amount of drugs fails during development due to toxicity and/or the lack of efficacy (Kola, I, 2008). Several attempts are being made to improve this, such as obtaining a better understanding of the pathophysiology of diseases, development of robust animal models, the application of biomarkers, development of pharmacokinetic (PK) - pharmacodynamic (PD) models, and the application of non-invasive techniques such as positron emission tomography (PET) in an early stage of development.

2. Application of radiolabeled compounds

A variety of radioisotopes is used in the R&D of drugs, such as ^{3}H, ^{14}C, ^{32}P, ^{35}S and ^{131}I (Penner et al, 2009). Carbon-14 is the isotope of choice in most of the ADME studies. The labeled part of the drug molecule should not be lost in metabolite formation. Tritium-labeled drugs are also used commonly, but have the risk of tritium exchange. Data from non-clinical studies performed with radiolabeled drugs can provide information to make the choice of the radioisotope and its position in the drug molecule. The tritium- or carbon-14-labeled drug at the metabolically stable position should have a radiochemical purity of ≥98%, and in special cases ≥95% may be acceptable. stability of the radiolabeled drug under dosing conditions should be checked at predose and postdose.

The advantage of using radiolabeled drugs is that the radioactivity can easily be detected and quantified using liquid scintillation techniques and disposition of drugs can be assessed. The choice of the radioisotope, the position of the radiolabel in the drug compound, radiochemical purity, and the specific activity are important parameters. These parameters can have an effect on the metabolic, chemical, and radiochemical stability of the drug, metabolite formation and detection, and the recovery of radioactivity.

2.1 Non-clinical studies

Radioactive labeled compounds are used in non-clinical studies to assess drug absorption, distribution, metabolism, and excretion (ADME). They can provide information on a) drug absorption in cell lines, b) P-glycoprotein (P-gp) transport or inhibition, c) metabolite profiling, d) drug transport (uptake, efflux), and e) drug binding to receptors (Marathe et al., 2004).

Other applications of radiolabeled compounds are to investigate apoptosis (Glaser et al., 2003), DNA replication and cell cycle progression (Hoy et al., 1990) in oncology. In the development of gemcitabine, a pyrimidine nucleoside anticancer drug, radiotracers were used to better understand the pharmacology and toxicology of this compound (Heinemann et al., 1988; Mackey et al., 1988). With the use of radiolabeled [^{3}H]-gemcitabine, its synthesized metabolite [^{3}H]-dFdU, and [^{3}H]-thymidine, novel active metabolites of gemcitabine were revealed, which turned out to be incorporated into DNA, formed *in vivo* in mice and accumulated into the liver following multiple oral dosing (Veltkamp et al.,2008a, 2008b). These findings gave new insights in the metabolism, pharmacology and toxicology of the drug.

2.2 Clinical studies

Administration of radiolabeled compounds to humans is not generally done in the first-in-man study. Considering the amount and type of information that human ADME studies can provide, it should be considered to conduct these studies early during clinical development. Possibilities are just after or in parallel with the multiple ascending dose (MAD) study. Radiolabeled drugs are used in a variety of clinical studies, such as in mass balance (MB), regional drug absorption, and microdose studies.

Mass balance studies provide valuable information on the absorption, metabolism, and elimination of a drug. Identification and quantification of metabolites is important in MB studies (Penner et al., 2009).

Regional drug absorption studies can be performed to determine the PK of a drug from a new modified release (MR) compared to a more conventional immediate release (IR) formulation and to measure absorption from specific areas of the gastrointestinal (GI) tract, such as the distal small bowel, ascending colon, and transverse colon. In this way, it can be determined whether the drug is sufficiently absorbed from the GI tract and together with other PK and its safety it can be determined whether development of a MR formulation would be beneficial.

Microdose studies can give important information on the distribution and metabolism of a new drug. It can also help to reduce the number of animals used in non-clinical studies (Combesa et al., 2003). There has been a growing interest in the safety of drug metabolites, particularly those not produced in experimental animals. In order to avoid unpredictable toxicity caused by such metabolites, the US Food and Drug Administration (FDA) issued the Guidance for Safety Testing of Drug Metabolites (MIST) in 2008 (FDA, 2008). It recommends that, before commencing phase III trials of a drug, a safety report must be prepared relating to drug metabolites that have a systemic exposure >10% of the parent drug and those that occur in significantly greater quantities in humans than in experimental animals.

Subsequently, the International Conference on Harmonisation, ICHM3 guidance, recommends safety assessment of metabolites whose exposure is >10% of the total exposure, not of the parent compound (ICHM3, 2009). This new guideline prompted the pharmaceutical industry to identify and quantify drug metabolites in humans in the early phases of drug development.

In microdose studies PK data are obtained after administration of a trace subpharmacologic quantity to human subjects (Lappin et al., 2006). The ultrasensitive analytic technique of accelerator mass spectrometry (AMS) has been used to quantify the low plasma concentrations anticipated after microdose administration. These studies do not provide information about the safety and tolerability of the drug. Requirements for microdose studies have been summarized by the EMEA, FDA and others (EMEA, 2003; FDA, 2006; Bergstrom et al., 2003; Marchetti et al., 2007).

3. PET imaging

3.1 Principles of PET

PET is a nuclear imaging technique used to map biological and physiological processes in living subjects following the administration of positron emitting radiopharmaceuticals. The technique is based on the detection of photons released by annihilation of positrons emitted by radioisotopes. Positron-emitting radionuclides are produced in a cyclotron by bombarding target material with accelerated protons. In the body, these radionuclides emit positrons that undergo annihilation with nearby electrons, resulting in the release of two photons. These so-called annihilation photons are detected by imaging and the resulting data can be used to reveal the distribution of the radiotracer in the body.

Unlike conventional imaging modalities, such as magnetic resonance imaging (MRI) or computed tomography (CT), which mainly provide detailed anatomical images, PET can measure biochemical and physiological aberrations that occur prior to macroscopic anatomical signs of a disease, such as cancer (Chen et al., 2011). Currently, many positron emitting isotopes are available with different characteristics (see Table 1). Fluorine-18 has the advantage of having a long half-life ($t_{1/2}$), which enables time-consuming radiosyntheses and imaging procedures.

Isotope	$t_{1/2}$
^{18}F	109.8 min
^{11}C	20.4 min
^{15}O	2.04 min
^{13}N	9.97 min
^{64}Cu	12.7 hours
^{68}Ga	68.1 min
^{124}I	4.2 days

Table 1. Positron emitters commonly used in PET studies.

3.2 Application of PET imaging

PET imaging has become an important tool in the process of drug development for a variety of compounds, such as for CNS targeted and anticancer drugs. PET imaging has the advantage of having high sensitivity and high specificity when using the appropriate PET probes. Therefore, it can be used for examination of the physiology of tissues and for evaluation of the distribution of a drug in specific organs or regions in a quantitive manner. PET imaging is used to examine the PK of drugs in tissues using a positron-labeled drug candidate or the PD (e.g. target expression, occupancy) using the radiolabeled drug itself or a different target-specific tracer. PET radiotracers used in R&D of drugs in the field of CNS, oncology and diabetes mellitus are listed in Table 2.

a.   CNS

PET tracer	Target	Purpose
[11C]-PIB (Rostomian et al., 2011)	Amyloid plaque	Diagnosis of Alzheimer's disease and efficacy on Aβ plaques
[18F]-AV45 (Wong et al., 2010)	Amyloid plaque	Diagnosis of Alzheimer's disease and efficacy measurement of Aβ plaques
[18F]-FK960 (Noda et al., 2003a)	Hippocampus (exact target not identified yet)	Dose setting Dementia
[11C]-SCH442416 (Mihara et al., 2008)	Adenosine A$_{2A}$R*	Determine receptor binding of a therapeutic drug Parkinson's disease
[18F]-FDG (Asai et al., 2009)	Glucose metabolism (rCMRglu**)	Diagnosis of Alzheimer's disease

*Adenosine A$_{2A}$R, adenosine A$_{2A}$ receptor; **rCMRglu, regional cerebral metabolic rate of glucose.

b.   Oncology

PET tracer	Target	Purpose
[18F]-FDG	Glucose metabolism	Tumor response/disease staging
[18F]-FLT	Cell proliferation	Tumor response
[11C]-Gly-Sar (Mitsuoka et al., 2008)	Peptide transport	Cancer detection (distinction between cancer and inflammatory tissue)
[11C]-methionine (Narayanan et al., 2002)	Amino acid metabolism	Tumor response (brain tumors)
[18F]-FMISO (Bruehlmeier et al., 2004)	Hypoxia	Diagnosis of hypoxic state of cancer
[18F]-FAZA (Piert et al., 2005)	Hypoxia	Diagnosis of hypoxic state of cancer

FDG; fluorodeoxyglucose; FLT; fluorothymidine; Gly-Sar, glycylsarcosine; FMT, fluoromethyltyrosine; FMISO, fluoromisonidazole; FAZA, fluoroazomycin-arabinofuranoside.

c.   Diabetes mellitus (see also sections 3.3.3 and 3.4.3)

PET tracer	Target	Purpose
[11C]-DTBZ	VMAT2 β-cells pancreas	Assessment of β-cell mass
[18F]-FP-(+)-DTBZ (AV133)	VMAT2 β-cells pancreas	Assessment of β-cell mass
[18F]-FP-(+)-epoxy-DTBZ	VMAT2 β-cells pancreas	Assessment of β-cell mass

DTBZ, dihydrotetrabenazine; VMAT2, vesicular monoamine transporter 2; FP-DTBZ, fluoropropyl-dihydrotetrabenazine.

Table 2. PET radiotracers in R&D of drugs in the field of CNS (a), oncology (b) and diabetes mellitus (c).

Amyloid plaque is a major feature of Alzheimer's disease (AD), for which several amyloid-imaging tracers have been developed (Kadir et al., 2010). Among these tracers, [11C]-PIB and [18F]-AV45 were examined as diagnostic agents for *in vivo* imaging of amyloid deposition in humans. According to the National Institute on Aging and Alzheimer's Association Lead Effort to Update Diagnostic Criteria for Alzheimer's Disease, biomarkers for AD have been developed and are being validated. These fall into several categories and include biomarkers for a) beta amyloid pathology, including amyloid PET imaging and levels of beta amyloid in cerebrospinal fluid (CSF), b) neuronal injury, including levels of CSF τ and phospho-τ, c) neuronal dysfunction, including decreased uptake of FDG on PET scans, and d) neurodegeneration, including brain atrophy on structural MRI scans.

PET imaging in oncology is used to assess tracer distribution and extent of uptake to identify the disease, for disease staging, and for monitoring therapeutic response. The most widely used PET tracer in oncology is [18F]-fluorodeoxyglucose ([18F]-FDG), a glucose analog that enters the cell after uptake by glucose transporter 1 (GLUT1). FDG is subsequently phosphorylated by hexokinase-II and accumulates in tumor cells, a mechanism also called metabolic trapping. In fact, increased glucose transport is associated with elevated glycolysis of the cancer cell and a corresponding increase in hexokinase activity. Tissues that metabolize gucose faster will accumulate more [18F]-FDG. Therefore, cancer cells can be differentiated from benign tissues by their increased metabolism. Correspondingly, this uptake can be semi-quantified on PET.

However, [18F]-FDG is not a target-specific tracer and it cannot differentiate between cells that have a high metabolic rate associated with neoplasia, and those for which the increased metabolic rate is associated with other etiologies, such as infection or inflammation. In addition, many malignancies do not exhibit high metabolic rates and, thus, are not properly diagnosed by [18F]-FDG (Chen et al., 2011).

A novel PET tracer, [11C]-glycylsarcosine ([11C]-Gly-Sar), targeted to H+/peptide transporters (PEPTs) was investigated for its specificity as compared to that of FDG in distinguishing between tumor and inflammatory tissues (Mitsuoka et al., 2008; Fig. 1). After i.v. administration of [11C]-Gly-Sar to mice, it was possible to visualize prostate, pancreatic and gastric tumor xenografts all expressing PEPTs. Accumulation of [11C]-Gly-Sar occured in kidneys and bladder, and was low in other tissues (e.g. brain, heart, lung, and liver), which had restricted functional expression of PEPTs. Accumulation of [18F]-FDG occured in brain, heart, kidneys and bladder. The detection of tumors in mice was improved using [11C]-Gly-Sar compared to [18F]-FDG (Fig. 1). Whereas 18F-FDG accumulated also in inflammatory tissue, uptake of [11C]-Gly-Sar was absent in inflammatory tissues. [11C]-Gly-

Sar is a promising tumor-imaging agent and appears superior to FDG for distinguishing between tumor and inflammatory tissue.

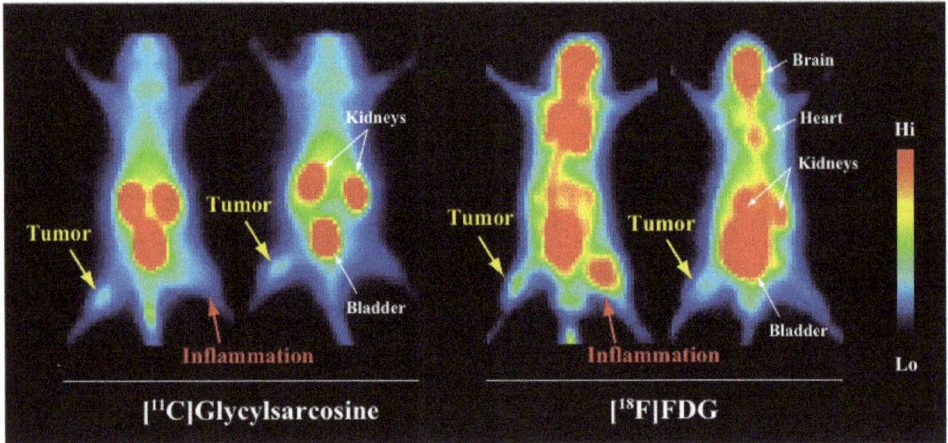

Fig. 1. Planer images of mice bearing human pancreatic tumors after injection of AsPC-1 pancreatic adenocarcinoma cells in their right hind limb. [11C]-Gly-Sar (~ 5 MBq) or [18F]-FDG (~ 5 MBq) were administered i.v. Turpentine oil was administered subcutaneously into the left hind limb of control mice (left mice for [11C]-Gly-Sar and [18F]-FDG) to induce inflammation. Yellow arrowhead, tumor; red arrowhead, inflammation.

Another promising radiopharmaceutical is 3'-deoxy-3'-[18F]-fluoro-L-thymidine (FLT), a marker for cell proliferation. Accumulation of FLT in tumor cells has been shown to be dependent on the activity of cellular thymidine kinase-1 (TK1). TK1 is the key enzyme and limiting step in the pyrimidine salvage pathway of DNA synthesis and is overexpressed in most tumor types. Intracellarly, FLT is phosporylated to FLT-monophosphate, but unlike thymidine, FLT is not incorporated into DNA as it lacks the 3'-hydroxyl group. As TK1 is functional only in the late G_1- and S-phase of the cell cycle, FLT uptake closely correlates with the amount of proliferating cells. L-[methyl-11C]-methionine, or [11C]-methionine, is a PET tracer for amino acid metabolism used in neuro-oncology, which makes it possible to assess the characteristics of lesions in the brain, impossible with FDG-PET or MRI. The uptake of [11C]-methionine provides information on the malignancy of lesions. Sodium-independent amino acid transporters (LATs), which mediate transport of large branched and aromatic amino acids, has attracted special interest, because system L is commonly upregulated in many tumors and correlates with tumor growth and prognosis. Other amino acid tracers have been developed to detect cancers. On the other hand, the prevalence of hypoxic areas is a characteristic feature of locally advanced solid tumours and has been described in a wide range of human malignancies. Evidence from experimental and clinical studies point to a role for tumor hypoxia in tumor propagation, resistance to therapy and malignant progression. To monitor the hypoxia, nitroimidazole compounds such as [18F]-fluoromisonidazole (FMISO)or [18F]-fluoroazomycin-arabinofuranoside (FAZA) have been described. These compounds are degraded into reactive intermediate metabolites by intracellular reductases in a process which is directly related to the level of oxygenation/hypoxia. This causes a gradient which is favorable for detection of hypoxic cells. Subsequently, these metabolites covalently bind to thiol groups of intracellular proteins and thereby accumulate within viable hypoxic cells.

PET images from orthotopically implanted pulmonary human tumor xenografts in mice are shown in Figure 2. The micro-PET images were coupled to micro-CT images in the lung. By using different radiotracers, these PET images provided information on the tumor condition, such as on glucose metabolism ([18F]-FDG), cell proliferation ([18F]-FLT], and hypoxia [18F]-FMISO]. Micro-PET/CT imaging could be a robust surrogate biomarker for antitumor activity in preclinical studies.

Assessment of treatment response is essential for disease management. Anatomic imaging alone using the Response Evaluation Criteria in Solid Tumors (RECIST) does not reflect physiological changes of the tumors, and therefore still has limitations in response evaluation. Functional imaging is very useful in the evaluation of the efficacy of novel anticancer drugs.

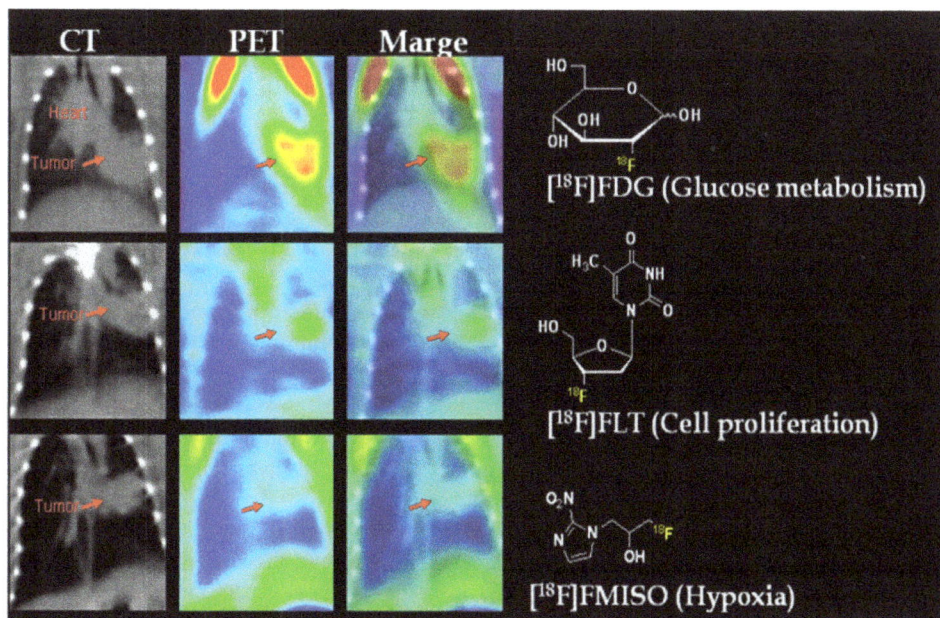

Fig. 2. PET, CT, and combined PET/CT-images of orthotopic human lung tumor xenografts in mice. Mice were administrated via the tail vein 15 MBq of [18F]-FDG for detection of glucose metabolism (top row), [18F]-FLT for detection of cell proliferation (middle row) and [18F]-FMISO for detection of hypoxia (bottom row). PET images were acquired for 50 to 60 min after tracer administration using three-dimensional ordered subset expectation maximization (3D-OSEM) reconstruction.

3.3 Examples of PET in non-clinical studies

Several examples of the application of PET in non-clinical studies for the development of CNS drugs, anticancer drugs and antidiabetic drugs is described in this Section.

3.3.1 Determination of the PK and PD of CNS drugs

The brain anatomy and function can be visualized using PET scanning in conjunction with a very small amount of a radiolabeled compound. It is also a useful technique to assess the PK of radioactive labeled therapeutic agents, and is commonly used in the development of CNS targeted drugs to determine the extent of brain penetration of the drug. After administration of the radioactive labeled therapeutic drug, the drug exposure in different areas of the brain can easily be determined noninvasively by measurement of the radioactivity with PET. In preclinical PET studies, the radioactive labeled drug is often evaluated in well-trained rhesus macaques in conscious condition without anxiety and stress. PET results in conscious monkeys are considered preferable above sedated monkeys for predicting brain penetration of a CNS drug in human, since anesthesia does depress the neuronal activity, regional cerebral blood flow (rCBF), and regional cerebral metabolic rate for glucose (rCMRGlu) in animal experiments.

For successful drug development, optimal dose setting in clinical trials is important. Because each drug has its corresponding target site, achievement of the most suitable drug concentration in the target organ or tissue becomes critical to allow the drug to exhibit its maximal effect. Although PK data in animal experiments are usually used to estimate the clinical dose, discrepancies in absorption, distribution, and metabolism of drugs between experimental animals and humans still remains in most cases, and these make the dose estimation difficult. Especially when a drug acts in the brain, further difficulty is caused by the presence of the blood-brain barrier.

FK960, (N-(4-acetyl-1-piperazinyl)-p-fluorobenzamide monohydrate) is a novel drug in development for Dementia. Its proposed mechanism of action is enhancement of somatostatin release in the hippocampus. Since somatostatin also enhances the release of acetylcholine, FK960, could indirectly activate the release of acetylcholine. FK960 improved memory impairment in several kinds of rodent and non human primate animal models. However, the dose-response relationships of FK960 in memory improvement in animal models of amnesia and in the enhancement of long-term potentiation (LTP) in hippocampal slices were bell shaped. Therefore, the optimal dose setting to exhibit efficacy in patients remains to be exactly determined. Because PET can measure radioactivity concentrations dynamically in a target organ in a living subject with minimal invasion, the acquisition of bridging data between animals and humans may be expected.

[18F]-FK960 was synthesized and administered orally in combination with non radioactive labeled drug to conscious rhesus monkeys. Drug concentration versus time profiles in plasma and the brain were obtained using PET imaging (Noda et al., 2003a; Fig. 3). Dynamic PET images were acquired for 4 hours from 5 min after drug administration. Arterial blood samples were drawn during the PET scan and analyzed by an auto well γ-counter and thin layer chromatography to determine [18F]-FK960 activity in plasma. FK960 concentrations (mol/L) were calculated using the specific activity of FK960.

The study demonstrated that [18F]-FK960 penetrated the blood-brain barrier and distributed dose-dependently into the entire brain. Maximal FK960 brain concentrations were comparable with the levels showing enhancement of LTP in hippocampal slices. The results suggested that this PET imaging method could be used to measure FK960 brain concentrations in humans. A different study showed that FK960 significantly improved regional cerebral blood flow and regional cerebral metabolic rate of glucose in conscious aged macaques (Noda et al., 2003b).

(A)

(B)

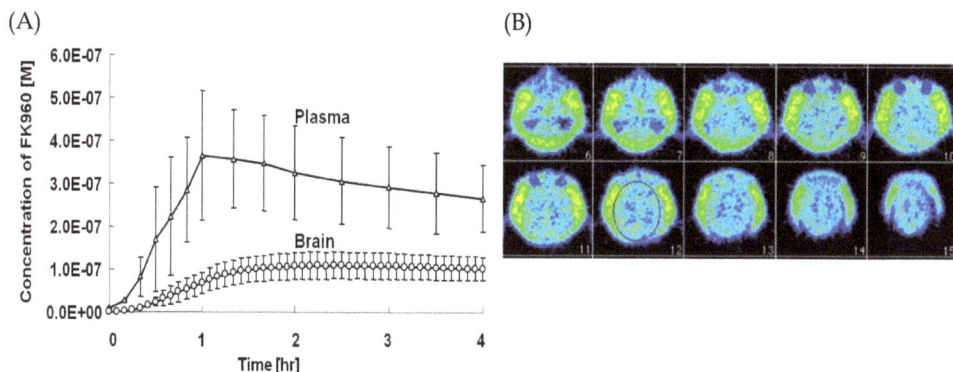

Fig. 3. (A) Brain and plasma concentrations of FK960 versus time in conscious male monkeys (n=3) after oral administration of a mixture of 0.1 mg/kg of FK960 and 370 MBq tracer of [18F]-]FK960 in saline. Concentrations of the drug into the entire brain were obtained from PET data. (B) Transverse PET images showing distribution of [18F]-FK960 into the brain. PET images were taken 90-120 min after drug administration as 3.6 mm transverse slices from the lower brain (top left) to the upper brain (bottom right). FK960 reached higher concentrations in the lower brain compared to the upper brain. The circle as presented in image 12 indicates the brain area in the skull.

Besides determination of the PK of a drug, PET imaging is important to assess the PD of drugs and is used in receptor and transporter occupancy studies, e.g. to investigate binding of the PET tracer to the target (receptor/transporter) and inhibition of this PET tracer binding by the drug of interest. These occupancy studies can reveal information on the binding potency and durability of binding *in vivo*, which can be used for dose selection.

In this way, PET imaging can be used to determine both the PK and PD of a drug and to establish the relationships between plasma and brain levels with the extent of receptor occupancy by the drug.

The use of PET to determine PD drug effects were gently shown for ASP5854 (5-[5-amino-3-(4fluorophenyl)pyrazin-2-yl]-1-isopropylpyridine-2(1H)-one), a novel adenosine A_{2A} receptor (A_{2A}R) antagonist, in development for Parkinson's disease. Adenosine A_{2A}Rs are abundantly expressed in the striatum of several species. They are co-expressed with dopamine D_2 receptors in the GABAergic striatopallidal neurons. Stimulation of adenosine A_{2A}Rs decreases the binding affinity of D_2 receptors and elicits effects opposite to the ones shown by D_2 receptor activation. These observations suggested that antagonistic adenosine-dopamine interactions can be important in the regulation of the activity of the basal ganglia and could explain the stimulating effects of adenosine A_{2A}R antagonists on motor behavior. Therefore, ASP5854 was considered of potential interest in the treatment of movement disorders and may reduce the symptoms in Parkinson's disease.

ASP5854 and [11C]-SCH442416, an adenosine A_{2A}R-specific radiotracer, were administered i.v. to conscious rhesus monkeys and adenosine A_{2A}R occupancy in the brain was examined using PET. ASP5854 dose-dependently increased adenosine A_{2A}R occupancy in the striatum (Mihara et al., 2008; Fig. 4) and showed long-lasting occupancy even at decreasing drug concentrations in plasma.

(A)   (B)

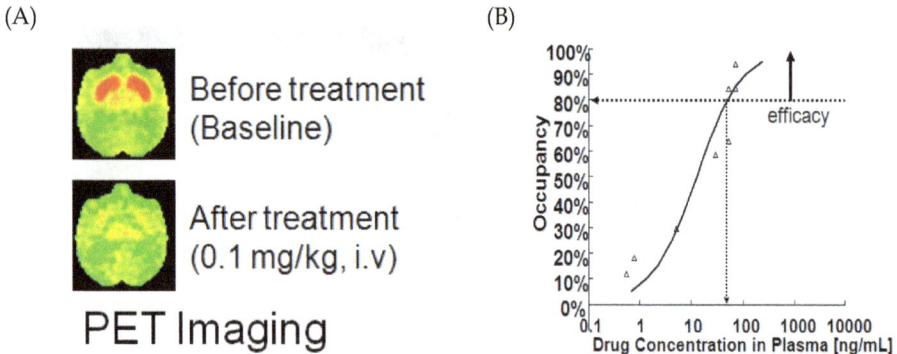

PET Imaging

Fig. 4. (A) PET images of [^{11}C]-SCH442416 before and after treatment of adenosine A$_{2A}$ receptor antagonist, ASP5854, in conscious monkeys. At baseline, the tracer accumulated in the striatum, which is rich of A$_{2A}$ receptors (20 mCi, i.v. injection; mean of six animals). At t=1h after drug administration (0.1 mg/kg, i.v.), the accumulation of the tracer in the striatum was lower compared to baseline, due to decreased binding of the PET tracer to the receptor as a result of ASP5854 binding to the receptor. (B) Relationship between the dose and receptor occupancy at 1 h after drug administration. Sigmoidal dose-response curve demonstrating an increase in adenosine A$_{2A}$R occupancy by ASP5854 with an increase in ASP5854 plasma concentration at t=1 h after drug administration (dose-levels: 0.001 to 0.1 mg/kg, i.v.); 80% receptor occupancy correlated with efficacy (catalepsy) in the monkeys.

Donepezil, an acetylcholine esterase inhibitor (AChEI), has been recommended as a treatment option for patients with AD. [^{18}F]-fluoro-2-deoxyglucose (FDG)-PET was used to measure the regional cerebral metabolic rate of glucose (rCMRglu), an index of neuronal activity, in rhesus monkeys (Asai et al., 2009). The effects on rCMRglu were measured following intramuscular (i.m.) administration of donepezil (500 μg/kg) or the non-selective muscarinic ACh receptor antagonist scopolamine (30 μg/kg, i.m.), or co-administration of both drugs. This FDG-PET study showed that administration of donepezil or scopolamine alone increased rCMRglu in conscious rhesus monkeys. The donepezil-induced increase in rCMRglu was abolished by simultaneous administration of scopolamine, suggesting that muscarinic ACh receptor function plays an important role in the effect of donepezil (Fig. 5).

3.3.2 Determination of early tumor response to anticancer drugs

Oncology is a main therapeutic area of many pharmaceutical companies. PET imaging is helpful to measure early tumor response to anticancer treatment in early phases of research and development.

YM155 is a small molecule survivin suppressant. Survivin is a member of the inhibitor of apoptosis protein family, acting as an inhibitor of caspase activation, and has been implicated in both cell survival and regulation of mitosis in cancer. Although survivin is expressed in a variety of normal fetal tissues, expression is absent in most adult tissues. In contrast, survivin is highly expressed in most tumors. Survivin overexpression in cancer patients is associated with resistance to cytotoxics and is correlated with poor survival. YM155 was shown to have nanomolar antitumor activity in a wide variety of human cancer cell lines. The continuous infusion of YM155 induced tumor regression in mice xenograft

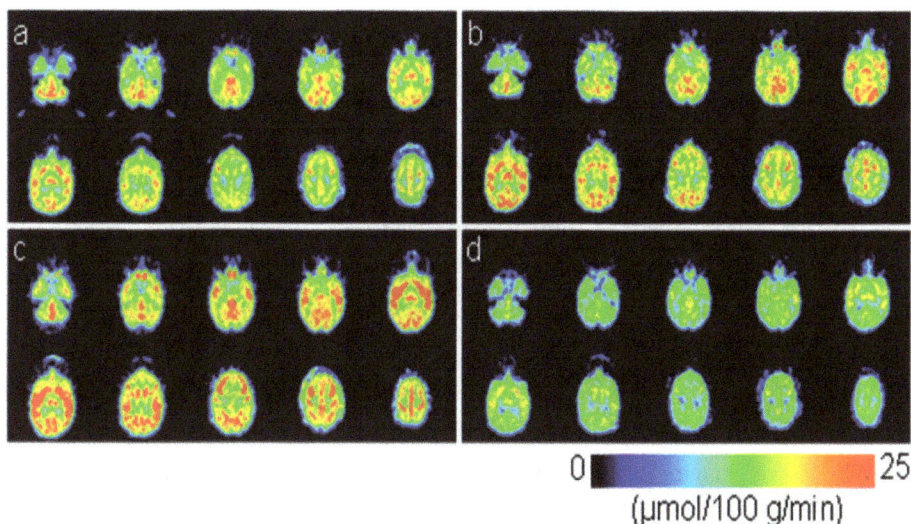

Fig. 5. PET images of the brain of consious monkeys demonstrating rCMRglu using FDG-PET after treatment with empty vehicle (a), donepezil 500 µg/kg (b), scopolamine 30 µg/kg (c), and donepezil+scopolamine (d). Images are orbitomeatal transverse slices (OM) + 0 to OM + 32.4 mm, 3.6 mm thick.

models (Nakahara et al., 2011a). Other results suggested that YM155 sensitized tumor cells to radiation (Iwasa et al., 2008) and platinum compounds both *in vitro* and *in vivo*, and that the effect was likely attributable to the inhibition of DNA repair and consequent enhancement of apoptosis (Iwasa et al., 2010). Non-clinical studies using radioactive ^{14}C-labeled YM155 demonstrated that the organic cation transporter 1 (OCT1) was the predominant transporter for the hepatic uptake of YM155 (Iwai et al., 2009). YM155, administered as 168 hours continuous infusion in 21-day cycles, appeared to be safe and well-tolerated, with a maximum tolerated dose of 8.0 mg/m^2/day in Phase I studies in patients with advanced refractory solid tumors (Satoh et al., 2009), and advanced solid tumors or lymphoma (Tolcher et al., 2008). Multi-center Phase II trials demonstrated the safety and tolerability of YM155 in patients with unresectable stage III or IV melanoma (Lewis et al., 2011) and safety and modest activity in patients with advanced refractory non-small cell lung cancer (NSCLC) (Giaccone et al., 2009).

PET imaging has been used for the development of YM155 in preclinical studies in mice. [18F]-FDG-PET has been used to assess early treatment response in animals with diffuse large B-cell lymphoma (DLBCL) and non-small cell lung cancer (NSCLC) xenografts, and [18F]-FLT-PET was used to determine the effect on cell proliferation. The combined effect of YM155 and docetaxel in human NSCLC xenografts was determined using whole-body imaging. Combination treatment of YM155 and docetaxel for 21 days resulted in increased inhibition of tumor growth (Nakahara et al., 2011b), accompanied by inhibition of tumor uptake of FDG, compared to monotherapy with either YM155 or docetaxel (Fig. 6).

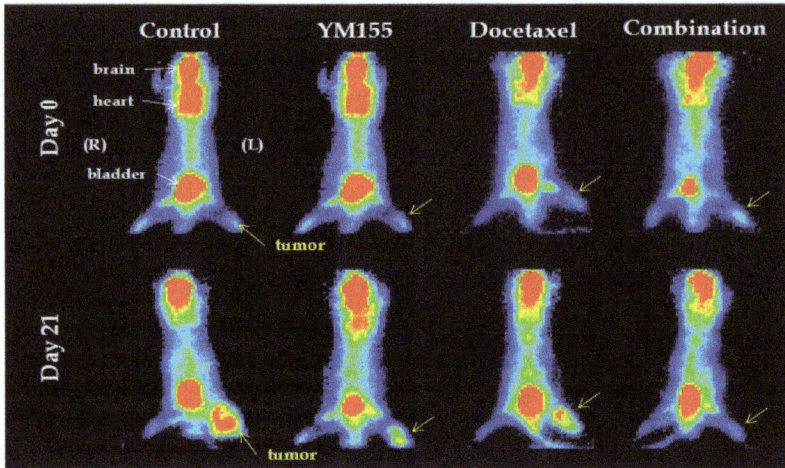

Fig. 6. Whole-body imaging of mice with [18F]-FDG. Human NSCLC xenografts were grown after injection of Calu-6 human tumor cells (3×10⁶ cells/100 μL) into the right hind limb of male Balb/c nude mice, aged 6-7 weeks. YM155 (2 mg/kg/day) was administred as s.c. infusion for 7 days. Docetaxel (10 mg/kg) was administered as i.v. bolus injection on days 0, 4 and 8 after the initiation of YM155 treatment. Mice were administerd 5 MBq of [18F]-FDG i.v. via the tail vein. After the appropriate time had elapsed, mice were anesthetized by isoflurane, placed in a prone position, and imaged by using a planar positron imaging system PPIS-4800 (Hamamatsu Photonics) to obtain the whole-body distribution of [18F]-FDG at 50–60 min from tracer injection.

3.3.3 Measurement of pancreatic β-cell mass in the development of antidiabetic drugs

Pancreatic β-cell mass (BCM) is known to decrease as a result of progression of disease in animals with diabetes mellitus. Measurement of BCM *in vivo* is an important tool for early diagnosis of diabetes mellitus and to monitor the efficacy of the treatment. Historically, the pancreas has been a difficult organ to image because of its retroperitoneal location and overlying visceral organs. Over the past 2 decades, the advance of CT and MRI has allowed to correlate disease processes of the pancreas, such as pancreatitis with anatomical structures. Unfortunately, these imaging modalities do not describe organ function or easily separate the endocrine from the exocrine pancreas. The ability to quantify and monitor β-cell mass (BCM) is of particular interest for diabetes mellitus.

Recently, it was reported that vesicular monoamine transporter 2 (VMAT2) is expressed mainly on the pancreatic β-cells as well as on dopaminergic neurons. A VMAT2 ligand, dihydrotetrabenazine (DTBZ), was developed previously for dopaminergic neuron imaging in the striatum and it has been successfully applied for PET studies of Parkinson's disease. DTBZ can also be used to monitor BCM quantitatively for diabetes. PET tracers that have been used to quantify VMAT2 are [11C]-dihydrotetrabenazine ([11C]-DTBZ) (Souza et al., 2006(a); Simpson et al., 2006; Murthy et al., 2008; Goland et al., 2009), [18F]-fluoropropyl-DTBZ ([18F]-FP-DTBZ or [18F]-AV-133) (Kung et al., 2007; Kung et al., 2008(a); Tsao et al., 2010) and [18F]-fluoropropyl-DTBZ epoxide ([18F]-FP-epoxy-DTBZ) (Kung et al., 2008(b)), which were shown to have high potency and selectivity for binding to VMAT2 in rats.

Longitudinal preclinical studies have been performed in induced T1DM and obese rat models (Freeby et al., 2008) and in a spontaneous T2DM diabetic rat model showing a decrease in [^{11}C]-DTBZ uptake that anticipated loss of glycemic control (Souza et al., 2006(a)). These preclinical *in vivo* PET studies in non-human primates are used to evaluate the between-species differences in PK of the tracer, and to establish the most likely optimal tracer for PET studies to establish BCM in humans.

3.4 Examples of PET in clinical studies

Several examples of the application of PET in clinical studies for the development of CNS drugs, anticancer drugs and antidiabetic drugs is described in this Section.

3.4.1 Biodistribution of CNS drugs in humans

Establishing the therapeutic dose and dose regimen of a new drug is challenging, especially for drugs with CNS activity. The major challenge is the blood-brain barrier (BBB), which limits the access of drugs to the brain tissue. It is important to select the right dose range of a drug in early phase clinical trials to establish the PK, PD and safety of a drug. PET is used to determine biodistribution and concentration of a drug in the brain non-invasively, *in vivo*. In such clinical investigation, Good Manufacturing Practice (GMP) and Good Clinical Practice (GCP) regulations are required. As described earlier in this Chapter, PET was used in non-clinical studies to predict the optimal clinical dose for FK960, a novel drug in development for the treatment of patients with Dementia. Subsequently, the drug was investigated in humans. FK960 has a bell-shaped dose–response relationship. Determining the proper clinical starting dose and dose range is difficult because of species differences between animals and humans. For example, differences in brain concentrations of the drug at comparable doses may be expected as a result of discrepancies in absorption, distribution, metabolism, and permeability of the drug through the BBB between animals and humans. This may lead to incorrect estimations of the therapeutic dose in humans from the animal data. Therefore it could be useful to determine drug concentrations in the brain, in addition to plasma, by PET in clinical studies.

In a clinical study in healthy male subjects, concentration versus time profiles of FK-960 in plasma and different sections of the brain were determined following oral administration of FK-960 (Fig. 7). A mixture of [^{18}F]-FK960 and FK960 (6 mg/man, male, n=3, p.o.) was administered to healthy subjects. In addition, whole-body PET imaging was carried out between 120 and 130 min after injection, showing that [^{18}F]-FK960 distributed to liver and kidneys, but not to the stomach and small intestine. The whole body scan suggested hepatic and renal clearance of FK960.

3.4.2 PET imaging in clinical oncology

In clinical oncology PET is used in the diagnosis of cancer, detection of metastasis and disease staging. It is also under investigation for the prediction of patient response to treatment (West et al., 2004). Early evaluation of treatment response can be of value to avoid unnecessary toxicity or ineffective treatment and to terminate development of uneffective drugs at an early stage, thereby saving costs of expensive late phase clinical studies. Currently, only two PET tracers have been described in a guidance by the FDA to assist applicants in preparing New Drug Applications (NDAs) (Guidance PET Drug

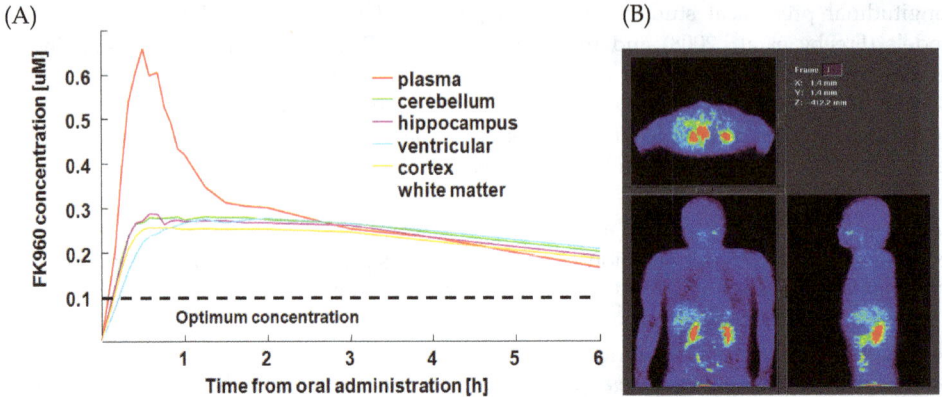

Fig. 7. (A) Concentration versus time curves of FK-960 in plasma and different area's of the brain in healthy subjects following oral administration of a mixture of [18F]-FK960 and FK960 (6mg, n=3). (B) Biodistribution of [18F]-FK960 in human whole-body using PET imaging. Images were acquired 120-130 min after administration.

Applications - Content and Format for NDAs and ANDAs, February 2011). There has been evidence that nuclear medicine imaging techniques could provide unique, biologically relevant, and prognostically important information unavailable through anatomic imaging such as CT or MRI. Multiple quantitative measurements can be performed by PET, which enables to monitor the effects of the treatment at an early stage, before those changes are detectable with conventional imaging modalities. The most important PET tracer in clinical oncology is FDG, which reflects glucose metabolism in the tumor. FDG-PET has shown to be of value in the differentiation of benign and malignant tissues, preoperative staging, detection of recurrent disease, and, more recently, in the identification of early tumor response to therapy. FLT-PET has attracted attention as a biomarker to evaluate tumor cell proliferation. The FDA has approved an investigational new drug (IND) application for [18F]-FLT, sponsored by the Society of Nuclear Medicine (SNM).

Early response evaluation using FDG and FLT has been used for example for erlotinib (Tarceva©), an inhibitor of the epidermal growth factor receptor (EGFR), used in the treatment of solid tumors, such as non-small cell lung cancer (NSCLC) and pancreatic cancer. [18F]-FDG PET has been recognized as an adequate staging tool in patients with NSCLC, and several studies suggested that the standardized uptake value (SUV) has a prognostic value in NSCLC. Furthermore, it has been reported that at an early stage during erlotinib therapy, [18F]-FDG PET/CT can predict treatment in NSCLC patients (Aukema et al., 2010).

The accuracy of FDG-PET and FLT-PET were evaluated in patients with advanced NSCLC for early prediction of progression after 6 weeks of therapy wit erlotinib (Zander et al., 2011). FDG-PET predicted progression-free survival (PFS), overall survival (OS), and absence of progression after 6 weeks of therapy with erlotinib in treatment-naive patients with advanced NSCLC independent from the EGFR mutational status. Moreover, it was reported that early FDG-PET response on day 14 after initiation of erlotinib therapy, was associated with improved PFS and OS, even in the absence of subsequent RECIST response (Mileshkin et al., 2011).

PET is also useful to assess the PK of the anticancer drug in tumor tissue. This is of particular interest for anticancer drugs, which have their target in the tumor cell, for example drugs that target mRNA of a specific cancer gene. Tumor cells generally overexpress efflux pumps, such as P-gp, which prevent anticancer drugs that are P-gp substrate, to enter the tumor cell, thereby causing tumor resistance to the treatment. This can lead to continuation of tumor cell proliferation, experience of unncessary adverse events of the chemotherapy, and delay in determination of the effective dose regimen or decision to switch to other potential effective treatment. The delivery of antisense oligonucleotide directed against survivin mRNA in tumor tissue was determined using [11C]-LY2181308 PET imaging in a phase 1 clinical study (Talbot et al., 2010).

3.4.3 PET imaging to determine the effect of antidiabetic drugs on β-cell mass
It is considered important to evaluate the effect of glucose lowering drugs on preservation of pancreatic BCM in patients with T2DM, since BCM is affected by the disease. Overall pancreatic BCM reflects the balance between the dynamic processes of β-cell expansion, through proliferation and neogenesis, and β-cell loss via apoptosis. Given that BCM can be modified significantly by altering the rate of any of these mechanisms, therapies that modulate β-cell expansion and loss have garnered recent interest. Therefore, it is also used as a tool to differentiate a novel antidiabetic drug in development from existing/marketed glucose lowering drugs with regard to its pharmacological activity. In fact, pioglitazone treatment preserved pancreatic β-cell morphology and β-cell function in obese diabetic *db/db* mice (Kawasaki et al., 2005). Furthermore, pioglitazone protects human β-cells against apoptosis or loss of function under exposure to interleukin-1β or high-glucose concentrations *in vitro* (Zeender et al., 2004). In obese Zucker rats, rosiglitazone maintained β-cell proliferation and prevented loss of β-cells (Finegood et al., 2001). Recently, treatment with thiazolidinediones was reported to improve β-cell function, which is strongly correlated with glycemic control, in patients with T2DM (Gastaldelli et al., 2007). DPP-IV inhibitors and incretins such as glucagon-like peptide-1 (GLP-1) and glucose-dependent insulinotropic polypeptide (GIP) stimulate not only insulin secretion but also augment BCM via β-cell proliferation and neogenesis. Also, GLP-1 receptor signaling modifies the cellular susceptibility to apoptosis. Unfortunately, most studies measuring BCM have relied on postmortem examination of the pancreas, because until recently it was impossible to prospectively measure BCM *in vivo*.

A PET study with [11C]-DTBZ quantified VMAT2 expression in the pancreas as a non-invasive measurement of pancreatic BCM, and demonstrated differences in BCM between healthy subjects and patients with type 1 diabetes patients (T1DM) (Goland et al., 2009). Others demonstrated that [18F]-FP-DTBZ compared to [11C]-DTBZ substantially improved, both qualitatively and quantitatively, the ability to image pancreatic BCM in T1DM (Normandin et al, 2010). Futhermore, [18F]-FP-DTBZ was suggested to be safe and to be used for biodistribution and radiation dosimetry for imaging in humans and possibly to be used repetitively in longitudinal studies (Lin et al., 2010). Up to date, no PET studies for BCM determination have been published in T2DM patients. Others showed that since there is a significant BCM reserve, T2DM symptoms related to unstable glucose homeostasis are not obvious until BCM has been reduced by more than 50–60% (Souza et al., 2006(b)). Determination of β-cell mass in T2DM patients can become relevant in the future (Rhodes et al., 2005).

4. Translation of PET imaging from animals to humans

As mentioned before, PET has been applied to determine the PK study as a brain tissue PK measurement of [18F]-FK962 in the brain in monkeys and in humans. A mixture of a tracer amount of [18F]-FK962 and cold FK962 was administrated orally to conscious monkeys (n=3, doses: 0.0032 mg/kg and 0.032 mg/kg) and orally to healthy subjects (n=5, doses: 0.2 mg/man and 2 mg/man). PET data showed a good linearity between the brain concentration and plasma concentration in both monkeys and humans (Fig. 8). The brain concentration in humans was found to be about 2-fold higher than that in monkeys at comparable plasma drug concentrations. In addition, the optimum dosage for oral administration was estimated at 2 mg in humans based on both human brain concentrations by PET and rat brain concentrations in a preclinical efficacy study (data not shown).

(A)

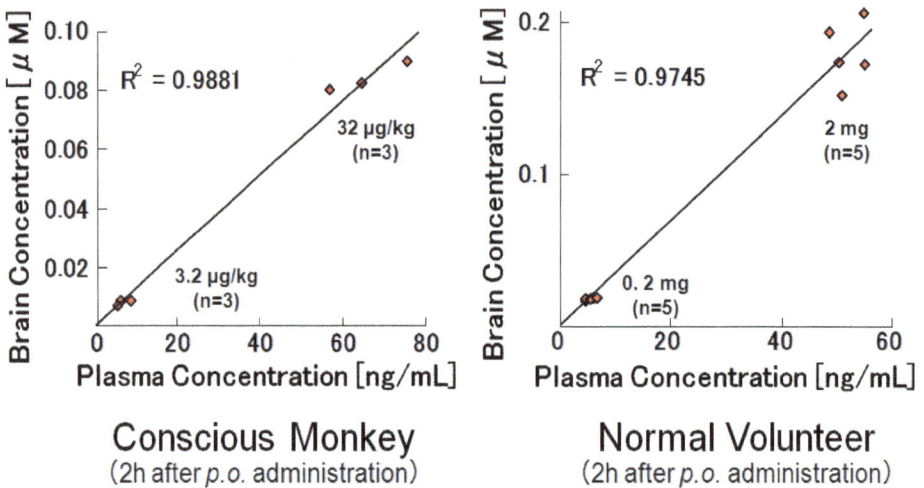

Conscious Monkey
(2h after *p.o.* administration)

Normal Volunteer
(2h after *p.o.* administration)

(B)

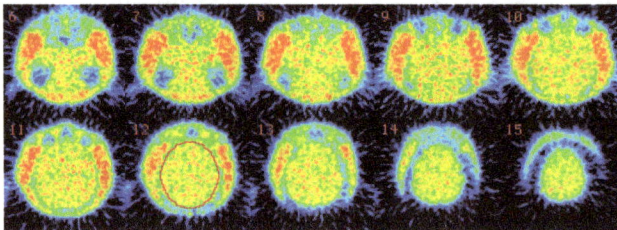

Fig. 8. (A) Correlations between FK-962 drug concentrations in plasma and the brain in monkeys and humans using PET imaging. (B) Transverse PET images showing distribution of [18F]-FK9602 in the brain of conscious monkeys. PET images were taken 90-120 min after drug administration as 3.6 mm transverse slices from the lower brain (top left) to the upper brain (bottom right). The circle as presented in image 12 indicates the brain area in the skull.

Thus, non-invasive imaging techniques such as PET have become available for assessment of drug distribution *in vivo*, including determination of drug brain concentrations in humans. Although PET enables monitoring of regional drug concentration differences with a spatial resolution of a few millimetres, discrimination between bound and unbound drug or parent compound and metabolite is difficult. Furthermore, labeling of a PET tracer is time consuming and expensive and requires special expertise on radiation exposure. PET studies in monkeys can be useful to establish the methodology for clinical PET studies. Complementary use of MB study results (see session 2.2) with PET imaging can provide more extensive drug distribution data than MB studies alone. The use of PET imaging could play an important role in future drug research and development with the potential to serve as translational tool for clinical decision making (Brunner et al., 2006).

5. Conclusions

Radiolabeled compounds are commonly used in the R&D of drugs. In ADME studies radiosotopes are relevant to a) evaluate the exposures of the parent compound and its metabolites in animals and humans for validation of toxicological species, b) identify the major metabolic pathways in humans to support drug–drug interaction studies, c) establish the rate and route of excretion of a drug candidate, and d) provide metabolism data of drugs for regulatory filing. Using radioactive materials in ADME studies in animals and humans helps to identify and quantify metabolites, and reveal the major metabolite(s) and clearance pathways.

In clinical development, radiolabeled compounds are used in MB studies to obtain information on the absorption, metabolism, and elimination of a drug. Other applications are the use in regional drug absorption studies to determine the PK of a drug in a novel modified release formulation and in microdose studies to obtain PK data after administration of a trace subpharmacologic quantity to human subjects.

PET is a powerful non-invasive technique, which is used to evaluate the PK (e.g. drug exposure), PD (e.g. receptor binding) in animals and humans. For CNS targeted drugs, it is used to assess drug penetration and distribution into different areas of the brain and to determine the target (receptor/transporter) occupancy by the drug, which provides an estimation of the binding potency and durability of the drug *in vivo*. In oncology PET is used in the diagnosis of cancer, detection of metastasis, disease staging, and for the prediction of patient response to treatment. In the field of diabetes mellitus, PET imaging could be considered a relevant tool to evaluate the effects of glucose lowering drugs on preservation of pancreatic β-cells in diabetes patients.

In summary, PET imaging can be used a) to estimate the pharmacological activity of a drug, b) to select the starting dose in first-in-man studies, c) as a surrogate biomarker for efficacy, and d) to proof mechanism and/or concept of a compound. This helps to determine at an early phase of development whether or not the drug is a good candidate to take into further clinical development. Thereby, PET imaging can also help to terminate the development of unfavourable drugs at an early stage, thereby reducing unnecessary development costs.

6. Acknowledgements

We thank our colleagues at the Bioimaging Research Laboratories, Astellas Pharma, Inc., Tsukuba, Japan. Especially we thank Dr. Akihiro Noda, Dr. Yoshihiro Murakami, Dr. Makoto Asai, Aya Kita, Kentaro Yamanaka, Takahito Nakahara and Dr. Ichiro Matsunari.

7. References

Kola, I. The state of innovation in drug development. Clin Pharmacol Ther 2008;83:227-30.

Penner N, Klunk L.J., Prakash C. Human Radiolabeled Mass Balance Studies: Objectives, Utilities and Limitations. Biopharm Drug Dispos 2009;30:185-203.

Marathe P.H., Shyu W.C., Humphreys W.G. The use of radiolabeled compounds of ADME studies in discovery and exploratory development. Current Pharmaceutical Design 2004;10:2991-3008.

Glaser M, Collingridge DR, Aboagye EO, Bouchier-Hayes L, Clyde Hutchinson O, Martin SJ, Price P, Brady F, Luthra SK. Iodine-124 labelled Annexin-V as a potential radiotracer to study apoptosis using positron emission tomography. Applied Radiation and Isotopes 2003;58:55-62.

Hoy CA, Lewis ED, Schimke RT. Perturbation of DNA Replication and Cell Cycle Progression by Commonly Used [³H]Thymidine Labeling Protocols. Molecular and Cellular Biology 1990;10:1584-1592.

Heinemann V, Hertel LW, Grindey GB, Plunkett W. Comparison of the cellular pharmacokinetics and toxicity of 2',2'-difluorodeoxycytidine and 1-h-D-arabinofuranosylcytosine. Cancer Res 1988;48:4024–31.

Mackey JR, Yao SY, Smith KM, et al. Gemcitabine transport in xenopus oocytes expressing recombinant plasma membrane mammalian nucleoside transporters. J Natl Cancer Inst 1999;91:1876–81.

Veltkamp SA, Pluim D, van Tellingen O, Beijnen JH, Schellens JH. Extensive metabolism and hepatic accumulation of gemcitabine after multiple oral and intravenous administration in mice. Drug Metab Dispos 2008;36:1606-15(a).

Veltkamp SA, Pluim D, van Eijndhoven MA, Bolijn MJ, Ong FH, Govindarajan R, Unadkat JD, Beijnen JH, Schellens JH. New insights into the pharmacology and cytotoxicity of gemcitabine and 2',2'-difluorodeoxyuridine. Mol Cancer Ther 2008;7:2415-25(b).

U.S. Food and Drug Administration, Department of Health and Human Services, Center for Drug Evaluation and Research: Guidance for Industry: Safety Testing of Drug metabolites, February 2008.

Combesa R.D., Berridge T., Connelly J., Eve M.D., Garner R.C., Toon S., Wilcox P. Early microdose drug studies in human volunteers can minimise animal testing: Proceedings of a workshop organised by volunteers in Research and Testing. European Journal of Pharmaceutical Sciences 2003;19:1–11.

International Conference on Harmonisation. ICH Harmonized Tripartite Guideline: Guidance on M3 (R2) Nonclinical Safety Studies for the Conduct of Human Clinical Trials and Marketing Authorization for Pharmaceuticals, June 2009.

Lappin G, Kuhnz W, Jochemsen R, Kneer J, Chaudhary A, Oosterhuis B, Drijfhout WJ, Rowland M, Garner RC. Use of microdosing to predict pharmacokinetics at the therapeutic dose: experience with 5 drugs. Clin Pharmacol Ther 2006;80:203-15.

The European Agency for the Evaluation of Medicinal Products. Evaluation of Medicines for Human Use. Position Paper on Non-Clinical Safety Studies to Support Clinical Trials With a Single Microdose. London: EMEA 2003:1-4.

U.S. Food and Drug Administration, Department of Health and Human Services, Center for Drug Evaluation and Research. Guidance for Industry, Investigators, and Reviewers. Exploratory IND Studies, 2006: 1-16.

Bergstrom M, Grahnen A, Langstrom B. Positron emission tomography microdosing: A new concept with application in tracer and early clinical drug development. Eur J Clin Pharmacol 2003;59:357–366.

Marchetti S, Schellens JH. The impact of FDA and EMEA guidelines on drug development in relation to Phase 0 trials. Br J Cancer 2007;97:577–581.

Chen K., et al. Positron Emission Tomography imaging cancer biology: Current status and future prospects. Seminars in Oncology 2011;38:70-86.

Rostomian AH, Madison C, Rabinovici GD, Jagust WJ. Early 11C-PIB frames and 18F-FDG PET measures are comparable: a study validated in a cohort of AD and FTLD patients. J Nucl Med 2011;52:173-9.

Wong DF, Rosenberg PB, Zhou Y, Kumar A, Raymont V, Ravert HT, Dannals RF, Nandi A, Brasić JR, Ye W, Hilton J, Lyketsos C, Kung HF, Joshi AD, Skovronsky DM, Pontecorvo MJ. In vivo imaging of amyloid deposition in Alzheimer disease using the radioligand 18F-AV-45 (Flobetapir F 18). J Nucl Med 2010;51:913-20.

Noda A, Takamatsu H, Murakami Y, Yajima K, Tatsumi M, Ichise R, Nishimura S. Measurement of brain concentration of FK960 for development of a novel antidementia drug: a PET study in conscious rhesus monkeys. J Nucl Med 2003;44:105-8(a).

Mihara T, Noda A, Arai H, Mihara K, Iwashita A, Murakami Y, Matsuya T, Miyoshi S, Nishimura S, Matsuoka N. Brain adenosine A_{2A} receptor occupancy by a novel A_1/A_{2A} receptor antagonist ASP5854 in rhesus monkey: Relationship to anti-cataleptic effect. J Nucl Med 2008;49:615-22.

Asai M, Fujikawa A, Noda A, Miyoshi S, Matsuoka N, Nishimura S. Donepezil- and scopolamine-induced rCMRglu changes assessed by PET in conscious rhesus monkeys. Ann Nucl Med 2009;23:877-82.

Mitsuoka K, Miyoshi S, Kato Y, Murakami Y, Utsumi R, Kubo Y, Noda A, Nakamura Y, Nishimura S, Tsuji A. Cancer detection using a PET tracer, 11C-glycylsarcosine, targeted to H+/peptide transporter. J Nucl Med 2008;49:615-22.

Narayanan TK, Said S, Mukherjee J, Christian B, Satter M, Dunigan K, Shi B, Jacobs M, Bernstein T, Padma M, Mantil J. A comparative study on the uptake and incorporation of radiolabeled methionine, choline and fluorodeoxyglucose in human astrocytoma. Mol Imaging Biol 2002;4:147–156.

Bruehlmeier M, Roelcke U, Schubiger PA, Ametamey SM. Assessment of hypoxia and perfusion in human brain tumors using PET with 18F-fluoromisonidazole and 15O-H2O. J Nucl Med 2004;45:1851–1859.

Piert M, Machulla HJ, Picchio M, Reischl G, Ziegler S, Kumar P, Wester HJ, Beck R, McEwan AJ, Wiebe LI, Schwaiger M. Hypoxia-specific tumor imaging with 18F-fluoroazomycin arabinoside. J Nucl Med 2005;46:106–113.

Kadir A, Nordberg A. Target-specific PET probes for neurodegenerative disorders related to dementia. J Nucl Med 2010;51:1418-30.

Noda A, Takamatsu H, Matsuoka N, Koyama S, Tsukada H, Nishimura S. Effect of N-(4-Acetyl-1-piperazinyl)-p-fluorobenzamide monohydrate (FK960), an antidementia drug with a novel mechanism of action, on regional cerebral blood flow and glucose metabolism in aged rhesus macaques studied with positron emission tomography. J Pharmacol Exp Ther 2003;306:213-7(b).

Nakahara T, Kita A, Yamanaka K, Mori M, Amino N, Takeuchi M, Tominaga F, Kinoyama I, Matsuhisa A, Kudou M, Sasamata M. Broad spectrum and potent antitumor activities of YM155, a novel small-molecule survivin suppressant, in a wide variety of human cancer cell lines and xenograft models. Cancer Science 2011;102:614-621(a).

Iwasa T, Okamoto I, Suzuki M, Nakahara T, Yamanaka K, Hatashita E, Yamada Y, Fukuoka M, Ono K, Nakagawa K. Radiosensitizing effect of YM155, a novel small-molecule survivin suppressant, in non-small cell lung cancer cell lines. Clin Cancer Res 2008;14:6496-504.

Iwasa T, Okamoto I, Takezawa K Yamanaka K, Nakahara T, Kita A, Koutoku H, Sasamata M, Hatashita E, Yamada Y, Kuwata K, Fukuoka M and Nakagawa K. Marked anti-tumour activity of the combination of YM155, a novel survivin suppressant, and platinum-based drugs. British Journal of Cancer 2010;103:36-42.

Iwai M, Minematsu T, Narikawa S, Usui T, Kamimura H. Involvement of Human Organic Cation Transporter 1 in the Hepatic Uptake of 1-(2-Methoxyethyl)-2-methyl-4,9-dioxo-3-(pyrazin-2-ylmethyl)-4,9-dihydro-1H-naphtho[2,3-d]imidazolium Bromide (YM155 Monobromide), a Novel, Small Molecule Survivin Suppressant. Drug Metabolism & Disposition 2009;37:1856-1863.

Satoh T, Okamoto I, Miyazaki M, Morinaga R, Tsuya A, Hasegawa Y, Terashima M, Ueda S, Fukuoka M, Ariyoshi Y, Saito T, Masuda N, Watanabe H, Taguchi T, Kakihara T, Aoyama Y, Hashimoto Y, Nakagawa K. Phase I study of YM155, a novel survivin suppressant, in patients with advanced solid tumors. Clinical Cancer Research 2009;15:3872-3880.

Tolcher AW, Mita A, Lewis LD, Garrett CR, Till E, Daud AI, Patnaik A, Papadopoulos K, Takimoto C, Bartels P, Keating A, and Antonia S. Phase I and Pharmacokinetic Study of YM155, a Small-Molecule Inhibitor of Survivin. Journal of Clinical Oncology 2008;26:5198-5203.

Lewis KD, Samlowski W, Ward J, Catlett J, Cranmer L, Kirkwood J, Lawson D, Whitman E, Gonzalez R. A multi-center phase II evaluation of the small molecule surviving suppressor YM155 in patients with unresectable stage III or IV melanoma. Invest New Drugs 2011; 29:161–166.

Giaccone G, Zatloukal P, Roubec J, Floor K, Musil J, Kuta M, van Klaveren R.J., Chaudhary S, Gunther A, and Shamsili S. Multicenter Phase II Trial of YM155, a Small-Molecule Suppressor of Survivin, in Patients With advanced, Refractory, Non–Small-Cell Lung Cancer. Journal of Clinical Oncology 2009;27:4481-4486.

Nakahara T, Yamanaka K, Hatakeyama S, Kita A, Takeuchi M, Kinoyama I, Matsuhisa A, Nakano K, Shishido T, Koutoku H, Sasamata M. YM155, a novel survivin suppressant, enhances taxane-induced apoptosis and tumor regression in a human Calu 6 lung cancer xenograft model. Anticancer Drugs 2011;22:454-462 (b).

Souza F, Simpson N, Raffo A, Saxena C, Maffei A, Hardy M, Kilbourn M, Goland R, Leibel R, Mann JJ, Van Heertum R, Harris PE. Longitudinal noninvasive PET-based beta cell mass estimates in a spontaneous diabetes rat model. J Clin Invest 2006;116:1506-13(a).

Simpson NR, Souza F, Witkowski P, Maffei A, Raffo A, Herron A, Kilbourn M, Jurewicz A, Herold K, Liu E, Hardy MA, Van Heertum R, Harris PE. Visualizing pancreatic beta-cell mass with [11C]DTBZ. Nucl Med Biol 2006;33:855-64.

Murthy R, Harris P, Simpson N, Van Heertum R, Leibel R, Mann JJ, Parsey R. Whole body [¹¹C]-dihydrotetrabenazine imaging of baboons: biodistribution and human radiation dosimetry estimates. Eur J Nucl Med Mol Imaging 2008;35:790-7.

Goland R, Freeby M, Parsey R, Saisho Y, Kumar D, Simpson N, Hirsch J, Prince M, Maffei A, Mann JJ, Butler PC, Van Heertum R, Leibel RL, Ichise M, Harris PE. ¹¹C-dihydrotetrabenazine PET of the pancreas in subjects with long-standing type 1 diabetes and in healthy controls. J Nucl Med, 2009;50:382-9.

Kung MP, Hou C, Goswami R, Ponde DE, Kilbourn MR, Kung HF. Characterization of optically resolved 9-fluoropropyl-dihydrotetrabenazine as a potential PET imaging agent targeting vesicular monoamine transporters. Nucl Med Biol 2007;34:239-46.

Kung MP, Hou C, Lieberman BP, Oya S, Ponde DE, Blankemeyer E, Skovronsky D, Kilbourn MR, Kung HF. In vivo imaging of beta-cell mass in rats using ¹⁸F-FP-(+)-DTBZ: a potential PET ligand for studying diabetes mellitus. J Nucl Med 2008;49:1171-6(a).

Tsao HH, Lin KJ, Juang JH, Skovronsky DM, Yen TC, Wey SP, Kung MP. Binding characteristics of 9-fluoropropyl-(+)-dihydrotetrabenzazine (AV-133) to the vesicular monoamine transporter type 2 in rats. Nucl Med Biol. 2010;37:413-9.

Kung HF, Lieberman BP, Zhuang Z-P, Oya S, Kung M-P, Choi SR, Poessl K, Blankemeyer E, Hou C, Skovronsky D, and Kilbourn M. In vivo imaging of VMAT2 in pancreas using a ¹⁸F epoxide derivative of tetrabenazine. Nucl Med Biol 2008;35:825-837(b).

Freeby M, Goland R, Ichise M, Maffei A, Leibe R and Harris P. VMAT2 quantitation by PET as a biomarker for β-cell mass in health and disease. Diabetes, Obesity and Metabolism 2008;10 (Suppl. 4):98–108.

West CML, Jones T, Price P. The potential of positron-emission tomography to study anticancer-drug resistance. Nature Reviews 2004;457-469.

U.S. Food and Drug Administration, Department of Health and Human Services; Center for Drug Evaluation and Research: Guidance PET Drug Applications - Content and Format for NDAs and ANDAs, February 2011.

Aukema TS, Vogel WV, Hoefnagel CA, and Vald´es Olmos RA. Prevention of Brown Adipose Tissue Activation in ¹⁸F-FDG PET/CT of Breast Cancer Patients Receiving Neoadjuvant Systemic Therapy. J Nucl Med Technol 2010; 38:12–15.

Zander T, Scheffler M, Nogova L, Kobe C, Engel-Riedel W, Hellmich M, Papachristou I, Toepelt K, Draube A, Heukamp L, Buettner R, Ko YD, Ullrich RT, Smit E, Boellaard R, Lammertsma AA, Hallek M, Jacobs AH, Schlesinger A, Schulte K, Querings S, Stoelben E, Neumaier B, Thomas RK, Dietlein M, and Wolf J. Early Prediction of Nonprogression in Advanced Non–Small-Cell Lung Cancer Treated With Erlotinib By Using [¹⁸F]-Fluorodeoxyglucose and [¹⁸F]-Fluorothymidine Positron Emission Tomography 2011;29:1701-1708.

Mileshkin L, Hicks RJ, Hughes BGM, Mitchell P, Charu V, Gitlitz BJ, Macfarlane D, Solomon B, Amler L, Yu W, Pirzkall A, and Fine BM. Changes in ¹⁸F-Fluorodeoxyglucose and ¹⁸F-Fluorodeoxythymidine Imaging in Patients with Non-Small Cell Lung Cancer Treated with Erlotinib. Clin Cancer Res 2011;17:3304-3315.

Talbot D, Ranson M, Davies J, Lahn MM, Callies S, Andre VAM, Kadam S, Burgess M, Slapak CA, Olsen AL, McHugh P, de Bono JS, Matthews J, Saleem A, and Price PM. Tumor Survivin is Downregulated by the Antisense Oligonucleotide

LY2181308: A Proof of Concept, First-in-Human Dose Study. Clinical Cancer Resrearch 2010;16:6150.

Kawasaki F, Matsuda M, Kanda Y, Inoue H, Kaku K. Structural and functional analysis of pancreatic islets preserved by pioglitazone in db/db mice. Am J Physiol Endocrinol Metab 2005;288:E510-8.

Zeender E, Maedler K, Bosco D, Berney T, Donath MY, Halban PA. Pioglitazone and sodium salicylate protect human beta-cells against apoptosis and impaired function induced by glucose and interleukin-1beta. J Clin Endocrinol Metab 2004;89:5059-66.

Finegood DT, McArthur MD, Kojwang D, Thomas MJ, Topp BG, Leonard T, Buckingham RE. Beta-cell mass dynamics in Zucker diabetic fatty rats. Rosiglitazone prevents the rise in net cell death. Diabetes 2001;50:1021-9.

Gastaldelli A, Ferrannini E, Miyazaki Y, Matsuda M, Mari A, DeFronzo RA. Thiazolidinediones improve beta-cell function in type 2 diabetic patients. Am J Physiol Endocrinol Metab 2007;292:E871-83.

Lin KJ, Weng YH, Wey SP, Hsiao IT, Lu CS, Skovronsky D, Chang HP, Kung MP, Yen TC. Whole-body biodistribution and radiation dosimetry of ^{18}F-FP-(+)-DTBZ (18F-AV-133): a novel vesicular monoamine transporter 2 imaging agent. J Nucl Med, 2010;51:1480-5.

Souza F, Freeby M, Hultman K, Simpson N, Herron A, Witkowsky P, Liu E, Maffei A, Harris PE. Current progress in non-invasive imaging of beta cell mass of the endocrine pancreas. Curr Med Chem. 2006;13:2761-73(b).

Rhodes CJ. Type 2 Diabetes-a Matter of β-Cell Life and Death? Science 2005;307:380-384.

Normandin M, Skaddan M, Petersen K, Calle R, Weinzimmer D, Skovronsky D, Treadway J, Carson R, Ding Y-S, and Cline G. PET imaging of pancreatic beta cell mass with [^{18}F]FP-DTBZ. J Nucl Med 2010;51(Suppl. 2):130.

Brunner M, Langer O. Microdialysis versus other techniques for the clinical assessment of in vivo tissue drug distribution. AAPS J 2006;14;8:E263-71.

4

Medical Radioisotopes Production: A Comprehensive Cross-Section Study for the Production of Mo and Tc Radioisotopes Via Proton Induced Nuclear Reactions on ^{nat}Mo

A. A. Alharbi[1,2] et al.[*]
¹Faculty of Sciences, Physics Department,
Princess Nora University Riyadh,
²Cyclotron institute, Texas A&M University,
College Station, TX,
¹Saudi Arabia,
²USA

1. Introduction

1.1 Radioisotopes in nuclear medicine

Nowadays, many different stable and radioactive isotopes, each with unique physical and chemical properties, play significant roles in technological applications of importance to our modern society and are substantial to scientific research. One of the most common applications is the use of the radioisotopes in medicine. Medical radioisotopes are used to label some special chemical compounds to form radiopharmaceuticals.

Radiopharmaceuticals are used extensively in the field of nuclear medicine in three main branches. The largest and the most common type involve diagnostic procedures in which a radionuclide in a chemically suitable form is administered to the patient, and the distribution of the radioactivity in the body is determined by an external radiation detector (Qaim, 2008). The results are in the form of image of the involved organ, which provides information about the functioning of person's specific organs via emission tomography. The second branch of nuclear medicine deals with radionuclide techniques that are used for the analysis of concentration of hormones, antibodies, drugs and other important substances in samples of blood or tissues. The third branch is radiation therapy, which is the ultimate aim of all diagnostic investigations. Here the tissues or organs are treated with radiation and restored to the normal functions in the human body (Loveland, et al., 2006).

[*] A. Azzam[1,3], M. McCleskey[2], B. Roeder[2], A. Spiridon[2],
E. Simmons[2], V.Z. Goldberg[2], A. Banu[2], L. Trache[2] and R. E. Tribble[2]
¹Faculty of Sciences, Physics Department, Princess Nora University Riyadh, Saudi Arabia,
²Cyclotron institute, Texas A&M University, College Station, TX, USA
³Nuclear Physics Department., Nuclear Research Center, AEA, Cairo, Egypt

The two fundamental considerations in the administration of radioactivity to the human body are (Krane, 1987):

- Efficient detection of the radiation from outside the body,
- Radiation dose caused to the patient.

Diagnostic techniques in nuclear medicine use radioactive tracers which are easily detectable and which help to investigate various physiological and metabolic functions of the human body. Diagnosis is usually conducted by short-lived radionuclides, generally attached to a suitable chemical compound. Depending on the nature of the radiopharmaceutical, it may be inhaled, ingested, or injected intravenously (Stöcklin, et al., 1995). The radiation emitted by the radionuclide provides different kinds of information, as required for diagnosis. Radionuclides are powerful tools for diagnosis due to three reasons:

1. The mass of the sample is infinitesimally small, as low as 10^{-10} g of radioactive material, so it does not disturb the biological equilibrium.
2. The radioactive form of an element behaves exactly the same way as the non-radioactive element.
3. Each radioactive material spontaneously decays into some other form with emission of radiation. This radiation can be detected from outside the body.

Depending upon the nature of radionuclide, today two different tomographic procedures are available for imaging:

- Single photon emission computed tomography (SPECT)
- Positron emission tomography (PET)

In SPECT, a single or a dominant photon is detected by a gamma camera, which can view organs from many different angles (Khan, 2003). The camera makes an image from the points where the radiation is emitted; this image achieved by the camera is enhanced on a computer and can be viewed by a physician.

Positron Emission Tomography (PET) is a more modern technique in which a positron-emitting radionuclide, attached to a proper chemical compound, is introduced in the body, usually by injection, where it accumulates in the target tissue. As it decays it emits a positron, which at first loses its kinetic energy in the tissue and then promptly combines with a nearby electron resulting in the simultaneous emission of two identifiable photons in opposite directions (180°). These are detected by two detectors in coincidence. An array of such detectors is known as a PET camera, it gives very precise and sophisticated information on the place of annihilation. The most important clinical role of PET is in oncology, with a suitable fluorine-18 labelled compound as the tracer, since it has been found to be the best non-invasive method of detecting and evaluating most cancers. It is also well used in cardiac and brain imaging (Qaim, et al., 1993).

The radiation therapy is often done by using external beams of protons, neutrons, electrons, or photons (Wolf & Jones, 1983). As far as radionuclides are concerned, there are many possibilities to utilize them in therapy. One such possibility is to use the radiation emitted by the radionuclides, e.g. electrons and high-energy γ-rays as in the case of ^{60}Co. However, in recent years internal radiotherapy has also been gaining enhanced attention. Internal radiotherapy involves the use of radionuclides of suitable decay characteristics (Qaim, 2003). When a therapeutic radionuclide is delivered to a specific organ by using a biochemical pathway, it is known as open source therapy or endoradiotherapy (Qaim, 2003; Krane, 1987; Wolf & Barclay Jones, 1983). This type of radiotherapy is a unique cancer treatment modality. It is systemic and non-invasive. The uptake and retention in the tumour can be assessed with a tracer study before administering a therapeutic dose to the patient.

The major criteria for the choice of a radionuclide for endotherapeutic use are suitable decay characteristics and suitable biochemical reactivity. Concerning the decay properties, the desired half-life is between 6 hours and 7 days and the emitted corpuscular radiation should have a suitable linear energy transfer (LET) value and range in the tissue (Qaim, 2003; Sharp, et al., 2005). The ratio of non-penetrating to penetrating radiation should be high. The daughter should be short-lived or stable. The stability of the therapeutically pharmaceutical is demanded over a much longer period than that in the case of a diagnostic pharmaceutical. Thus, the choice falls on about 30 radionuclides. Most of them are β- emitters but several of them are α emitters and Auger electron emitters.

1.2 Medical radioisotopes production

The main processes to produce the medical radioisotopes are neutron activation, nuclear fission, charged particles induced reactions and radionuclide generators. Mostly, chemical separation is needed to separate the required isotope from targets and any produced impurities before using in the labeling process.

The medical radioisotopes can be produced using nuclear reactors either by neutron activation or by nuclear fission. The first procedure depends mostly on the thermal neutron capture process (n,γ). These isotopes will decay by means of β- emission accompanied with some gamma rays and could be used in treatment or Single Photon Emission Computed Tomography (SPECT). The second procedure based on the fission of a heavy nucleus, from the fuel after thermal neutron absorption. Some of the produced fission fragments have found medical applications such as ^{99}Mo (used as ^{99}Mo/^{99m}Tc generator), ^{131}I, and ^{133}Xe (Qaim, 2004).

Charged particle accelerators are another tool for producing medical radioisotopes using charged particle induced reactions on some stable isotopes. The accelerators used for this purpose should deliver ion beam with enough energy suitable for the used nuclear reaction and high beam intensity for production of reasonable radioactive yield in a reasonable irradiation time. Usually cyclotron accelerators with energies in the range 10 to 50 MeV are suitable for this purpose.

Cyclotron radionuclide production involves various constraints. First, a target has to be prepared, quite often from isotopically enriched material and energy should be carefully chosen to reduce, as much as possible, the impurities level. Second, the target should be stable in respect to ionizing radiation and heat generated by slowing down of the charged particles. Therefore, targets should be as thin as possible, just enough to degrade the incident energy to the required threshold energy, and they should display good heat conductivity to allow efficient cooling. After irradiation, the target is dissolved and various radiochemical operations are performed to isolate and purify the radionuclide.

The produced isotopes will usually be neutron deficient. This type of isotopes decay with β+ and/or EC accompanied with specific gamma rays and can be used for Positron Emission Tomography (PET) such as ^{11}C, ^{15}O, ^{13}N, and ^{18}F or SPECT such as ^{111}I, ^{67}Ga and ^{201}Tl (Lamberecht, 1979; Qaim, 2001). A number of isotopes as shown in Table 1 are technically available for use in medical applications (Troyer & Schenter, 2009).

Purpose	Accelerator-produced	Reactor-produced
Therapeutic Isotopes	^{64}Cu, ^{67}Cu, ^{77}Br, ^{88m}Br, ^{88}Y, ^{89}Zr, ^{103}Pd, ^{111}In, ^{124}I, ^{186}Re, ^{211}At	^{32}P, ^{47}Sc, ^{60}Co, ^{64}Cu, ^{67}Cu, ^{89}Sr, ^{90}Sr, ^{90}Y, ^{103}Pd, ^{103}Ru, ^{106}Ru, ^{109}Cd, ^{109}Pd, ^{117m}Sn, ^{115}Cd, ^{125}I, ^{131}I, ^{137}Cs, ^{145}Sm, ^{153}Sm, ^{165}Dy, ^{166}Dy, ^{166}Ho, ^{169}Er, ^{169}Yb, ^{180}Tm, ^{175}Yb, ^{177}Lu, ^{186}Re, ^{188}Re, ^{192}Ir, ^{195m}Pt, ^{198}Au, ^{199}Au, ^{211}At, ^{213}Bi, ^{225}Ac, ^{241}Am
Diagnostic Isotopes	^{11}C, ^{13}N, ^{15}O, ^{18}F, ^{55}Fe, ^{57}Co, ^{61}Cu, ^{64}Cu, ^{67}Ga, ^{74}As, ^{76}Br, ^{81m}Kr, ^{82m}Rb, ^{94m}Tc, ^{97}Ru, ^{111}In, ^{123}I, ^{124}I, ^{179}Ta, ^{201}Tl	^{3}H, ^{14}C, ^{51}Cr, ^{64}Cu, ^{97}Ru, ^{99m}Tc, ^{123}I, ^{131}I, ^{133}Xe, ^{153}Gd, ^{195m}Pt

Table 1. Common medical isotopes sorted by use category and production method (Troyer & Schenter, 2009)

1.3 Molybdenum and technetium in nuclear medicine

Molybdenum is used as a target material for the production of medically important radioisotopes, such as ^{99m}Tc/^{99}Mo, $^{96(m+g)}$Tc and ^{94m}Tc.

^{94m}Tc (52min), has shown its applicability as a PET isotope (Rösch and Qaim, 1993; Nickles, et al., 1993; Sajjad and Lambrecht, 1993; Rösch, et al., 1994; Fabbender, et al., 1994; Qaim, 2000; Hohn, et al., 2008). ^{96}Tc (4.28d) has been proposed for the use in prevention of coronary restenosis by Fox (2001). Despite of favorable moderate half-life, other isotopes of technetium, like, ^{93}Tc (2.75h), ^{94}Tc (4.883h) and ^{95}Tc (20.0h) are seldom discussed. Specially, radiological half-life of ^{94}Tc is ideal for diagnostic purposes. ^{95}Tc (20.0h), due to its comparatively longer half-life is also promising for tracking long processes, like, metabolic pathways for brain and heart, studies with proteins, anti bodies, etc. Among short-lived radionuclides, ^{93}Tc (2.75 h) is another promising isotope for imaging as suggested by (Lambrecht and Montner, 1982).

One of the most important medical radioisotopes is ^{99m}Tc ($T_{1/2}$= 6.01 h), which has a gamma ray energy of about 140 keV. The fact that both its physical half-life and its biological half-life are very short, as seen in Table 2, leads to a very fast clearing from the body after an imaging process. A further advantage is that the gamma is a single energy, not accompanied by beta emission, and that permits a more precise alignment of imaging detectors.

Isotope	Half-lives in days		
	$T_{Physical}$	$T_{Biological}$	$T_{Effective}$
^{99m}Tc	0.25	1	0.20

Table 2. The physical, biological and effective half lives for ^{99m}Tc

^{99m}Tc is a vital part of diagnostic tests for heart diseases and cancers; It accounts for over 80% of all diagnostic nuclear medicine procedures worldwide. According to the latest survey, the world demand for production of ^{99}Mo/^{99m}Tc is estimated to be around 7 kCi/week and further growth is predicted (Takács, et al., 2003). Currently, only five nuclear reactors produce ^{99}Mo/^{99m}Tc leading to a predicted shortage in covering the world demand. Consequently, many studies nowadays concentrate on producing ^{99}Mo generators with an alternative method using cyclotron accelerators (Van der Marck, 2010; Gull, 2001).

^{99m}Tc is obtained from the decay of its parent isotope ^{99}Mo. It was discovered in 1937, and the first ^{99}Mo/^{99m}Tc generator was invented at the Brookhaven National Laboratory in the U.S. in 1957. General usage of ^{99m}Tc began in the early seventies when the Chalk River Laboratory established routine production of ^{99}Mo, its parent isotope (Tammemagi and Jackson, 2009; Ullyett, 1997). ^{99m}Tc is versatile and can be used to produce some 20 different compounds of radiopharmaceuticals. There are various technological options for the production of ^{99m}Tc/^{99}Mo listed in Table 3.

Reactors	Fission of ^{235}U	$n + ^{235}U \rightarrow ^{99}Mo + xn +$ other fission products
	Neutron activation of ^{98}Mo	$n + ^{98}Mo \rightarrow ^{99}Mo$
Accelerators	Photo-fission of ^{238}U	$Photon + ^{238}U \rightarrow ^{99}Mo + xn +$ other fission products
	^{100}Mo transmutation	$Photon + ^{100}Mo \rightarrow ^{99}Mo + n$
	Direct ^{99m}Tc production	$P + ^{100}Mo \rightarrow ^{99m}Tc + 2n$

Table 3. The various technological options for the production of ^{99m}Tc/^{99}Mo

The usual production of ^{99}Mo for nuclear medicine depends on:
1. The neutron induced fission of ^{235}U, which results in expensive but high specific activity ^{99}Mo (IAEA-TECDOC-1065, 1999), or
2. The (n,γ) nuclear reaction with ^{98}Mo, 24% using natural Molybdenum, resulting in inexpensive but low-specific activity ^{99}Mo.
Thus, for either method, at least one neutron is required for the reaction.

Neutrons can be produced from accelerator reactions where the charged particles strike heavy atoms, also from alpha or gamma reactions with light atoms, such as beryllium or lithium. However, to produce the large quantities of neutrons needed for production of useful quantities of ^{99}Mo, the most effective source is a critical nuclear reactor operating at powers in the range of megawatts. Each fission process of an atom of ^{235}U produces an average of about 2.5 neutrons. In an operating reactor, these neutrons either are absorbed by materials in the reactor or escape from the boundaries of the reactor. One neutron must cause fission in another ^{235}U atom. Of the remaining 1.5 neutrons from each fission process in a critical reactor, some small fractions are available for production. The most appropriate target material for low specific activity ^{99}Mo production is molybdenum trioxide (MoO_3); neutron activation occurs via the reaction $^{98}Mo(n,\gamma)^{99}Mo$.

The potential use of accelerators for these purposes is another issue of current scientific and technological interest. Recently, a matter of concern has been the availability and supply of ^{99}Mo for the manufacturing of generators. These concerns arose from several factors including, amongst others, the shutdown of some nuclear reactors, uncertainty of reliable operating condition for radioisotope production and easy availability of enriched ^{235}U target materials.

More recently, the utilization of charged particle accelerators, either LINAC's or cyclotrons, has been discussed as a potential alternative technology to the fission route. These discussions have been prompted by basic research concerns as well as the need to explore new production routes to offset the perceived situation of future problems with the availability of ^{99}Mo if no new dedicated reactors are licensed.

The production of ^{99}Mo via the ^{100}Mo(p,pn) reaction was evaluated. A good agreement was found among the different excitation functions available. However, because of the rather low cross-section values found in these measurements, the production of ^{99}Mo via this potential process was found to be largely impractical. A significant limiting factor of this approach appears to be the need for a large inventory (tens of kg quantities) of enriched ^{100}Mo, the logistical considerations of its distribution and recovery, and the cost (2 US $/mg). Furthermore, proton accelerators delivering mA beam on target would be required including the development of high power targets.

The production of ^{99m}Tc via the ^{100}Mo(p,2n) reaction was also evaluated, and the cross section data available were found to be consistent and in good agreement. Extrapolating ^{99m}Tc yields obtained from this data, using the operational conditions of the existing 30 MeV accelerator technologies, suggest that large-scale (kCi) production of ^{99m}Tc is possible (Glenn, et al., 1997).

1.4 Nuclear data needs

The excitation function measurements of charged particle induced reactions are needed to improve and study the ideal way for medical radioisotope production. The optimization of nuclear reaction for the production of radioisotope at a cyclotron involves a selection of the projectile energy range that will maximize the yield of the product and minimize that of radionuclide impurities. The IAEA Coordinated Research program (CRP) which deals with all aspects of the production of medical radioisotopes that can be used for diagnostic and therapeutic purposes, requires a reliable database for production cross sections, not only for the main and the monitor reactions but also for the associated producing impurity reactions (IAEA-TECDOC-468, 2009). The program includes targetry (preparation, cooling and chemistry), yields, radionuclidic impurities, radiation dose from targets and target backings. By revising the database situation for ^{99}Mo & $^{94,95g,95m,96(m+g),96g,99m}$Tc production, it could be seen that the status of the present information is still not satisfactory for a detailed optimization of the production processes. Several authors (Kormali, et al., 1976; Takács, et al., 2002; Bonardi, et al., 2002; Uddin, et al., 2004; Khandaker, et al., 2006; Khandaker, et al., 2007; Uddin, et al., 2008) have reported a variety data for proton-induced reaction cross-sections on molybdenum in the medium-energy range, but large discrepancies can be found among them. These discrepancies limit the reliability of data evaluations.

2. Experimental techniques

The reaction cross-section of the proton-induced reactions on molybdenum were measured, in this work, as a function of proton energy in the range from the respective threshold for each contributing reaction (E_{thr}) to about 40 MeV using the activation method and the well-established stacked foil technique combined with high resolution gamma-ray spectroscopy.

2.1 Stacked foil technique

By this method a series of thin target foils are put together to form the target as in Figure 1. Each target foil (Mo in this study) is followed by another material (mainly Al in our case) to catch the ejected product nuclides (recoils) from the preceding Mo foil. This catcher foil is selected so that it does not produce any radioactive product by the given bombarding particle at the energy range used. The catchers should be also as low Z- material as possible to decrease the gamma attenuation during the activity measurements. Therefore, a pair of foils (Mo+Al catcher) will contain the total produced radioactive isotopes from the given Mo foil after the irradiation. The catcher Al foil contains only the ejected atoms (radionuclides) from the Mo implanted into it. The advantage of the stacked foil method is that one can get a whole excitation function curve using a lower number of irradiations. Another advantage of this method is that each target of the stack is irradiated with the same integrated beam charge. The main conceptual disadvantage of the staked foil technique is concerned with the energy straggling that is induced in the beam by passing through the stack of thin foils, recoil catchers and energy degraders (Zeigler, J.F., 1995). The inaccuracy of the foil thickness and surface roughness, which cause the accumulation of the error in energy calculations from the first to the last foil of the stack, which can be corrected by inserting some beam current monitor foils in different regions over the stack.

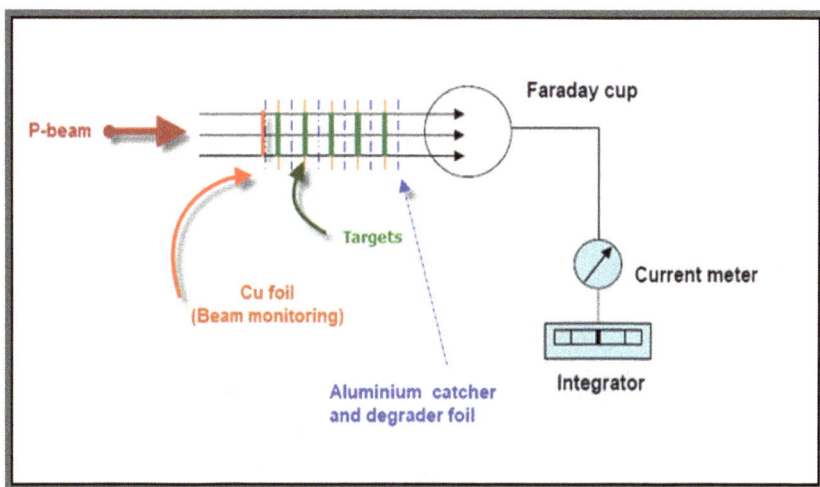

Fig. 1. Schematic diagram of the stacked foil arrangements

2.2 Target holder and experimental setup
An aluminum target holder (12 mm aperture) was designed as shown in Figure 2. It also acts as a Faraday cup equipped with secondary electron suppressor by applying -300 Volts to an electrically isolated cylinder attached to the target holder. An earthed collimator ring (10 mm diameter) was placed in front of the holder facing the beam. This target holder was attached to a reaction chamber shown in Figure 3, which adapted for the activation purpose. The total charges collected by the Faraday cup have been integrated using current integrator circuit with good linearity at low current values. The target foils of 10 mm diameter were sufficiently larger than the proton beam diameter. Care was taken to ensure that equal areas of the monitor and the target foils intercepted the beam. The irradiation geometry used guaranteed that practically the whole beam passed through every foil. The secondary effect

of the interactions of the secondary produce neutrons with the molybdenum targets was checked by placing some foils in the end of the stack far behind the range of the fully stopped proton beam followed by the measurement of its activities.

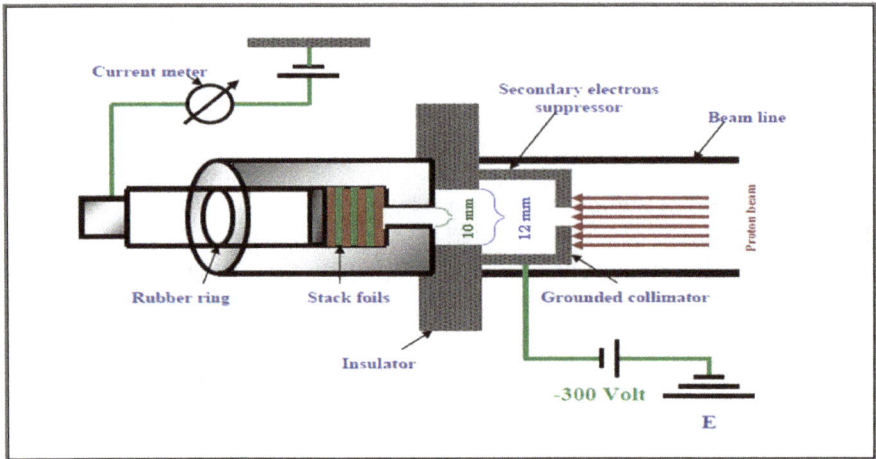

Fig. 2. Schematic diagram of the target holder and the Faraday cup

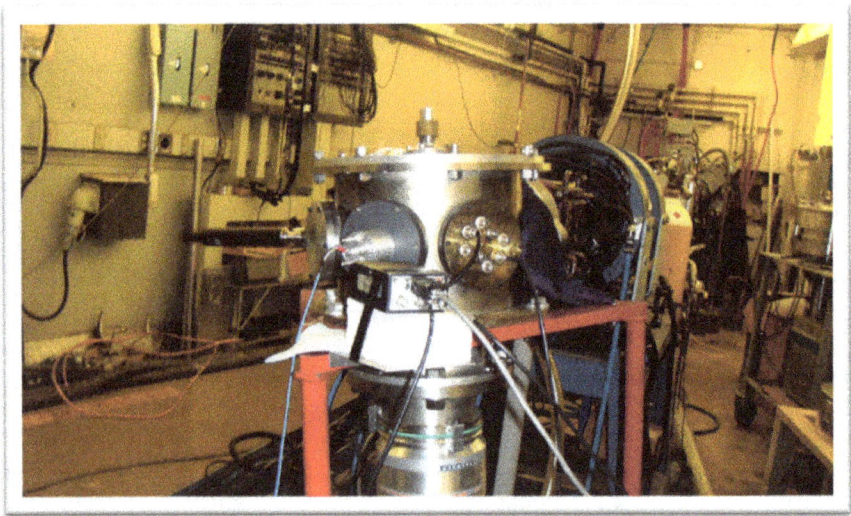

Fig. 3. A photograph of the experimental setup

2.3 Targets and irradiations

Thin foils of molybdenum with natural isotopic composition were used as our main targets. There are 35 known isotopes of molybdenum ranging in atomic mass from 83 to 117, as well as four metastable nuclear isomers. The seven stable isotopes are listed in Table 2 (Audi, et al., 2003). All unstable isotopes of molybdenum decay into isotopes of niobium, technetium, and ruthenium.

Isotope	Natural abundance (%)
^{92}Mo	14.84
^{94}Mo	9.25
^{95}Mo	15.92
^{96}Mo	16.68
^{97}Mo	9.55
^{98}Mo	24.13
^{100}Mo	9.63

Table 4. Most stable radioisotopes of molybdenum

The irradiations were performed using an external beam of accelerated protons with energy of about 40 MeV provided by K500 superconducting cyclotron at Texas A&M University, Cyclotron institute, USA. Two different sets of stacks were irradiated to cover the energy range from the respective threshold for each reaction up to 40 MeV. Each stack was made of several groups of targets; ^{nat}Mo (99.999% and 50 μm thickness) as the main target foils, ^{nat}Cu (99.98% and 125 μm thickness) were used as monitor foils that acted also as beam degraders and ^{nat}Al (99.999% and 50,100 μm thickness) as catcher foils, all foils were supplied by Goodfellow, Cambridge, UK. The set of foils was pressed together to avoid air gaps between them, which could have influence on the vacuum and particles stopping. The proton energy degradation along the stack was determined using the computer program SRIM-2003 assuming the incident energy was 40 MeV (Ziegler, et al., 1985). The irradiation conditions for each stack are shown in Table 5.

Stack number	Incident energy (MeV)	Energy range (MeV)	Irradiation time (hour)	Beam current (nA)
Stack 1	39.4 ± 0.4	39.4 - 19	30 min	27
Stack 2	20.3 ± 0.8	20.3 - 0	50 min	24

Table 5. Irradiation conditions in the experiments relevant to cross-section measurements

2.4 Monitor reactions

To confirm the cyclotron beam intensity and energy, a thin copper monitor foil (50 μm) was placed in the front of the stack (Al-Saleh, et al., 2006). Copper is an ideal target material with respect to its availability, physical, mechanical and chemical properties to be used in monitoring process. This Cu foil was irradiated simultaneously with the main target foils and then analyzed with the same gamma ray spectrometer in a comparable geometry. Thus, the ratio (α_{exp}) between the measured cross section values for the $^{63}Cu(p,n)^{63}Zn$ and $^{63}Cu(p,2n)^{62}Zn$ nuclear reactions can be calculated using equation (1) (Piel, et al., 1992):

$$\alpha_{exp} = \frac{\sigma_{62_{Zn}}}{\sigma_{63_{Zn}}} = \frac{A_{62}\left(1 - e^{-t_b \lambda_{63}}\right)}{A_{63}\left(1 - e^{-t_b \lambda_{62}}\right)} \tag{1}$$

where, t_b is the irradiation time, A_{62} and A_{63} are the measured decay activities for both ^{62}Zn and ^{63}Zn, respectively. By comparing the determined ratio which found to be (0.0118) with the ratios obtained from the recommended cross-section values by the IAEA (Tárkányi, et al., 2001) and plotted, in dotted line, as a function of the proton energy in Figure 4. The energy value of the accelerated protons was estimated to be $E_p = 39.4 \pm 0.4$ MeV.

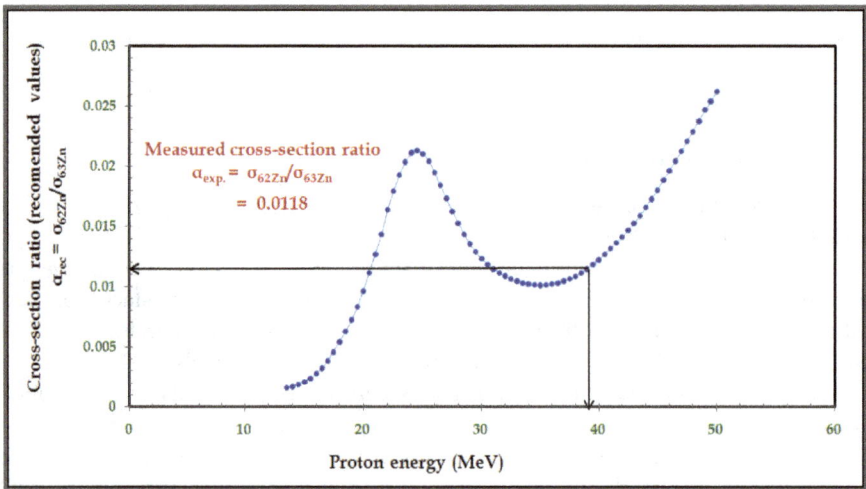

Fig. 4. The energy calibration for the Proton beam using the α_{exp} ratio for σ_{62Zn} and σ_{63Zn}

The measured Cu monitoring reactions were also used for beam intensity calculations, using the reverse relation to the well-known reaction cross section values. The charge collected in the Faraday cup was registered, from which the average beam current was deduced. The two results generally agreed within 10%. The uncertainty of the proton energy along the stack was checked by inserting Al and Cu monitor foils into different points of the stack then by comparing the measured excitation functions for $^{nat}Al(p,x)^{22,24}Na$ and $^{nat}Cu(p,x)^{62,63,65}Zn$ monitor reactions with their recommended values (Tárkányi, et al., 2001), as shown in Figure 5. The individual uncertainties of the contributing reactions were taken into account considering the cumulative effects. The total uncertainty for each energy point depends on the irradiation circumstances and the position of each foil in the stack. These are the uncertainties of the target homogeneity and thickness, the incident beam energy and the beam straggling. Typical uncertainty in the energy was (± 0.3 MeV) at the beginning of the stack and (± 1.2 MeV) at the end. Furthermore, the very good agreement with the recommended values for the measured cross-sections of the studied monitoring reactions confirms the reliability of our experimental setup.

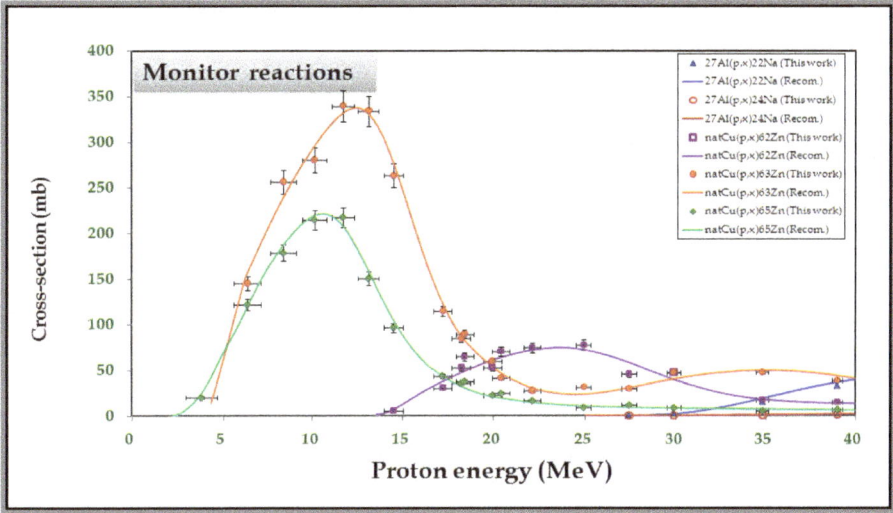

Fig. 5. Excitation functions of the monitor reactions compared with the recommended cross-sections by the IAEA.

2.5 Radioactivity measurements

The radioactivity of the residual nuclei in the activated foils was measured nondestructively using a HPGe γ-ray detector with 70% efficiency relative to a (3"x3") NaI detector, and energy resolution of 2.2 keV for the 1.332 MeV γ-line of the ^{60}Co standard source, a peak to Compton ratio of 58: 1. The detector absolute efficiencies for various source-detector distances and photon energies were determined experimentally by using a selected set of γ-ray standard sources (^{60}Co, ^{137}Cs, ^{133}Ba and ^{152}Eu), of known activities, to cover the whole energy range of the studied γ-rays. The detector-sample distance was kept large enough to ensure the point source geometry and to keep the dead time within 8% or less. In addition to the main characteristic γ-lines for each studied radioisotope, some other weaker γ-lines were also considered to minimize the relative errors due to counting statistics, wherever possible. In the cases of the longer-lived radionuclides, activity measurements were carried out after sufficient cooling time, which is enough for the complete decay of most of the undesired short-lived isotopes, to avoid any possible interference of nearly equal energies γ-lines. The stack was dismantled and each foil was counted 2-3 times after different cooling times following the end of bombardment EOB to avoid disturbance by overlapping γ-lines from undesired sources and to evaluate accurately the cross-sections for cumulative formation of the corresponding longer-lived daughter radionuclide.

Figure 6 presents an example of the calibrated measured γ-ray spectrum with identified γ-lines covering the energy range up to 1350 keV. Table 6 shows the contributing reactions and the decay data of all the investigated radionuclides, which were taken from the Table of Isotopes (Firestone, 1998 and T-16, Nuclear Physics Group, LANL 1997).

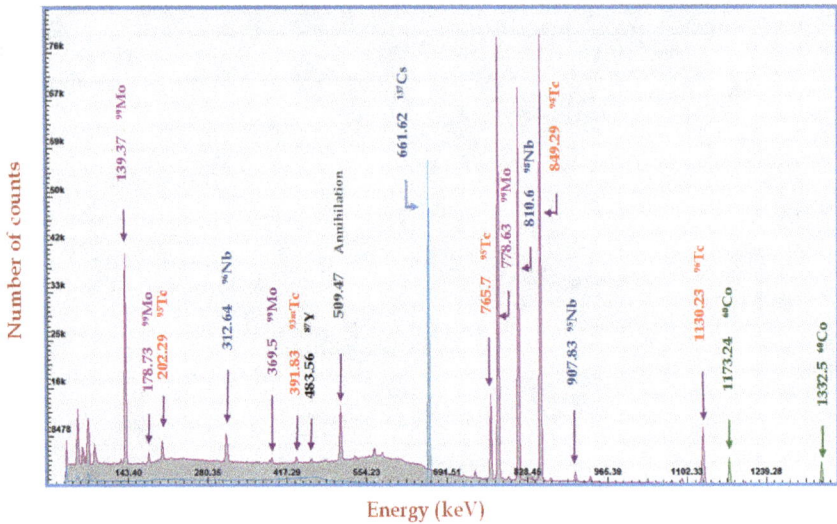

Fig. 6. A calibrated Gamma ray spectrum with identified γ-lines

Nuclide	Half life	Principal contributing reactions	Q-value MeV	Decay mode	E_γ keV	$I\gamma$ %
^{99}Mo	2.75 d	^{100}Mo(p,pn) ^{99}Nb→decay	-8.30 -11.14	β⁻ (100)	140.51 181.07 739.5	89.43 5.99 12.13
^{94g}Tc	4.88 h	^{94}Mo(p,n) ^{95}Mo(p,2n) ^{96}Mo(p,3n)	-5.03 -12.41 -21.56	EC (87.94%) β⁺ (11.71%)	702.63 849.92 871.08	99.6 95.7 100
^{95g}Tc	20 h	^{95}Mo(p,n) ^{96}Mo(p,2n) ^{97}Mo(p,3n) ^{96m}Tc→ decay	-02.47 -11.63 -18.45	EC (100%)	765.79 947.67 1073.71	93.82 01.95 03.74
^{96g}Tc	4.28 d	^{96}Mo(p,n) ^{97}Mo(p,2n) ^{98}Mo(p,3n)	-03.76 -10.58 -19.22	EC (100%)	778.22 812.58 849.92	99.76 82.0 98.0
^{96m}Tc	51.50 min	^{96}Mo(p,n) ^{97}Mo(p,2n) ^{98}Mo(p,3n)	-03.76 -10.58 -19.22	IT (98%) EC (2%)	34.28 778.22 1200	100.0 01.90 01.08
^{99m}Tc	6.01 h	^{100}Mo(p,2n) ^{99}Mo→decay	-7.60	IT +β⁻ (100)	140.51	89.06

Table 6. The contributing reactions and the decay data of the investigated radioisotopes

2.5.1 Separation of interfered γ-lines

Some investigated radionuclides emit γ-rays that have very close energies, which were difficult to be separated using the HPGe spectrometer.

The individual activities of those overlapped γ-rays were analyzed using the difference in half-lives of the contributing nuclides by plotting the γ-ray emission rate as a function of time. Figure 7 shows the radioactive decay curve for the 140.5 keV γ-peak which resulted from the decay of the directly produced ^{99}Mo (65.94 h, 140.51 keV), the directly and indirectly produced ^{99m}Tc (6.01 h, 140.51 keV), and ^{90}Nb (14.6 h, 141.2 keV). The radionuclides decay completely in the order of their half-lives, ^{99}Mo the longest-lived nuclide is the last to decay. After more than 14 days, the remaining activity was due to decay of the daughter nuclide ^{99m}Tc in transient equilibrium with the parent ^{99}Mo radionuclide. The activities of the radionuclide; ^{99}Mo(A_2) $\rightarrow$^{99m}Tc(A_1) at the end of bombardment (EOB) were estimated by using equation (2) (Uddin, et al., 2004):

$$A_{1(EOB)} = \frac{A_{1+2}(\lambda_2 - \lambda_1)}{\exp(-\lambda_1 t_c) + \lambda_2[\exp(-\lambda_1 t_c) - \exp(-\lambda_2 t_c)]} \tag{2}$$

where t_c is the respective cooling time, λ_1 and λ_2 are the decay constants of ^{99}Mo and ^{99m}Tc, respectively, and $A_{1(EOB)}$ is the activity of ^{99}Mo at the EOB. To separate the activities after the EOB of ^{90}Nb$(A_{3(EOB)})$ and ^{99m}Tc$(A_{4(EOB)})$, we used the following equation (3):

$$A_{3+4} = A_{3(EOB)} \exp(-\lambda_3 t_c) + A_{4(EOB)} \exp(-\lambda_4 t_c) \tag{3}$$

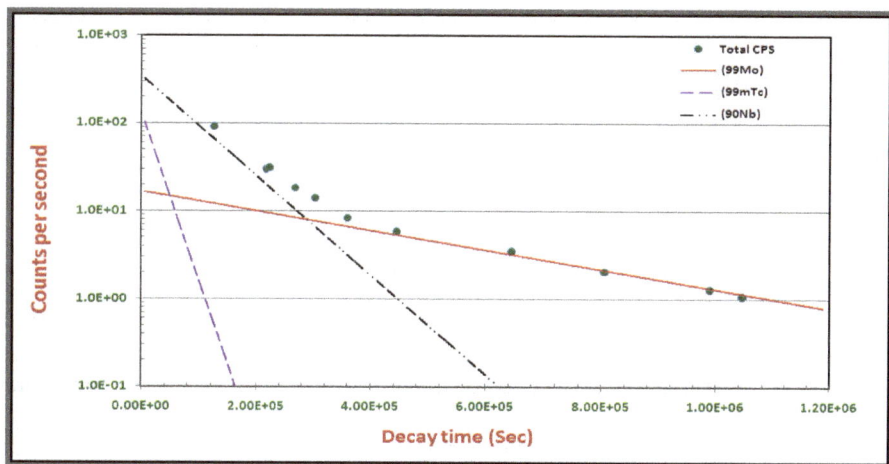

Fig. 7. Resolving the 140 keV γ-line which produced from three different radioisotopes ^{99}Mo, ^{99m}Tc and ^{90}Nb

The daughter ^{99m}Tc activity decreases from the maximum at a constant rate, which depends on the decay rate of ^{99}M. Then the directly produced ^{99m}Tc completely decayed out before the measurement. The measured activity for the 140.5 keV γ-line was the sum of the γ-line from the daughter ^{99m}Tc and from ^{90}Nb. We deduced the activities of 140.5

and 141.2keV γ-lines from the independent γ-lines of ^{99}Mo and ^{90}Nb, respectively; an excellent agreement was obtained when compared with the results of radioactive decay curve.

2.6 Cross section calculations and uncertainty

The reaction cross sections for the nuclear reactions ^{nat}Mo(p,x) were calculated using the activation formula as in equation 4 considering the decay data and rates of the radioactive isotopes produced, the detector absolute efficiency, and the measured beam intensity (Helus & Colombetti, 1980).

$$\bar{\sigma} = \frac{M\,Z\,e\,\lambda\,T_\gamma}{I_\gamma \Delta x\, N_A\, f\, \rho\, I\, \varepsilon_{abs}\, (1 - e^{-\lambda\, t_b})\, e^{-\lambda\, t_c}\, (1 - e^{-\lambda\, t_m})} \tag{4}$$

Whereas; M is the target molecular weight, Ze is the projectile charge, λ is the decay constant, T_γ is the net area under each γ-peak, I_γ is the gamma line intensity, Δx is the thickness of each target foil, N_A is the Avogadro's number, f is the abundance of the isotope, ρ is the target density, I is the beam intensity, ε_{abs} is the detector efficiency corresponding to each γ-line energy, t_c is the cooling time and t_m is the measuring time.

The total experimental error was calculated by combining the individual errors as a square root of the sum of squares of the contributing relative errors, which are the lack of precision in: measuring the absolute detector efficiency of 3-6%, the calculation of the area under the photoelectric peak 1-4%, measuring the current intensity 4-7%, the calculation of irradiation time 2 %, determining the foil thicknesses and composition 1-4% and the nuclear decay data of 3%. The total experimental errors were obtained to be (8-12%). The total uncertainty in each energy point depends on the irradiation circumstances and the position of the foil in the stack.

3. Nuclear model calculations

All the measured cross sections over the whole energy range were simulated using TALYS (Koning, et al., 2008) code. A short description for the codes is given in the following:

3.1 TALYS code

We calculated the independent formation cross sections for both the ground and/or the isomeric states by using the TALYS code, which is a computer program that integrates all types of nuclear reactions in the energy range of 1 keV-200 MeV. TALYS incorporates modern nuclear models for the optical model, level densities, direct reactions, compound reactions, pre-equilibrium reactions, fission reactions, and a large nuclear structure database (Koning, et al., 2008). The database of this code is derived from the (Reference Input Parameter Library, http://www-nds.iaea.org/ripl2/). The pre-equilibrium particle emission is described using the two-component exciton model. The model implements new expressions for internal transition rates and new parameterization of the average squared matrix element for the residual interaction obtained using the optical model potential. The phenomenological model is used for the description of the pre-equilibrium complex particle emission. The contribution of direct processes in inelastic scattering is calculated using the

ECIS-94 code (Raynal, 1994) incorporated in TALYS (Raynal, 1994). The equilibrium particle emission is described using the Hauser-Feshbach model. The default optical model potentials (OMP) which used in TALYS are the local and the global parameterizations for neutrons and protons. These parameters can be adjusted in some cases by the user. The present results of all the calculated excitation functions were evaluated using the default values of the code.

4. Result and discussion

The experimentally constructed excitation functions for the main investigated ^{nat}Mo(p,x)^{99}Mo,$^{94g,95g,96(m+g),99m}$Tc nuclear reactions are shown in Figures 8-12 together with the results of the theoretical calculation using TALYS code and the previously published data. The numerical values of the present experimental cross-sections and their estimated uncertainties are presented in Table 7.

Proton Energy MeV	Reaction Cross-section (mb)				
	^{nat}Mo(p,xn)^{99}Mo	^{nat}Mo(p,xn)^{94g}Tc	^{nat}Mo(p,xn)^{95g}Tc	^{nat}Mo(p,xn)$^{96(m+g)}$Tc	^{nat}Mo(p,xn)^{99m}Tc
39 ± 0.3	159 ± 10	62 ± 6	113 ± 11	73 ± 7	17 ± 2
35 ± 0.3	166 ± 11	75 ± 7	109 ± 11	122 ± 12	22 ± 2
30 ± 0.4	165 ± 11	77 ± 6	84 ± 9	184 ± 11	20 ± 2
27 ± 0.4	159 ± 10	75 ± 6	90 ± 9	192 ± 11	28 ± 2
25 ± 0.4	141 ± 10	77 ± 7	120 ± 11	173 ± 12	35 ± 3
22 ± 0.5	122 ± 9	72 ± 7	158 ± 12	115 ± 12	84 ± 8
20 ± 0.5	103 ± 9	69 ± 7	146 ± 12	100 ± 10	120 ± 10
20 ± 0.5	95 ± 9	74 ± 7	140 ± 11	97 ± 9	152 ± 11
18 ± 0.5	79 ± 8	70 ± 7	122 ± 11	100 ± 9	182 ± 12
18 ± 0.5	71 ± 8	72 ± 7	120 ± 11	95 ± 9	202 ± 15
17 ± 0.6	51 ± 5	73 ± 7	115 ± 11	110 ± 9	220 ± 18
15 ± 0.6	19 ± 10	60 ± 6	120 ± 11	127 ± 10	222 ± 16
13 ± 0.6	10 ± 1	43 ± 4	106 ± 10	153 ± 13	196 ± 16
12 ± 0.7	4 ± 1	28 ± 2	84 ± 8	165 ± 11	170 ± 14
10 ± 0.7	2 ± 0.3	9 ± 0.7	77 ± 8	125 ± 10	116 ± 11
8 ± 0.8			66 ± 7	89 ± 8	6 ± 0.9
6 ± 0.9				53 ± 6	

Table 7. Measured cross-sections for the proton-induced nuclear reactions on ^{nat}Mo.

4.1 Excitation functions
4.1.1 ^{nat}Mo(p,xn)^{99}Mo

^{99}Mo is produced by proton activation on ^{nat}Mo target via the contribution of two reaction channels ^{100}Mo(p,pn)^{99}Mo (Q= 8.3 MeV) and ^{100}Mo(p,2p)^{99}Nb (Q= 11.14 MeV) through the β-decay of the parent isotope ^{99}Nb(15 s). The highest cross-section value of about 160 mb corresponds to E_p= 30 MeV.

A comparison between our measured cross-sections and the previously reported data together with the theoretical calculations using TALYS code is presented in Figure 8. (Takács, et al., 2003) reported cross-section data up to 37 MeV and (Levkovskij, 1991) reported up to 29 MeV for ^{99}Mo production on the enriched ^{100}Mo isotope. Our measured values are consistent with the data presented by (Uddin, et al., 2004). The data reported by (Scholten, et al.,1999) are consist with our data in energy range lower than 22 MeV, although his results at the higher energies are scattered. Our results showed agreement with (Takács, et al., 2003) in low energy region. The data presented by (Levkovskij, 1991) are about 25% higher than our data. (Lagunas-solar, et al., 1991) reported numerical cross-section data that are much lower than our measured data and the other published data as well in the energy region above 20 MeV. A good agreement exists between the measured cross-sections and the TALYS code calculations within the experimental error and that fact confirms the reliability of our measured data.

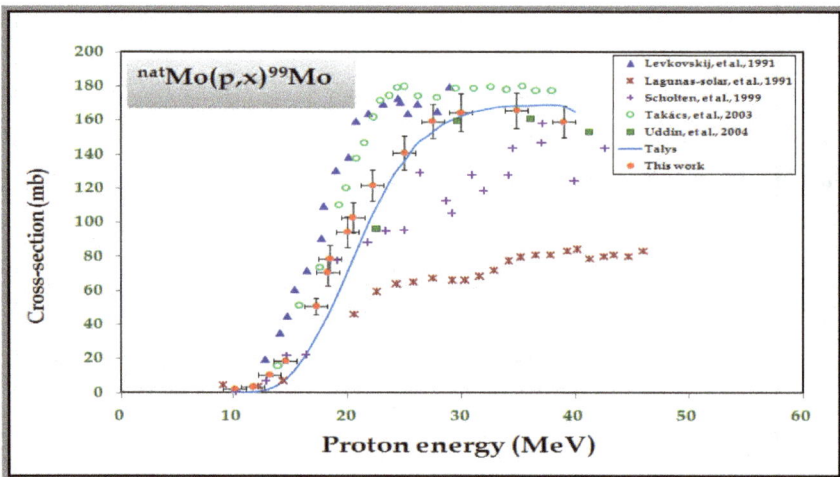

Fig. 8. Excitation function of the ^{nat}Mo(p,x) reaction (full red dots with vertical and horizontal error bars) compared to some previously published results and the TALYS code calculations (curve).

4.1.2 ^{nat}Mo(p,xn)^{94g}Tc

^{94}Tc has two isomeric states, metastable state ^{94m}Tc (T½ = 52 min, 2$^+$) and ground state (T½ = 4.86 h, 7$^+$). We studied the excitation function for the ground state only due to the relatively short half-life of the metastable state. The contribution of the isomeric transition (IT< 0.1) for ^{94m}Tc is small enough to be neglected. Therefore, we can study each state separately by eliminating the interfering gamma rays from the measurements, such as 849.92 keV and

871.08 keV as listed in Table 6. Mainly we used the 702.63 keV γ-line, which has no interference with any other γ-lines from any other produced isotopes in a cooling time of about 5 hours, to determine the cross section for ^{94g}Tc production.

The present experimental excitation function for the reaction ^{nat}Mo(p,xn)^{94g}Tc is presented in Figure 9 together with the previously published results and the calculated cross sections by the used nuclear model code TALYS.

A good agreement is found between our measured cross sections and the ones reported by (Bonardi, et al., 2002 and Uddin, et al., 2004) over the entire energy range. There is a remarkable difference between the present results and the reported data by (Khandaker, et al., 2007) especially for the energies lower than 20 MeV and above 30 MeV. The measured cross sections by (Kormali, et al., 1976) show about 40% lower values than our data in the energy range from 11-20 MeV. The TALYS code calculation is about 50% higher than our measured data and higher than all the previously reported data sets.

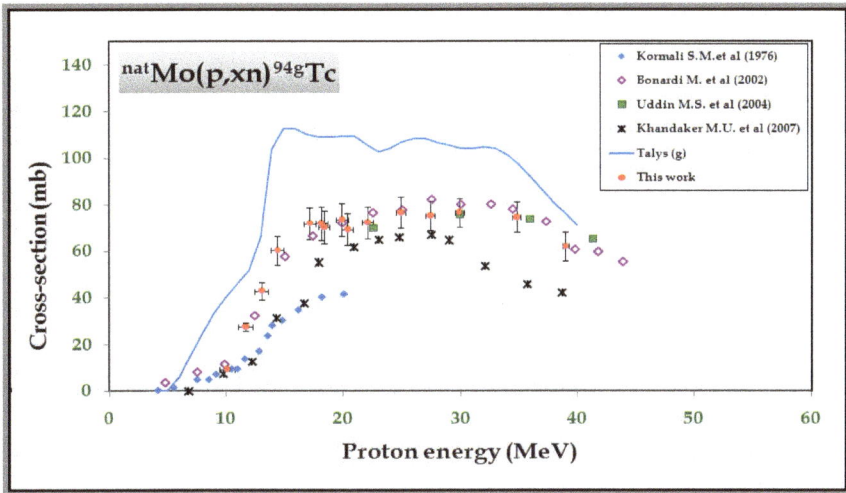

Fig. 9. Excitation function of the ^{nat}Mo(p,x) reaction (full red dots) compared to some previously published results and the TALYS code calculations (curve).

4.1.3 ^{nat}Mo(p,xn)^{95g}Tc

^{95}Tc is formed in two different states: the longer lived isomeric state ^{95m}Tc (T½ = 61 d, 1/2⁻) and the shorter lived ground state ^{95g}Tc (T½ = 20 h, 9/2⁺). In this study, we report only the measured cross sections for ^{95g}Tc due to the difficulty in measuring the interfering characteristic γ-rays for ^{95m}Tc as shown in Table 6. The ^{95g}Tc activity measurement was based on detecting the main γ-line at 765.79 keV. A comparison of the present measured data with some previously reported data and the TALYS code calculations is shown in Figure 10.

The cross section is only measurable at 8 MeV, then increases gradually due to the ^{95}Mo (p,n) reaction. The contribution of the ^{96}Mo (p,2n) reaction appears as a little plateau starting at about 12 MeV, while the ^{97}Mo (p,3n) reaction contribution starts at about 20 MeV, creating another small peak. There is a good agreement between our experimental excitation function and the previously published data by (Bonardi, et al., 2002 and Khandaker, et al., 2007) within the experimental error, while the earlier presented study by (Khandaker, et al., 2006) shows 35% higher value than our experimental data at energies above 26 MeV. However, the data reported by (Birattari, et al., 2002) shows higher cross-section values in the proton energy range > 10 MeV. The presented data by (Uddin, et al., 2004) shows inconsistency with most of the other experimental data, , especially for the point at about 22 MeV.

The TALYS code calculation results are in good consistency with our experimental data within the experimental error, but there exists a small drop in the measured cross section values in the higher values of the energy range.

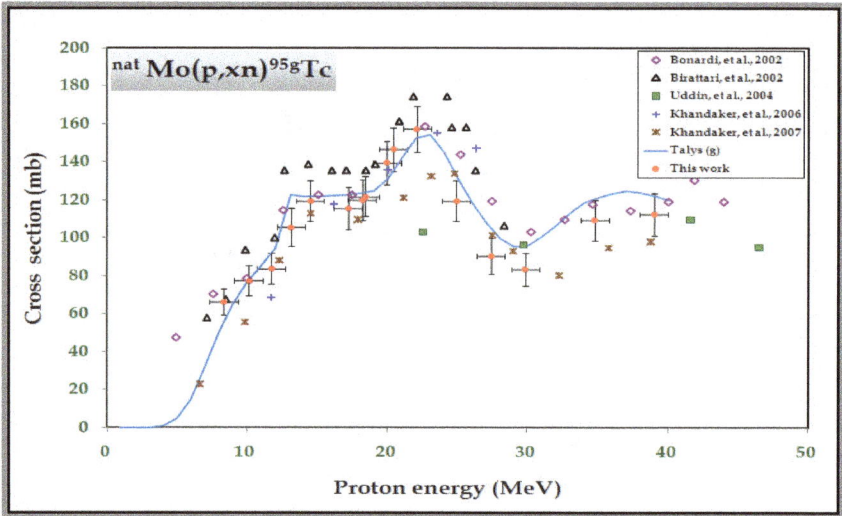

Fig. 10. Excitation function of the ^{nat}Mo(p,x)^{95g}Tc reaction compared to some previously published results and the TALYS code calculations.

4.1.4 ^{nat}Mo(p,xn)$^{96(m+g)}$Tc

^{96}Tc is formed in two energy states: ^{96m}Tc (T½ = 51.5 min, 4⁺) that decays by 98% isomeric transition to the ground state ^{96g}Tc (T½ = 4.28 d, 7⁺). In this study we measured the cross-section of ^{96g}Tc using the main characteristic γ-line 778.2 keV, while it was not possible to measure the characteristic isomeric transition 34.28 keV of the metastable state due to the intensive interfering of the X-rays. According to the short half-life and the high IT decay rate of the metastable state, we can consider the measured cross section as the total cross section of $^{96(m+g)}$Tc without measuring the metastable state independently. Figure 11 illustrates a comparison between our measured cross sections and the available published data together with the TALYS code calculations. Some findings can be summarized from this figure as follows:

- The first part of the curve is due to ^{96}Mo(p,n) reaction. It starts to increase rapidly to form a peak at 12 MeV. Then it decreases slowly and forms a plateau in the range 16-21 MeV due to the contribution of the ^{97}Mo(p,2n) and ^{98}Mo(p,3n) reactions which start at

11 and 19 MeV, respectively. The rapid increase in the cross-section values at energies higher than 22 MeV indicates the increasing contribution of the (p,3n) reaction.

- Very good agreement is found in the energy range above 9MeV between the present data and those reported by (Takács, et al., 2002; Uddin, et al., 2004 & Khndaker, et al 2006,2007).
- The results by (Bonardi, et al., 2002) overestimate the cross-section value in the energy range < 10 and >26 MeV.
- The data by (Khandaker, et al., 2007) are somewhat low in the proton energy range below 10 MeV.

An overall good agreement is found between the present experimental excitation function for $^{96(m+g)}$Tc formation and the calculated theoretical results by TALYS code and the recommended data (Takács, et al., 2002), within the experimental error.

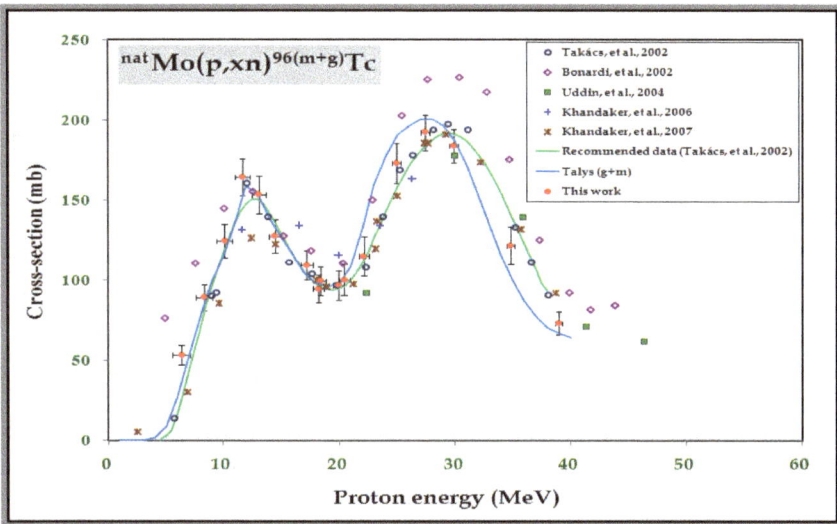

Fig. 11. Excitation function of the ^{nat}Mo(p,x)$^{96(m+g)}$Tc reaction compared to some previously published results and the TALYS code calculations.

4.1.5 ^{nat}Mo(p,xn)^{99m}Tc

Three reactions contribute to the production of ^{99m}Tc by direct way are ^{98}Mo(p,γ), ^{100}Mo(p,2n), and indirect way by ^{100}Mo(p,pn). Possibly, the highest contribution is from the ^{100}Mo(p,2n)^{99m}Tc reaction (on the 9.63% ^{100}Mo present in the highly chemically pure Mo sample). Activity of ^{99m}Tc was measured in this work by detecting the gamma peak at energy 140.5 MeV after the resolution of this peak as described before. The measured excitation function is compared with some earlier published data and the TALYS code calculations in Figurer 12. The data of (Takács, et al. 2003) and (Kandaker, et al. 2007) fit nicely our measured data specially in the low energy part up to 20 MeV. At higher energies (Kandakar, et al. 2007) data clearly over estimate our results. The results of (Challan, et al. 2007) agree with our results except the last two points. The cross section data for (Scholtan, et al. 1999) are clearly lower than our values over the hall energy range. The TALYS calculations over estimate the present results, especially in the energy range lower than 18 MeV , while they fit, within the experimental errors in the higher range.

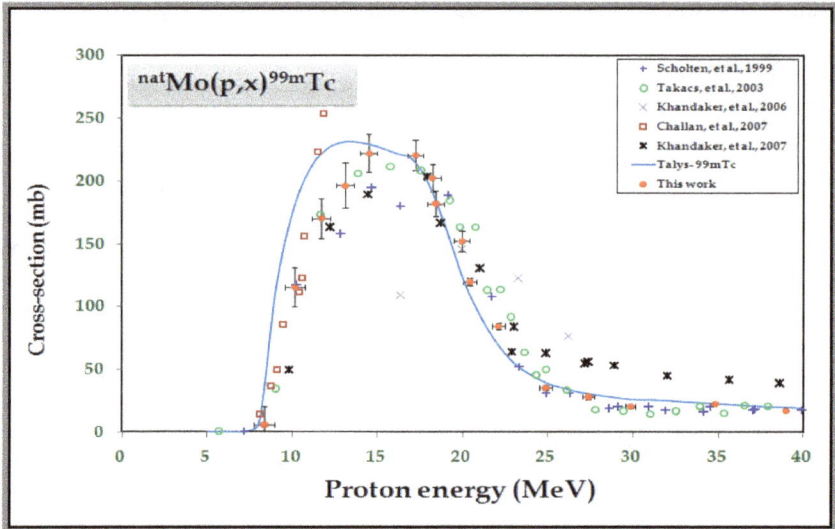

Fig. 12. Excitation function of the natMo(p,x)99mTc reaction compared to some previously published results and the TALYS code calculations.

4.2 Integral yield calculations

The integral yields, at the end of bombardment, for the production of the different isotopes were derived using the measured excitation functions for the production of these radioisotopes. The method was done by assuming the thick target as dividend to several thin targets each of an equivalent thickness of about 0.5 MeV. The cross section at each thin target is assumed constant, because of the small energy interval through the target. The number of target atoms/cm2 was calculated using the target thickness, which reduce the proton energy by 0.5 MeV. The differential yield produced in each thin target was calculated using the following equation (5):

$$Y(E)\left(\frac{MBq}{\mu A.h}\right) = N.P.\bar{\sigma}(E).10^{-30}.\left(1 - e^{-\lambda t_b}\right) \qquad (5)$$

Whereas, $\bar{\sigma}(E)$ (mb) is the average cross section at a specific energy; N is the number of target atoms/cm2; λ is the decay constant for the produced isotopes; P is the number of incident protons/sec for (1 µA) and the irradiation time (t_b= 1 h). We then calculated the integral target yield by summing up the differential yields.

Figure 13 represents the values of the integral target yield for the studied reactions as a function of the proton energies. Obviously, the yields of the investigated radioisotopes increase with the proton energy and start to saturate at energy of about 30 MeV. The nearly saturation values for 99Mo, 94gTc, 95gTc, 96(m+g)Tc, and 99mTc are equal to 110, 600, 310, 90 and 910 MBq/µA.h, respectively.

For the production of ^{99m}Tc via cyclotron, it is highly recommended to use an enriched target of ^{100}Mo to exclude all the other impurities by using the indirect ^{100}Mo(p,pn)^{99}Mo and the direct ^{100}Mo(p,2n)^{99m}Tc nuclear reactions. From the present data we conclude that the optimum energy range for the production of ^{99m}Tc directly and indirectly using protons is Ep= 35-18 MeV, the integral target yield amounting to to 412 MBq/µA.h to 1000 MBq/µA.h at saturation with respect to the half lives of both ^{99}Mo and ^{99m}Tc.

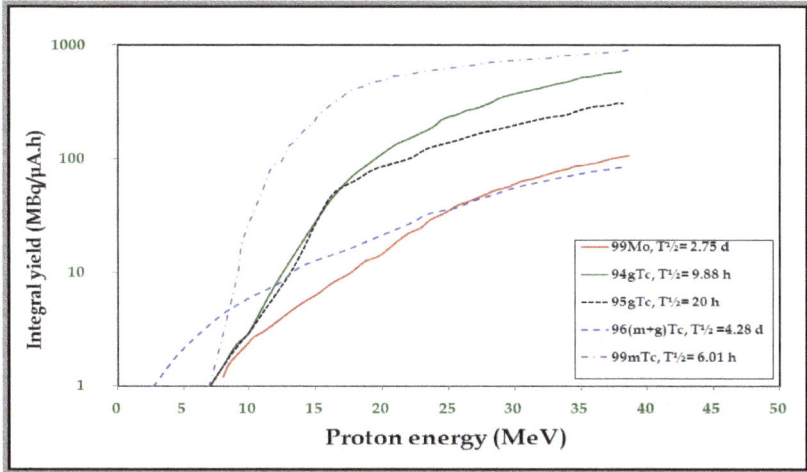

Fig. 13. Integral Yields for the ^{nat}Mo(p,x)^{99}Mo,$^{94g,95g,96(m+g),99m}$Tc nuclear reactions calculated from the excitation functions measured in this work.

5. Conclusion

^{99m}Tc radioisotope is a very important medical radioisotope for diagnostic tests. In this work an alternative root of producing this isotope, either directly or through the generator ^{99}Mo (^{99m}Tc) , namely using cyclotrons, is introduced and discussed. The excitation functions for the different proton-induced nuclear reactions on ^{nat}Mo target are measured and compared with some previously measured data. This study aims to resolve some contradictions between the existing data, and to give a reliable data set for the production of ^{99m}Tc and some other isotopes of importance in nuclear medicine beside some impurities. Monitoring reactions on Al and Cu targets are also measured and compared with the recommended IAEA data sets, in order to give high degree of consistency to our results. The present excitation functions confirm some previously measured sets, while contradict with others. Theoretical code calculations using TALYS code are performed and show a good consistency with the measured cross section values. The code calculations can be used for cross section estimations, when not enough experimental data exist. Furthermore, the integral or thick target yields are estimated based on the measured excitation functions for all the investigated reactions. Finally, it is well known that for medical uses, enriched targets have to be used in the production to avoid the secondary produced unwanted impurities. While the studies on natural targets, gives an idea about the suitable energy range for maximum production of the wanted isotope and minimum of the impurities.

6. Acknowledgment

We thank G. J. Kim, D. P. May, and the staff of the CI for delivering the stable beam of protons. One of the authors Dr. A. Alharbi wish to express her appreciation to the international Fulbright U.S. exchange Scholar Program. This work was supported in part by the United States Department of Energy Office of Nuclear Physics under award number DE-FG02-93ER40773, and the Texas A&M University Cyclotron Institute.

7. References

Al-Saleh, F.S.; Al-Harbi, A.A. & Azzam, A. (2006). Excitation functions of proton induced nuclear reactions on natural copper using a medium-sized cyclotron, Radiochim. Acta, 94, 391.

Audi, G.; Bersillon, O.; Blachot, J. & Wapstra, A. H. (2003). The NUBASE evaluation of nuclear and decay properties, *Nuclear Physics A*, 729, 3–128.

Birattari, C.; Bonardi, M.; Gini, L.; Groppi, F. & Menapace, E. (2002) J. *Nucl. Scien. Tech., Suppl.2*, 1302.

Bonardi, M.; Birattari, C.; Groppi, F. & Sabbioni, E. (2002). Thin-target excitation functions, cross-sections and optimised thick-target yields for $^{nat}Mo(p,xn)^{94g,95m,95g,96(m+g)}Tc$ nuclear reactions induced by protons from threshold up to 44MeV. No Carrier Added radiochemical separation and quality control, *Appl. Radiat. & Isot.*, 57, 617.

Challan, M.B.; Comsan, M.N.H. & Abou-Zeid, M.A. (2007) Thin Target Yields and Empire-II Predictions on the accelerator production of Technetium-99m, *Journal of Nuclear and Radiation Physics*, 2, 1, 1-12

Fabbender, M.; Novgorodov, A.F.; Rösch, F. & Qaim, S.M. (1994). Excitation functions of $^{93}Nb(3He,xn)^{93m,g,94m,g,95m,g}Tc$-processes from threshold up to 35 MeV :possibility of production of ^{94m}Tc in high radiochemical purity using a thermo chromatographic separation technique, *Radiochim. Acta*, 65, 215–221.

Firestone, R.B. (1998). Table of Isotopes, 8th edition, *John Wiley & Sons*, New York , USA.

Glenn, D.; Heger, S. & Hladik, W. (1997), Comparison of Characteristics of Solution and Conventional Reactors for Mo-99 Production, *Nuclear Technology*, Vol 118.

Gull, K.; Hermanne, A.; Mustafa, M.G.; Nortier, F.M.; Oblozinsky, P.; Qaim, S.M.; Scholten, B.; Shubin, Yu.; Takács, S.; Tárkányi, T.F. & Zhuang, Y. (2001). Charged particle cross-section database for medical radioisotope production: diagnostic radioisotopes and monitor reactions, *Final Report of a Co-ordinated Research Project*, IAEA-TECDOC-1211, IAEA, Vienna, Austria.

Helus, F. and Colombetti, L.G. (1980). Radionuclides Production. *CRC Press Inc.*, Boca Raton, Florida.

Hohn, A.; Zimmermann, K.; Schaub, E.; Hirzel, W.; Schubiger, P.A. & Schibli, R. (2008). Production and separation of "non-standard" PET nuclides at a large cyclotron facility: the experiences at the Paul Scherrer Institute in Switzerland, *Q. J., Nucl. Med. Mol. Imaging*, 52, 145–150.

IAEA-TECDOC-1065, (1999). Production technologies for molybdenum-99 and technetium-99m, *IAEA*, Vienna, Austria, ISSN 1011-289.

IAEA-TECDOC-468, (2009). Cyclotron produced radionuclides: physical characteristics and production methods, *IAEA*, Vienna, ISSN 0074–1914, ISBN 978–92–0–106908–5.

Khan, F.M. (2003).*The Physics of Radiation Therapy*, Third Edition, Lippincott Williams & Wilkins, USA.

Khandaker, M.U.; Meaze, A.K.M.M.H.; Kim, K.; Son, D.; Kim, G. & Lee, Y.S. (2006). Measurements of the proton-induced reaction cross-sections of ^{nat}Mo by using the MC50 cyclotron at the Korea Institute of Radiological and Medical Sciences, J. Korean Phys. Soc., 48, 821.

Khandaker, M.U.; Uddin, M.S.; Kim, K.S.; Lee, Y.S. & Kim, G.N. (2007). Measurement of cross-sections for the (p,xn) reactions in natural molybdenum, Nucl. Instrum. Methods Phys. Res., B 262, 171.

Koning, A. J.; Hilaire, S. & Duijvestijn, M.C. (2008). TALYS-1.0, Proceedings of the International Conference on Nuclear Data for Science and Technology - ND2007, April 22-27, 2007, Nice, France, eds. O. Bersillon, F. Gunsing, E. Bauge, R. Jacqmin and S. Leray, EDP Sciences, 211. Available online: http://www.talys.eu/home/

Kormali, S.M.; Swindle, D.L. & Schweikert, E.A. (1976). Charged particle activation of medium Z elements. II. Proton excitation functions, J. Radiat. Chem., 31, 437.

Krane; K.S. (1987). Introductory Nuclear Physics, John Wiley and Sons, Inc. New York, ISBN 978-812-6517-85-5, USA.

Lagunas-Solar, M.C.; Kiefer, P.M.; Carvacho, O.F.; Lagunas, C.A.; Ya Po Cha, (1991). J. Appl. Radiation Isotopes, 42, 463.

Lambrecht, R.M. & Montner, S.M. (1982). Production and radio chemical separation of ^{92}Tc and ^{93}Tc for PET, J. Labelled Compd. Radiopharm., 19, 1434–1435.

Levkovoskii, N. (1991). Middle mass nuclides (A=40-100) activation cross sections by medium energy (E+10-50 MeV) protons and α-particles (Experiment and Systematics). Inter-Vesti, Moscow.

Loveland, W.; Morrissey, D.J. & Seaburg, G.T. (2006). Modern Nuclear Chemistry, John Wiley & Sons, ISBN: 978-047-1115-32-0, USA.

Lamberecht, R.M. (1979). Positron emitting radionuclides-present and future status, Proceedings second international symposium on radiopharmaceuticals, 19-22 March, Seattle, Washington. The Society of Nuclear Medicine Inc., New York.

Nickles, R.J.; Nunn, A.D.; Stone, C.K. & Christian, B.T., (1993). Technetium-94m -teboroxime-synthesis, dosimetry and initial PET imaging studies, J. Nucl. Med., 34, 1058–1066.

Piel, H.; Qaim, S.M. & Stöcklin, G. (1992). Excitation functions of (p,xn) reactions on natNi and highly enriched 62Ni – possibility of production of medically important radioisotope 62Cu at a small cyclotron, Radiochem. Acta, 57, 1.

Qaim, S.M.; Clark, J.C.; Crouzel, C.; Guillaume, M.; Helmeke, H.J.; Nebeling, B.; Pike, V.W. & Stöcklin, G. (1993). PET radionuclide production. In: Radiopharmaceuticals for Positron Emission Tomography. (Stöcklin, G., Pike, V. W., Eds.), Kluwer Academic Publishers, Dordrecht, The Netherlands, pp. 1–42.

Qaim, S.M. (2000). Production of high purity ^{94m}Tc for positron emission tomographic studies, Nucl. Med. Biol., 27, 323–328.

Qaim, S.M. (2001). Nuclear data relevant to the production and application of diagnostic radionuclides, Radiochim. Acta, 89, pp. 223–232.

Qaim, S.M. (2003). Cyclotron production of medical radionuclides, Handbook of nuclear chemistry 4, Kluwer Academic Publishers, Dordrecht, Netherlands.

Qaim, S.M. (2004). Use of cyclotron in medicine, Radiation Physics and Chemistry, 71, pp. 917-926.

Raynal, J. (1994). Notes on ECIS-94, CEA Saclay Report No. CEA-N-2772.

Reference Input Parameter Library, Available online: http://www-nds.iaea.org/ripl2/

Rösch, F. & Qaim, S.M. (1993). Nuclear data relevant to the production of the positron emitting technetium isotope ^{94m}Tc via the ^{94}Mo (p,n) reaction, Radiochim. Acta, 62, 115–121.

Rösch, F.; Novgorodov, A.F. & Qaim, S.M. (1994). Thermo chromatographic separation of 94mTc from enriched molybdenum targets and its large scale production for nuclear medical application, *Radiochim. Acta*, 64, 113–120.

Sajjad, M. & Lambrecht, R.M. (1993). Cyclotron production of medical radionuclides, *Nucl. Instrum. Meth.*, B79, 911–915.

Scholten, B.; Lambrecht, R.M.; Cogneau, M.; Ruiz, H.V. & Qaim, S.M. (1999). Excitation functions for the cyclotron production of ^{99m}Tc and ^{99}Mo, *J. of Applied Radiation and Isotopes*, 51, 69-80

Sharp, P.F.; Germmell, H.G. & Murray, A.D. (3^{ed}.) (2005), Practical Nuclear Medicine, ISBN-13: 978-1852338756, *Springer-Verlag London Limited*, USA.

Stöcklin, G.; Qaim, S.M. & Rösch, F. (1995). The impact of radioactivity on medicine, *Radiochim. Acta*, 70/71, 249-272.

T-16, Nuclear Physics Group, Theoretical Division of the Los Alamos National Laboratory (1997). *Nuclear Information Service*, Los Alamos, USA. Available online: http://t2.lanl.gov/data/data.html

Takács, S.; Tárkányi, F.; Sonck, M. & Hermanne, A. (2002). New cross sections and intercomparison of proton monitor reactions on Ti, Ni and Cu, *Nucl. Instrum. Methods Phys. Res.*, B 188, 106.

Takács, S.; Szűcs, Z.; Tárkányi, F.; Hermanne, A. & Sonck, M. (2003). Evaluation of proton induced reactions on ^{100}Mo:New cross sections for production of ^{99m}Tc and ^{99}Mo, *Journal of Radio analytical and Nuclear Chemistry*, Vol. 257, No. 1, pp. 195.201

Tammemagi, H., Jackson, D. (2009). Half-Lives A Guide to Nuclear Technology in Canada, *Oxford University Press*, pp. 11-13, 156.

Tárkányi, F.; Takács, S.; Gul, K.; Hermanne, A.; Mustafa, M.G.; Nortier, M.; Obložinský, P.; Qaim, S.M.; Scholten, B.; Shubin, Yu.N. and Zhuang Y. (2001). Beam Monitor Reactions, IAEA-TECDOC-1211, IAEA, Vienna, p. 49, Updated version January, 2007. Available from: http://www-nds.iaea.org/medical

Troyer, G.L. & Schenter, R.E. (2009). Medical isotope development and supply opportunities in the 21st century, *J. Radioanal. Nucl. Chem.*, 282:243–246, DOI 10.1007/s10967-009-0267-4.

Uddin, M.S.; Hagiwara, M.; Tárkányi, F.; Ditrói, F. & Baba, M. (2004). Experimental studies on the proton-induced activation reactions of molybdenum in the energy range 22-67 MeV, *Appl. Radiat. & Isot.*, 60, 911.

Uddina, M.S. & Baba, M. (2008). Proton-induced activation cross-sections of the short-lived radionuclides formation on molybdenum, *Appl. Radiat. & Isot.*, 66,208–214.

Ullyett, B. (1997), Chapter Five - Canada Enters the Nuclear Age, published for Atomic Energy of Canada Limited, *McGill-Queen's University Press*.

Van der Marck, S. C., Koning, A. J. & Charlton, K. E., (2010). The options for the future production of the medical isotope ^{99}Mo, *Eur J Nucl Med Mol Imaging*, 37:1817–1820 DOI 10.1007/s00259-010-1500-7

Wolf, A.P.; Barclay Jones, W. (1983). Cyclotrons for biomedical radioisotope production, *Radiochim. Acta*, 34, 1.

Ziegler, J.F. (1995). TRIM 95.4 code: The Transport of Ions in Matter, *IBM-Research*, Yorktown, New York, USA.

Ziegler, J.F.; Biersack, J.P. & Littmark, U. (1985). The Stopping and Range of Ions in Solids. Vol. 1 of the Stopping and Ranges of Ions in Matter, *Pergamon Press*, New York.

Use of Radioactive Precursors for Biochemical Characterization the Biosynthesis of Isoprenoids in Intraerythrocytic Stages of *Plasmodium falciparum*

Emilia A. Kimura et al.[*]
Department of Parasitology, Institute of Biomedical Sciences,
University of São Paulo, São Paulo,
Brazil

1. Introduction

Malaria continues to be one of the major threats to human health, affecting 300-500 million people and causing the death of approximately 1 million individuals per year, mostly children under 5 years of age (WHO 2010b). Human malaria is caused by five species of the genus *Plasmodium*, namely *Plasmodium falciparum, Plasmodium vivax, Plasmodium ovale, Plasmodium malariae* and *Plasmodium knowlesi,* whereas the latter is found exclusively in the Southeast Asian region (Cox-Singh & Singh 2008). Current estimates suggest that approximately 2.4 billion people are at risk of stable or unstable *Plasmodium falciparum* transmission, similar global estimates are also available for *P. vivax*, and while there is considerably less mortality attributed to this species, its geographical reach is far greater. An estimated 2.9 billion people are at risk for vivax malaria, with an estimated 80 million to 300 million clinical cases annually (Guerra *et al.*, 2010). These global estimates are a direct result of an increasing ability to collate and assimilate large data sets that also allow the monitoring of trends in malaria incidence and parasite prevalence. *P. falciparum* is strongly associated with a potentially fatal form of the disease, although recent reports indicate an underestimation of the severity of *P. vivax* infections (Alexandre *et al.*, 2010). Efforts were made to eradicate malaria and although these were successful over large geographical areas, they did not succeed in tropical Africa or in many parts of Asia. In the past few years, malaria has once again attracted more attention partly because of increasing recognition that the malaria prevalence in sub-Saharan Africa has increased during the past decade. The main cause of the worsened malaria situation recorded in recent years has been the spread of drug-resistant parasites, which has led to rising malaria-associated mortality, especially in east Africa.

---

[*] Gerhard Wunderlich, Fabiana M. Jordão, Renata Tonhosolo, Heloisa B. Gabriel,
Rodrigo A. C. Sussmann, Alexandre Y. Saito and Alejandro M. Katzin
Department of Parasitology, Institute of Biomedical Sciences,
University of São Paulo, São Paulo, Brazil

The emergence of resistance occurs due to widespread and indiscriminate use of antimalarials. This fact exerts a strong selective pressure on malaria parasites to develop high levels of resistance. On the other hand, the spread of resistance is due to the existence of a sexual cycle in the invertebrate host where there is genetic exchange.

Antimalarial drug resistance is not the same as malaria treatment failure, which is the absence of success in clearing malarial parasitaemia and/or resolve clinical symptoms even with the administration of an antimalarial. While drug resistance may lead to treatment failure, not all treatment failures are caused by drug resistance. Treatment failure can also be the result of incorrect dosing, problems of treatment adherence, poor drug quality, interactions with other drugs, compromised drug absorption or misdiagnosis of the patient. Apart from leading to inappropriate case management, all these factors may also accelerate the spread of true drug resistance by exposure of the parasites to inadequate drug levels (WHO 2010b).

To assess if a strain is resistant to an antimalarial, the World Health Organization (WHO) recommended some methods: *in vivo* assessment of therapeutic efficacy; molecular genotyping to distinguish between re-infections and recrudescence; *in vitro* studies of parasite susceptibility to drugs in culture and identification of molecular markers.

Among the major antimalarial compounds recommended by WHO for treatment of malaria are the aminoquinolines (chloroquine, amodiaquine, primaquine, quinine, mefloquine), the antifolates (sulfadoxine), diaminopyrimidine (pyrimethamine), sesquiterpene lactones (artemisinin, artemether, artesunate) and some antibiotics (WHO 2010a). In counterpart, with the exception of artemisinin derivates, there is a widespread drug resistance confirmed to all these drugs in many malaria-endemic regions as shown in figure 1 (Ekland & Fidock 2008).

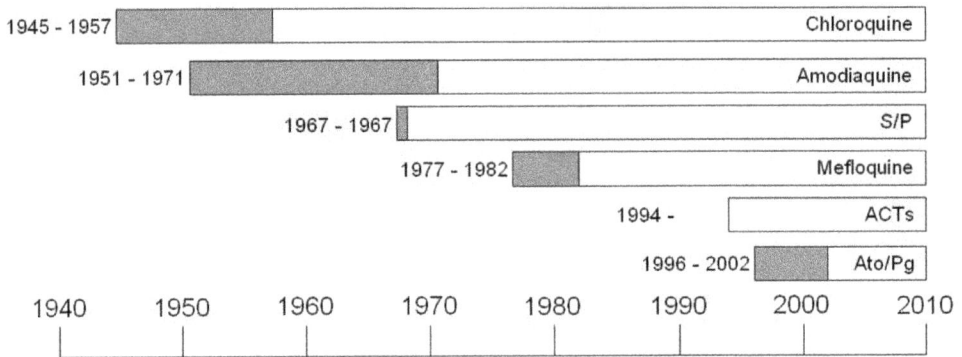

Fig. 1. Emergence of resistance to the principal antimalarials. Each bar represents an antimalarial monotherapy or combination. Years to the left of each bar represent the date the drug was introduced and the first reported instance of resistance. Chloroquine and sulfadoxine/pyrimethamine remained effective for considerable periods after the first reported instances of resistance. Artemisinin-based combination therapies (ACTs); atovaquone/proguanil (Ato/Pg); sulfadoxine/pyrimethamine (S/P), (Adapted from Ekland & Fidock., 2008).

Use of Radioactive Precursors for Biochemical Characterization the Biosynthesis of Isoprenoids in
Intraerythrocytic Stages of Plasmodium falciparum

97

Field trials of artemisinin (Qinghaosu) and its derivates were first implemented in China in the early 1970s. Artemisinin has a low radical cure rate when used alone in a short course, presumably due to its very short half-life in vivo. Since 1994, artemisinin and its derivates have been used in combination therapies (ACTs). More recently, in reports about the molecular marker SERCA-type PfATPase6 associated to artemisinin-derivate resistance was described as possible target of the drug (Eckstein-Ludwig *et al.*, 2003), but some groups do not agree about the correlation of mutation of this gene to the artemisinin (Dondorp *et al.*, 2009; Valderramos *et al.*, 2010).

Several strategies have been used to control malaria in the world, and these rely on the efficient and fast treatment of infected individuals, environmental measures including vector control programs and prevention by the stimulation of use of insecticide-treated bednets (WHO 2010b). Given the ever-looming surge of resistance of the parasite against the drugs currently in use, the development of an efficient vaccine or novel drugs are important issues.

The symptoms of malaria are linked to the stage of schizogony. After the invasion of erythrocyte the parasite consumes and destroys the intracellular proteins, especially hemoglobin which is polymerized into an inert substance denominated hemozoin or malarial pigment. According to classical symptomatology, fever coincides with lysis of red blood cells (RBCs) and is caused by the release into the bloodstream of the hemozoin and other toxic parasite products. Coincident with the rupture of erythrocyte occurs the increased expression of TNF-α and other cytokines (TNF-β and IL-6) which can also induce the release of other mediators (prostaglandin and TGF-β) that are responsible for the onset of symptoms (including fever) and tissue damage (Beeson & Brown 2002; Hemmer et al., 2006).

P. falciparum is strongly associated with a potentially fatal form of the disease, most affecting young children, non-immune adults and pregnant women, although recent reports indicate an underestimation of the severity of *P. vivax* infections (Alexandre *et al.*, 2010). The pathogenesis of human *P. falciparum* infection is a complex interaction of parasite induced RBCs alterations (Maier *et al.*, 2009) and microcirculatory anomaly, (Grau *et al.*, 2003) accompanied by local and systemic immune reactions, resulting in a accumulation or sequestration of parasite infected RBCs in various organs, such as the brain, lung and placenta, and together with other factors is important in the pathogenesis of severe forms of malaria (Marsh *et al.*, 1995; Buffet *et al.*, 2011). RBC infected with early parasite stages (rings) display mild modifications of adhesion and/or deformability properties and may circulate, whereas late parasite stages, called trophozoites and schizonts (mature forms), have substantial alterations of adhesion and deformability that favor their sequestration in small vessels, thus preventing their circulation in the peripheral blood. Sequestration of mature forms is induced by their adherence to endothelial cells, blood cells, platelets, and uninfected RBCs. These interactions are mediated by multiple host receptors recognized by parasite adhesins (Maier *et al.*, 2009). Cerebral malaria is the main clinical manifestation of severe *P. falciparum* infection and seems to be responsible for most deaths. It is characterized by coma, often with convulsions (Lalloo *et al.*, 1996). Severe anaemia is probably the second most common presentation of severe *P. falciparum* infection and probably results from RBCs destruction, indirect destruction of parasitized erythrocytes by immune mechanisms and reduced erythropoiesis associated with imbalances in cytokine concentrations (Menendez *et al.*, 2000; Ekvall 2003). Renal dysfunction or failure, circulatory collapse and shock, disseminated intravascular coagulation and spontaneous bleeding, and acidosis can also occur. Among adults with malaria, pregnant women are particularly susceptible to malaria, despite substantial immunity before pregnancy, and the risk is highest in first pregnancies. The major complications of infection are maternal anaemia, which in turn increases maternal deaths, and reduced infant birthweight from a combination of intrauterine growth retardation and premature delivery leading to excess infant mortality. In some settings maternal malaria may also cause spontaneous abortion or stillbirth (Granja *et al.*, 1998).

Severe clinical forms are uncommon in infections with P. vivax however in countries where this parasite is dominating, more and more frequently is becoming common severe cases and even deaths to P. vivax infection are reported (Anstey et al., 2009). In Brazil cases such as severe rhabdomyolysis (Siqueira et al., 2010) and immune thrombocytopenic purpura have been reported (Lacerda et al., 2004) and in south-east Asia, especially in India and Vietnam, cases of acute renal failure were documented (Sanghai & Shah 2010). A serious problem encountered in the P. vivax infection are hypnozoites, this liver stage that can cause relapses many months or even years after the initial infection, and these hypnozoites can only be eliminated by additional treatment with primaquine (Watkins & Sibley 2011).

2. The life cycle of plasmodium

Laveran was responsible for the discovery of the Plasmodium, observing them in human erythrocytes and was the first to describe it in 1880. The life cycle of parasites of humans Plasmodium genus is very similar between species, showing two distinct phases. The life cycle of malaria parasite is complex, and there are four critical points in the life cycle of *Plasmodium* parasites in which a small number of parasites rapidly multiply to generate much larger populations. These life cycle stages are male gamete development, sporozoite formation, liver stage development and blood stage asexual reproduction. The first two of these processes occur within the mosquito vector, and the second two processes take place in the vertebrate host.

Infective sporozoites from the salivary gland of the Anopheles mosquito are injected into the human host along with anticoagulant-containing saliva to ensure an even-flowing blood meal. Once entered in the human bloodstream, P. falciparum sporozoites reach the liver and penetrate the liver cells where they remain at for 9–16 days and undergo asexual replication known as exo-erythrocytic schizogony. Each sporozoite gives rise of thousands of merozoites inside the hepatocyte and each merozoite can invade a red blood cell (RBC) upon release from the liver. According to the Plasmodium species, the liver phase takes on average 6 days (P. falciparum), 10 days (P. vivax), or 15 days (P. ovale and P. *malariae*).

Merozoites enter erythrocytes by a complex invasion process, requiring a series of highly specific molecular interactions. Asexual division starts inside the erythrocyte and the parasites develop through different stages therein. The early trophozoite stage is often referred to as the "ring form". Trophozoite enlargement is accompanied by highly active metabolism. The end of this stage is marked by multiple rounds of nuclear division withouth cytokinesis resulting in the formation of schizonts. Each mature schizont contains up to 32 merozoites and these are released after lysis of the RBC to invade further uninfected RBCs. This release coincides with the sharp increases in body temperature during the progression of the disease. This cycle takes about 36-48 h in P. *falciparum*, 48 h in P. *vivax* and P. *ovale* and 72 h in P. *malariae*. A small proportion of the merozoites in the RBCs eventually differentiate to produce micro and macrogametocytes. These gametocytes are essential for transmitting the infection to new hosts through female *Anopheles* mosquitoes.

A mosquito taking a blood meal on an infected individual may ingest these gametocytes into its midgut, where macrogametocytes form macrogametes and exflagellation of microgametocytes produces microgametes. These gametes fuse, undergo fertilization and form a zygote. This transforms into an ookinete, which penetrates the wall of a cell in the midgut and develops into an oocyst. Inside the oocyst many nuclear divisions occur, resulting in thousands of sporozoites and they migrate to the salivary glands for onward transmission into another host.

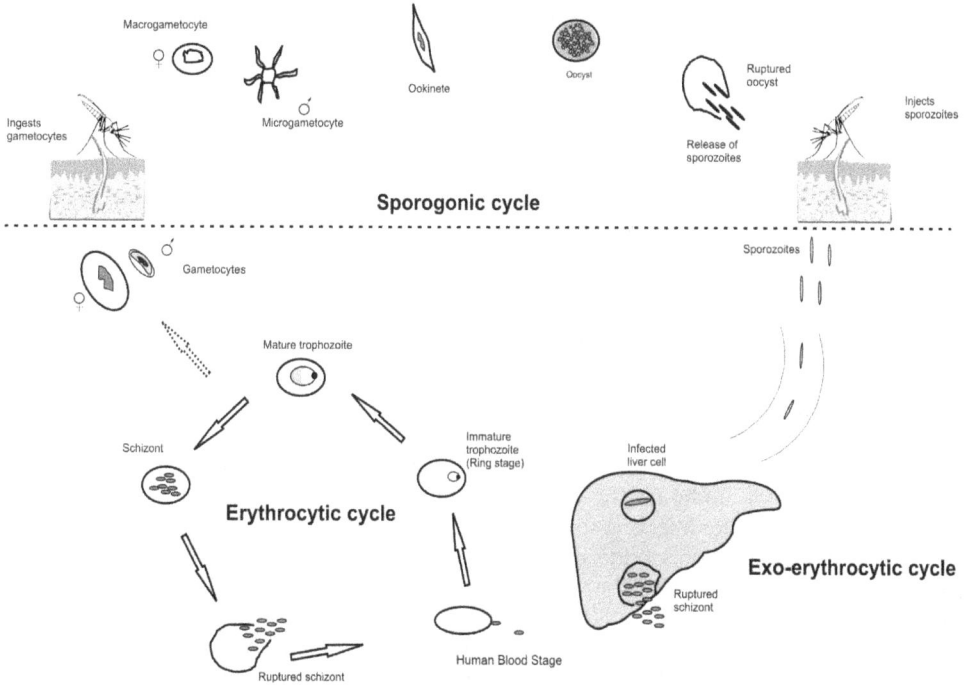

Fig. 2. During a blood meal, the mosquito inoculates sporozoites into the human host, infecting the liver cells initiating the exo-erythrocytic cycle. The parasites multiply, forming the schizonts, which then rupture releasing merozoites. The merozoites infect red blood cells, initiating the erythrocytic cycle (differentiating into ring, trophozoite and schizont); after the rupture of the red blood cells that contain them, merozoites are released and invade other red blood cells; some parasites differentiate into male (microgametocytes) and female (macrogametocytes) sexual forms the mosquitoes, by feeding off infected patients, ingest the gametocytes; the gametocytes evolve into gametes, and the microgametes penetrate the macrogametes generating zygotes, which in turn evolve into motile ookinetes; these invade the midgut wall of the mosquito, where they develop into oocysts; where sporogony takes place, releasing sporozoites; these migrate to the mosquito salivary gland.

The erythrocytic stages of the malaria parasite are responsible for the symptoms and pathology of the disease (Miller et al., 1994). Immunity against the erythrocytic stages is not well understood, although they are important from the vaccine development perspective (Good 2001). To a greater or lesser extent, all four species have been cultured or maintained in vitro; *P. falciparum*, however, is the only species for which all life cycle stages have been established in culture (Hollingdale 1992).

Cultivation of the intraerythrocytic stages of *P. falciparum* is vital for many aspects of malaria research. The *in vitro* continuous cultivation of the erythrocytic stages of *P. falciparum* achieved by (Trager & Jensen 1976) was a turning point in the history of malaria research. Since the establishment of this technique and refinements thereafter, the pace of research on malaria has increased significantly.

The methods for cultivation of the erythrocytic stages of *P. falciparum* reported have been usefully applied in nearly every aspect of research on malaria: chemotherapy, drug resistance, immunology and vaccine development, pathogenesis, gametocytogenesis and mosquito transmission, genetics, the basis for resistance of certain mutant red cells, cellular and molecular biology and biochemistry of the parasites and of their relationship with their host erythrocyte (Trager & Jensen 1997).

Although several methods have been developed, the technique used for the *in vitro* culture of the intraerythrocytic stages of *P. falciparum* remains essentially the same as that originally described by Trager and Jensen. The protocol described by these authors was based on the use of HEPES (4-(2-hydroxyethyl)-1-piperazineethanesulfonic acid)-buffered RPMI tissue culture medium supplemented with human serum, erythrocytes and sodium bicarbonate. Since then, efforts and time have been invested in trying to improve the *in vitro* growth of the asexual stages of the *Plasmodium* life cycle.

The availability of the technique has also spawned efforts to cultivate other malaria pathogens, both human and non-human. In most instances, it is more convenient working with malarial parasites from non-human hosts because they can be maintained *in vivo*, thereby allowing testing for infectivity of in vitro-cultured stages in the vertebrate host and providing an animal model for the study of the parasite in the human host, is as the case culture malarial parasites infecting simian, avian, and rode hosts.

Techniques for cultivation of the exoerythrocytic stages of avian malarial parasites *P. gallinaceum*, *P. lophurae*, and *P. fallax* were described by Huff (Huff 1964). Primary cultures of hepatocytes from rhesus monkeys (*Macaca mulatta*) were used to support growth of several simian malarias (*P. cynomolgi*, *P. knowlesi*, *P. coatneyi*, and *P. inui*) (Millet *et al.*, 1988). In a later study, the same group used rhesus hepatocytes for cultivation of developmental stages of *P. fieldi* and *P. simiovale*, two parasites that infect macaques (Millet *et al.*, 1994). Erythrocytic development of a number of simian malarial parasites has been studied, including that of P. *knowlesi*, *P. cynomolgi*, *P. fragile*, *P. gonderi*, *P. coatneyi*, *P. inui*, *P. fieldi*, and *P. simiovale*. Some of these aforementioned malarias are non human facsimiles of human malarias. Among the rodent malarias, the erythrocytic stages of *P. berghei* and *P. chabaudi* have been cultivated *in vitro* (Mons *et al.*, 1983; O'Donovan & Dalton 1993). The rodent malarial parasite *P. berghei* from in vivo-infected livers was maintained in primary cultures of rodent liver cells by Foley et al. (Foley *et al.*, 1978). Among the avian malarias, the erythrocytic stage of *P. lophurae* has been cultivated *in vitro* (Langreth & Trager 1973).

3. Characterization of Isoprenoid pathway by metabolic labeling

Widespread resistance to most anti-malarial drugs and the unavailability of an effective vaccine have fueled the emergence of malaria in recent years as a major global health and economic burden. Despite these hurdles, the field of malaria research has witnessed some extremely notable developments in the recent past including sequencing of the malaria genome (Gardner *et al.*, 2002) the application of proteomics to studying malaria life cycle (Florens *et al.*, 2002), the malaria transcriptome (Bozdech *et al.*, 2003), and several web resources such as Mal-Vac (database of malaria vaccine candidates), PlasmoDB (genome database of the Plasmodium genus), and VarDB (database for antigenic variation genes families) (Chaudhuri *et al.*, 2008; Hayes *et al.*, 2008; Aurrecoechea *et al.*, 2009).

A thorough knowledge of the biochemistry of *P. falciparum* is required in order to develop new drugs. This aim can be achieved by two means: either by focusing on validated targets in order to generate new drug candidates; or by identifying new potential targets for malaria chemotherapy. This last strategy will be commented and discussed in this chapter, focusing on the biosynthesis of isoprenoids. In the apicomplexan parasites of the genus *Plasmodium* the isoprenoid pathway is localized in the apicoplast which was acquired millions of years ago during an evolutionary event (Lim & McFadden 2010). At this occasion, the ancestor of the phylum apicomplexa gained a plastid by the secondary endosymbiosis of a photosynthetic eukaryote (McFadden *et al.*, 1996). The chloroplast of this photosynthetic eukaryote was retained and during evolution, many chloroplastid genes were transferred to the apicomplexan nucleus while others were lost (Funes *et al.*, 2002; Moore *et al.*, 2008). In the case of malaria parasites, especially its most virulent species *P. falciparum*, a series of new "plant-like" enzymes associated with this organelle were recently discovered (Luo *et al.*, 1999).

When testing for metabolic activities, usually the cultivable blood stage forms of *P. falciparum* are employed (Trager & Jensen 1976), that we will describe briefly below. Since the parasite is an obligate intracellular organism in this stage, several precautions must be taken to not confound host cell and parasite metabolic activities. Also, the chemical nature of an unlabeled or radiolabeled precursor plays an important role, since some substrates do not enter the red blood cell or do not cross the parasitophorous vacuole or the parasite cell membrane. Other substrates may require solvents which are toxic to either the parasite or the red blood cell. In the following, we describe the progress that was made in the detection of a number of metabolic activities of the parasites, many of which configure prime targets for drug intervention.

One of the strategies to identify each product of a metabolic pathway is the metabolic labeling using a radioactive specific precursor and a posterior analysis by an appropriate method. Due to the difficulty to obtain higher quantities of parasite biomass, the labeling with non-radioactive isotopes often does not lead to clear results, for this reason, most research groups rely on radiolabeled compounds when monitoring metabolic turnover. Several criteria must be evaluated before proceeding to experimental metabolic labeling of intraerythrocytic stages of *P. falciparum*: 1- determine type of study (structural or biosynthetic), 2- choose labeling protocol and select a radiolabeled precursor(s), 3- check incorporation of radioactivity into the parasite, 4- if the incorporation is sufficient for analysis, proceed with steady-state, pulse, pulse-chase or sequential transfer protocols for analysis of intermediates of the metabolic pathway to be studied. If incorporation is insufficient for analysis, it is recommended to try using more labeled precursor or decrease the level of unlabeled precursor in the culture medium. In the next step, it is advantageous to check effects on the parasite and molecule(s) of interest and effects on incorporation of labeled precursors. Afterwards, the optimal concentration of unlabeled precursor is selected and the experiment may proceed with steady-state, pulse, pulse-chase or sequential transfer protocols for the analysis of intermediates of the metabolic pathway that is studied.

For the characterization the composts isoprenics in parasite, *P. falciparum* clone 3D7 was cultured according to the protocol described by Trager and Jensen (Trager & Jensen 1976) where human sera was substituted by Albumax I (0.5%). Parasites were grown in tissue culture flasks (75 cm^2) a 40 ml volume with a gas mixture of 5% CO_2, 5% O_2, and 90% N_2. In asynchronous cultures we can to obtain the ring, trophozoite or schizont stages. In synchronous cultures (~ 15% parasitemia), the stage initially was in ring stage (1-20 h after reinvasion) after two treatments with 5% (w/v) D-sorbitol solution in water, for subsequent maintenance in culture until the differentiation to trophozoite (20-30 h after reinvasion) or schizont (30-35 h after reinvasion) stages. Parasite development and multiplication were monitored by microscopic evaluation of Giemsa-stained thin smears.

Cultures of *P. falciparum* with a parasitemia of approximately 10% were labeled with the different precursors in normal RPMI 1640 medium for 15 h and after each stage purified. The asynchronous parasites were purified on a 40%/70%/80% discontinuous Percoll® gradient (Braun-Breton *et al.*, 1986). The culture was centrifuged at 2000 rpm for 5 min at room temperature, the pellet resuspended in RPMI-1640 (1:1 vol/vol), and carefully placed on top of the gradient. The tubes were centrifuged at 10,000 x *g* for 20 min at 20°C. The cells containing schizonts, trophozoite and ring stages, respectively were collected, washed twice in RPMI-1640, the pellet of parasites were stored in liquid N_2 for subsequent analysis. The other hand, synchronous culture with schizont stages predominantly were purified with magnetic column separation (MACS Separation Columns "CS"). The pre-equilibration, washing, and elution the column were all carried out at room temperature with RPMI-1640. The culture was centrifuged, the pellet resuspends in RMPI-1640 (1:10 vol/vol), 10 ml of the 10% suspension of erythrocytes were applied to a CS column assembled in a magnetic unit to remove non-infected erythrocytes, ring-infected erythrocytes, and young trophozoites. After wash the column by filling from top with 50 ml of RMPI-1640, the column was removed from the magnetic field and its contents eluted with 50 ml of RMPI-1640, thus, the schizonts stages fraction were obtained, centrifuged at 2,000 rpm for 5 min at room temperature, the supernatant discarded, and the pellet of parasites were stored in liquid N_2 for subsequent analysis. Parasite form was monitored by microscopic evaluation of Giemsa-stained thin smears.

In the following, we specifically focus i) on the methylerythritol phosphate (MEP) pathway which leads to the isoprenoids isopentenyl pyrophosphate (IPP) /dimethylallyl pyrophosphate (DMAPP) and ii) on downstream reactions which result in dolichol, ubiquinone, menaquinone, tocopherol, carotenoids and other related compounds.

4. Mevalonate-independent methylerythritol phosphate (MEP) pathway in *P. falciparum*

The MEP pathway starts with the condensation of pyruvate and glyceraldehyde-3-phosphate, which yields 1-deoxy-D-xylulose-5-phosphate (DOXP) as a key metabolite (Rohmer *et al.*, 1993). Cassera and colleagues showed by metabolic labeling that the MEP pathway is functionally active in the intraerythrocytic stages of *P. falciparum*. Using different radioactive precursors such as [1-14C]sodium acetate, D-[U-14C]glucose and [2-14C]deoxy-D-xylulose, five intermediates of the MEP pathway were identified. The intermediates were isolated by high performance liquid chromatography (HPLC) and characterized by electrospray mass spectrometric analyses. All but one of the intermediates of the MEP pathway were characterized, including 1-deoxy-D-xylulose-5- phosphate, 2-C-methyl-D-erythritol-4-

phosphate, 4-(cytidine-5-diphospho)-2-C-methyl-D-erythritol, 4-(cytidine-5-diphospho)-2-C-methyl-D-erythritol-2-phosphate, and 2-C-methyl-D-erythritol-2,4-cyclodiphosphate. The effect of fosmidomycin on the levels of MEP pathway intermediates was found to be most prominent in ring stages followed by schizont stages of P. falciparum. It was also shown that the MEP pathway provides IPP precursors for the synthesis of higher isoprenic compounds like ubiquinone and dolichol, and this was demonstrated by the decrease in the ubiquinone and dolichol content in fosmidomycin-treated parasites (Cassera et al., 2004). To achieve efficient labeling, [1-^{14}C] sodium acetate and D-[U-^{14}C] glucose were employed instead of pyruvic acid, which is not incorporated by blood stage P. falciparum. These relatively simple compounds which enter in many biochemical pathways obviously require a refined analysis of synthesized molecules and this was achieved by demonstrating the chemical nature of the found molecules by different analytic methods.

5. Detection of N-linked glycoproteins in *P. falciparum*

Glycoconjugates have been shown to be important in the penetration, cellular growth, host immunity regulation and differentiation (Schwarz & Datema 1982). The presence of glycoproteins in P. falciparum has been demonstrated in several studies but remains controversial because there is little information addressing the function of P. falciparum glycoconjugated molecules. Also, the few available studies are inconsistent due to differing methodologies. N-linked glycosylation is a protein modification that occurs co-translationally in the endoplasmic reticulum. Kimura and colleagues described briefly the methodology used for detection of N-linked glycoproteins. For this purpose parasite culture containing ring stage parasites were labeled for 48 hs with D-[U-^{14}C]glucose or D-[U-^{14}C]mannose. The metabolic labeling was also done in the presence of 12 μM tunicamycin, the classic inhibitor of N-glycosylation. The total profile of glycoproteins was analyzed by SDS-PAGE of treated and untreated parasites. The N-linked glycoproteins were confirmed by 18 h radiolabeling with L-[^{35}S]methionine and affinity chromatography using Concanavalin A-Sepharose. In parallel, tunicamycin pre-treated parasites showed a differing band pattern, thus identifying protein species which carried N-glycosyl moieties. In order to increase L-[^{35}S]methionine or D-[U-^{14}C] glucose labeling, the cultures were starved in methionine- or glucose-deficient medium for 1 h before the addition of the radioactive substance (Kimura et al., 1996). By this approach, these authors identified N-linked glycoprotein when they used the radioactive precursor D-[U-^{14}C]glucose. As above, although this form of labeling is unspecific, other methodologies confirmed the nature the compounds where radioactive glucose became incorporated. On the other hand, Dieckmann-Schuppert et al. (1992) did not detect N-linked glycoproteins using the specific 2-[^{3}H]mannose, 6-[^{3}H]glucosamine, or 1-[^{14}C]mannosamine. In 1997 Gowda and colleagues confirmed that P. falciparum contains low levels of N-glycosylation activity. The amount of N-linked carbohydrates in whole parasite proteins is approximately 6% compared with the GPI anchors attached to proteins based on radioactive GlcN incorporated into the proteins (Gowda et al., 1997). Bushkin et al. (2010) suggested that the occupation of N-glycan sites is markedly reduced in apicoplast proteins versus some secreted proteins in Plasmodium. Clearly, more work has to be done in order to elucidate the nature and purpose of N-linked glycoproteins in Plasmodium. Also, the differences in the results published by distinct groups mentioned above may have occurred due to differences in basic aspects of each experiment, such as the duration of labeling, the tracing compound used and the parasite stage that was analyzed.

6. Characterization of dolichol in *P. falciparum*

The protein modification by N-linked glycosylation is dependent on the *de novo* synthesis of dolichyl-P, a long chain non-sterol isoprene which acts as a membrane-bound carrier of oligosaccharides in the assembly of glycoproteins (Leloir 1977). Couto *et al* (1999) demonstrated the presence of dolichol, dolichyl-P and dolichyl-PP species of 11 and 12 isoprenic units in parasites of *P. falciparum* cultivated *in vitro*.

In these experiments, cultures of *P. falciparum* were labeled for 15 h with [1-(n)-3H]geranylgeranyl pyrophosphate triammonium salt ([1-(n)-3H]GGPP) or with [1-(n)-3H]farnesyl pyrophosphate triammonium salt ([1-(n)-3H]FPP). Each stage was purified by Percoll gradient; the pellets were extracted and analyzed by C18 reverse-phase HPLC (RP-HPLC). From extracts labeled with [1-(n)-3H]GGPP, 3 major peaks were detected: at 8 min, coincident with an authentic sample of geraniol, at 21 min, coincident with a dolichol standard with 11 isoprenic units and at 23 min, coincident with a dodecaprenol standard. Parasites labeled with [1-(n)-3H]FPP showed a number of other labeled products, but the presence of dolichol of 11 and 12 isoprenic units was evident (Couto *et al.*, 1999).

7. Characterization of ubiquinone and carotenoids in *P. falciparum*

Coenzyme Q is a molecule composed of a benzoquinone ring with a side chain of several isoprenic units, and the number of which defines the type of coenzyme Q. A polyprenyl diphosphate synthase is involved in the elongation of the side chain (Ogura *et al.*, 1997). This isoprenic chain is then attached to p-hydroxybenzoic acid (PHBA), which is synthesized via the shikimate pathway. The isoprenic chain then allows the molecule to attach to the inner membrane of mitochondria, where it participates in many metabolic processes, like the electron transport chain (Ernster & Dallner 1995).

Macedo *et al.* (2002) had identified coenzyme Q8 and coenzyme Q9 by metabolic labeling of parasites with [1-14C]acetic acid, [1-14C]isopentenyl pyrophosphate triammonium salt ([1-14C]IPP), [1-(n)-3H]FPP, and [1-(n)-3H]GGPP in all intraerythrocytic stages. To our knowledge, this was the first report on the incorporation of [1-14C]IPP into *P. falciparum*. It is well known that the considerable increase in total lipid content associated with *P. falciparum* invasion is due to the existence of an intense lipid transport system in infected erythrocytes. The efficient uptake of [1-(n)-3H]FPP, [1-(n)-3H]GGPP and, to a lesser extent, [1-14C]IPP, may be ascribed to this transport mechanism for lipid-like components. The parasite is capable of synthesizing two different homologues of coenzyme Q, depending on the given radioactive intermediate. When labeling is performed with [1-(n)-3H]FPP, coenzyme Q with an isoprenic chain of 40 carbons (Q8) is detected; while [1-(n)-3H]GGPP labeling leads to Q9 (45 carbons) chains. These findings can be explained by the fact that both FPP and GGPP are substrates of the prenyltransferase involved in the biosynthetic pathway of the isoprenic chain of ubiquinone as shown in other systems (Ogura & Koyama 1998). This difference in the length of the isoprenic chain according to the precursor was also observed in the biosynthesis of dolichols in *P. falciparum*. In order to check whether this difference could be induced by the length of the isoprenoid intermediate, the basic isoprenic unit [1-14C]IPP was used as a metabolic marker. It would be expected that by labeling with [1-14C]IPP, both coenzymes Q would be detected. Surprisingly, HPLC analysis showed a single radioactive peak, which co-eluted with a coenzyme Q9 standard. The enzyme of *P. falciparum* (www.PlasmoDB.org, entry PfB0130w) that biosynthesizes the isoprenic chain attached to benzoquinone ring of coenzyme Q was cloned and expressed and showed octaprenyl pyrophosphate synthase activity. Enzymatic activity was measured by determination of the amount of [1-^{14}C]IPP incorporated into butanol-extractable polyprenyl diphosphates. The recombinant and native versions of the enzyme had similar Michaelis constants with the substrates [1-^{14}C]IPP and farnesyl pyrophosphate. The initial rate was calculated by determining the quantities of product formed or IPP consumed at each time point by counting the ^{14}C radioactivity in the butanol phase (product) and in the aqueous phase (IPP). The recombinant protein, as well as *P. falciparum* extracts, showed an octaprenyl pyrophosphate synthase activity, with the formation of a polyisoprenoid with eight isoprenic units, as detected by reverse-phase HPLC and reverse-phase TLC, and confirmed by electrospray ionization and tandem MS analysis. Additionally, the recombinant enzyme could be competitively inhibited in the presence of the terpene nerolidol. Since the *P. falciparum* enzyme shows quite low similarity to its human counterpart, decaprenyl pyrophosphate synthase, it was suggested that the identified enzyme and its recombinant version could be exploited in the screening of novel drugs (Tonhosolo *et al.*, 2005). The enzyme octaprenyl pyrophosphate synthase (PfB0130w) turned out to be a bi-functional enzyme with phytoene synthase activity, which was shown by *in vitro* enzymatic assays using [1-(n)-^{3}H]GGPP as a substrate, followed by HPLC analysis and confirmation by LC-APCI-MS/MS analysis. The identification of the enzyme phytoene synthase showed that intraerythrocytic stages of *P. falciparum* can perform the crucial step of the pathway that leads to the biosynthesis of carotenoids. Carotenoids are widespread lipophilic pigments synthesized by all photosynthetic organisms and some nonphotosynthetic fungi and bacteria. All carotenoids are derived from the C40 isoprenoid precursor geranylgeranyl pyrophosphate. HPLC analysis from extracts of intraerythrocytic stages of *P. falciparum* labeled with [1-(n)-^{3}H]GGPP, revealed several compounds in all intraerythrocytic stages, with retention times coincident with lutein, phytoene, phytofluene, *all-trans-β*-carotene, neurosporene and 6-*all-trans*-lycopene. Some of these compounds were structurally characterized by electrospray mass spectrometric analysis (Tonhosolo *et al.*, 2009).

Considering that carotenoid biosynthesis is absent in humans, and also that possibly other uncharacterized carotenoid synthesizing enzymes are present, in *Plasmodium*, it possible to speculate that this pathway could be exploited for the design of new antimalarial drugs. Indeed, sequence data from additional organisms, functional studies, improved bioinformatics screening approaches, together with biochemical evidence, may reveal whether other interesting targets and pathways are present in the phylum Apicomplexa.

8. Vitamin E biosynthesis

The biosynthesis of vitamin E depends on both the MEP and shikimate pathways. This compound consists of a polar chromanol head group attached to a hydrophobic phytyl (tocopherols) or geranylgeranyl (tocotrienols) tail, both of which are critical for their roles as lipid-soluble antioxidants (Schneider 2005).

Using [1-(n)-^{3}H]GGPP or [1-(n)-^{3}H]FPP as radiotracer precursors and HPLC systems for purification of vitamin E, the biosynthesis of α and γ-tocopherol was detected in the three intraerythrocytic stages from 5x10^{8} parasites. To confirm the chemical identity of these compounds, unlabeled extracts from 10^{10} parasites were purified by HPLC, the retention time of tocopherol isomers was collected and analyzed by gas chromatography coupled to a mass spectrometer (GC-MS). Although the two isomers were found in parasites as well as in erythrocytes and culture medium extracts, the parasite extracts showed higher concentrations than the other samples (Sussmann 2010).

Usnic acid is a secondary metabolite from lichens and capable to inhibit the 4-hydroxyphenylpyruvate dioxygenase, an enzyme from the shikimate pathway which is responsible for the biosynthesis of homogentisate from hydroxyphenylpyruvate. Homogentisate is the aromatic portion which receives the isoprenoid side chain from MEP pathway to form vitamin E isomers. When labeled parasites are treated with 25 μM of usnic acid, the biosynthesis of α and γ-tocopherol was inhibited in $53{,}5 \pm 7\%$ (Sussmann 2010).

9. Menaquinone biosynthesis

In plants and cyanobacteria the shikimate pathway and MEP provide the precursors for the biosynthesis of phylloquinone (PhQ) while in bacteria menaquinone is synthesized (Lombardo *et al.*, 2006). Tonhosolo *et. al.* (2010) showed by metabolic labeling with the precursor [1-(n)-^{3}H]GGPP that intraerythrocytic stages of *P. falciparum* biosynthesizes menaquinone (MQ-4), employing different chromatographic methods reported for this type of molecule and further confirmed the nature of the molecules by ESI-MS/MS analysis. Additionally, they showed that the mycobacterial inhibitor of menaquinone synthesis Ro 48-8071 also suppressed MQ biosynthesis and growth of parasites, pointing possibly again to an interesting drug target.

10. Posttranslational modification

10.1 Protein farnesylation and geranylgeranylation in *P. falciparum*

Post-translational modification of proteins with isoprenoids was first recognized as a general phenomenon in 1984 (Schmidt *et al.*, 1984). The isoprenyl group is linked post-translationally to cysteine residues at the C-terminus of the protein through a thioether bound (McTaggart 2006). Studies have shown that FPP (15 carbons) and GGPP (20 carbons) are the most common isoprenoids found attached to proteins. Several of the proteins that undergo these modifications have been identified and may participate in important cell regulatory functions, particularly signal transduction pathways (Zhang & Casey 1996). Protein prenylation is a general phenomenon in eukaryotic cells and has also been described for several protozoan parasites (Lujan *et al.*, 1995; Field *et al.*, 1996; Shen *et al.*, 1996; Ibrahim *et al.*, 2001) including *P. falciparum* (Chakrabarti *et al.*, 2002).

In order to investigate the presence of isoprenylated proteins in *P. falciparum*, the labeled intermediates [1-(n)-^{3}H]FPP and [1-(n)-^{3}H]GGPP were used. Parasites were incubated with radioactive for 18 h, purified on a Percoll gradient, lysed, and analyzed by SDS/PAGE and autoradiography. [1-(n)-^{3}H]GGPP labeled proteins appeared in the ring, trophozoite, and schizont stages. Non-infected red blood cells showed no incorporation of radioactivity under these conditions. The isoprenylated proteins were later identified as members of the Ras and Rab protein family (Rodrigues Goulart *et al.*, 2004).

Moura *et al.* (2001) and Rodrigues Goulart *et al.* (2004) showed that terpenes can inhibit protein isoprenylation in *P. falciparum*. The process of protein prenylation is a very attractive target for the development of new drugs for cancer and parasites (Docampo & Moreno 2001; Stresing *et al.*, 2007). One of the most potent bisphosphonates clinically used

Use of Radioactive Precursors for Biochemical Characterization the Biosynthesis of Isoprenoids in
Intraerythrocytic Stages of Plasmodium falciparum

107

to treat bone resorption diseases, risedronate, inhibited the protein isoprenylation in *P. falciparum* (Jordao *et al.*, 2011). In order to investigate the mechanism of action for risedronate in intraerythrocytic stages of *P. falciparum*, parasites were incubated with or without risedronate and with [1-(n)-^{3}H]FPP, [1-(n)-^{3}H]GGPP and [^{14}C] IPP, purified on a Percoll gradient, lysed, and analyzed by TLC and SDS-PAGE. The results showed that protein prenylation is inhibited by decreasing the biosynthesis of farnesyl pyrophosphate and geranylgeranyl pyrophosphate. Additionally, it was demonstrated that risedronate inhibits the transference of [1-(n)-^{3}H]FPP to proteins but not the [1-(n)-^{3}H]GGPP in *P. falciparum*.

10.2 Protein dolichylation in *P. falciparum*

Another type of protein modification is the attachment of a dolichyl group to proteins. This type of modification is characterized by covalently and post-translationally bound dolichyl groups to the C-terminal cysteine residues of proteins. Protein dolichylation was described in tumor cells and dolichylated proteins could be involved in the cell cycle control (Hjertman *et al.*, 1997).

Our group has previously shown that *P. falciparum* synthesizes dolichols of 11 and 12 isoprene units (Couto *et al.*, 1999) and that these compounds can be attached to a group of 21–24 kDa proteins of this parasite (Moura *et al.*, 2001). To confirm the existence of dolichyl groups attached to *P. falciparum* proteins, D'Alexandri *et al.* (2006) performed *in vitro* metabolic labeling of the parasites with [1-(n)-^{3}H]FPP or [1-(n)-^{3}H]GGPP. They used these precursors instead of [^{3}H]dolichol for analysis of protein dolichylation because commercially available [^{3}H]dolichol has dolichols of 16 and 21 isoprene units that are longer than those synthesized by *P. falciparum*. After metabolic labeling, the proteins were extracted, extensively delipidated and analyzed by SDS–PAGE. RP-TLC and RP-HPLC analysis of [1-(n)-^{3}H]FPP-labeled compounds released from the 21 to 28 kDa from *P. falciparum* proteins revealed that a dolichol of 11 isoprene units and a polyisoprenoid of 12 isoprene units can be attached to proteins of this parasite. The dolichol structure was confirmed by electrospray-ionization mass spectrometry analysis. Treatment with protein synthesis inhibitors and RP-HPLC analysis of the proteolytic digestion products from parasite proteins labeled with [^{35}S]cysteine and [1-(n)-^{3}H]FPP showed that the attachment of dolichol to protein is a post-translational event and probably occurs via a covalent bond to cysteine residues. This was the first demonstration of protein dolichylation in parasites, and also may represent a new potential target for anti-malarial drugs.

11. Concluding remarks

The use of radioactive tracers in metabolic labeling in cultures of *P. falciparum* allowed the identification of many steps of the isoprenoid biosynthesis. In figure 3 we summarize the intermediaries' biosynthesizing by the MEP pathway and the isoprenoids identified in *P. falciparum* until the moment. This pathway is different from the human host and we hypothesize that the identification of related enzymes may directly lead to the development of new antimalarial drugs. In many aspects, radiotracers are important, often indispensable tools for the identification of metabolic intermediates not only in *Plasmodium* but also in many other intracellular protozoa and may ultimately help to point to yet undetected novel drug targets.

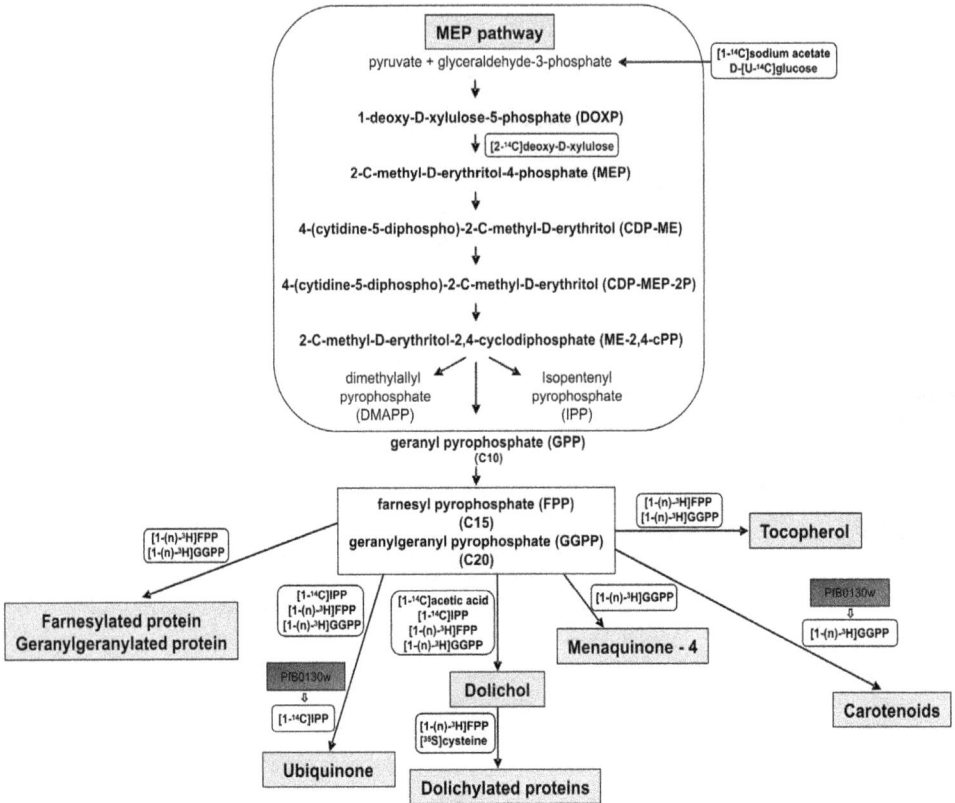

Fig. 3. Isoprenoids biosynthesized by *P. falciparum*. IPP and DMAPP are biosynthesized by the 2-C-methyl-D-erythritol (MEP)-4-phosphate pathway. Shaded boxes indicate presence the isoprenoids biosynthesized by malaria parasite identified until this moment. White boxes indicate radioactive tracers used for identification of products biosynthesized by *P. falciparum*. PfBO130w correspond to a bi-functional enzyme octaprenyl pyrophosphate synthase/phytoene synthase that use [1-¹⁴C]IPP or [1-(n)-³H]GGPP respectively as substrate.

12. Acknowledgments

This work was supported by grants from Brazilian Agencies CNPq and FAPESP (Brazil). F.M.J. and R.A.C.S. are the recipient of a post-graduate fellowship from CNPq. A.Y.S. and H.B.G. receive post-graduate fellowships from FAPESP. R.T. is the recipient of a post-doctoral fellowship from PNPD/CAPES.

13. References

Alexandre, M. A., C. O. Ferreira, et al. (2010). Severe Plasmodium vivax malaria, Brazilian Amazon. *Emerg Infect Dis*, Vol. 16, No. 10, (Oct 2010), pp. (1611-1614).

Anstey, N. M., B. Russell, et al. (2009). The pathophysiology of vivax malaria. *Trends Parasitol*, Vol. 25, No. 5, (May 2009), pp. (220-227).

Aurrecoechea, C., J. Brestelli, et al. (2009). PlasmoDB: a functional genomic database for malaria parasites. *Nucleic Acids Res*, Vol. 37, No. Database issue, (Jan 2009), pp. (D539-543).

Beeson, J. G. and G. V. Brown (2002). Pathogenesis of Plasmodium falciparum malaria: the roles of parasite adhesion and antigenic variation. *Cell Mol Life Sci*, Vol. 59, No. 2, (Feb 2002), pp. (258-271).

Bozdech, Z., M. Llinas, et al. (2003). The transcriptome of the intraerythrocytic developmental cycle of Plasmodium falciparum. *PLoS Biol*, Vol. 1, No. 1, (Oct 2003), pp. (E5).

Braun-Breton, C., M. Jendoubi, et al. (1986). In vivo time course of synthesis and processing of major schizont membrane polypeptides in Plasmodium falciparum. *Mol Biochem Parasitol*, Vol. 20, No. 1, (Jul 1986), pp. (33-43).

Buffet, P. A., I. Safeukui, et al. (2011). The pathogenesis of Plasmodium falciparum malaria in humans: insights from splenic physiology. *Blood*, Vol. 117, No. 2, (Jan 13 2011), pp. (381-392).

Bushkin, G. G., D. M. Ratner, et al. (2010). Suggestive evidence for Darwinian Selection against asparagine-linked glycans of Plasmodium falciparum and Toxoplasma gondii. *Eukaryot Cell*, Vol. 9, No. 2, (Feb 2010), pp. (228-241).

Cassera, M. B., F. C. Gozzo, et al. (2004). The methylerythritol phosphate pathway is functionally active in all intraerythrocytic stages of Plasmodium falciparum. *J Biol Chem*, Vol. 279, No. 50, (Dec 10 2004), pp. (51749-51759).

Chakrabarti, D., T. Da Silva, et al. (2002). Protein farnesyltransferase and protein prenylation in Plasmodium falciparum. *J Biol Chem*, Vol. 277, No. 44, (Nov 1 2002), pp. (42066-42073).

Chaudhuri, R., S. Ahmed, et al. (2008). MalVac: database of malarial vaccine candidates. *Malar J*, Vol. 7, No. 2008), pp. (184).

Couto, A. S., E. A. Kimura, et al. (1999). Active isoprenoid pathway in the intra-erythrocytic stages of Plasmodium falciparum: presence of dolichols of 11 and 12 isoprene units. *Biochem J*, Vol. 341 (Pt 3), No. (Aug 1 1999), pp. (629-637).

Cox-Singh, J. and B. Singh (2008). Knowlesi malaria: newly emergent and of public health importance? *Trends Parasitol*, Vol. 24, No. 9, (Sep 2008), pp. (406-410).

D'Alexandri, F. L., E. A. Kimura, et al. (2006). Protein dolichylation in Plasmodium falciparum. *FEBS Lett*, Vol. 580, No. 27, (Nov 27 2006), pp. (6343-6348).

de Macedo, C. S., M. L. Uhrig, et al. (2002). Characterization of the isoprenoid chain of coenzyme Q in Plasmodium falciparum. *FEMS Microbiol Lett*, Vol. 207, No. 1, (Jan 22 2002), pp. (13-20).

Dieckmann-Schuppert, A., S. Bender, et al. (1992). Apparent lack of N-glycosylation in the asexual intraerythrocytic stage of Plasmodium falciparum. *Eur J Biochem*, Vol. 205, No. 2, (Apr 15 1992), pp. (815-825).

Docampo, R. and S. N. Moreno (2001). Bisphosphonates as chemotherapeutic agents against trypanosomatid and apicomplexan parasites. *Curr Drug Targets Infect Disord*, Vol. 1, No. 1, (May 2001), pp. (51-61).

Dondorp, A. M., F. Nosten, et al. (2009). Artemisinin resistance in Plasmodium falciparum malaria. *N Engl J Med*, Vol. 361, No. 5, (Jul 30 2009), pp. (455-467).

Eckstein-Ludwig, U., R. J. Webb, et al. (2003). Artemisinins target the SERCA of Plasmodium falciparum. *Nature*, Vol. 424, No. 6951, (Aug 21 2003), pp. (957-961).

Ekland, E. H. and D. A. Fidock (2008). In vitro evaluations of antimalarial drugs and their relevance to clinical outcomes. *Int J Parasitol*, Vol. 38, No. 7, (Jun 2008), pp. (743-747).

Ekvall, H. (2003). Malaria and anemia. *Curr Opin Hematol*, Vol. 10, No. 2, (Mar 2003), pp. (108-114).

Ernster, L. and G. Dallner (1995). Biochemical, physiological and medical aspects of ubiquinone function. *Biochim Biophys Acta*, Vol. 1271, No. 1, (May 24 1995), pp. (195-204).

Field, H., I. Blench, et al. (1996). Characterisation of protein isoprenylation in procyclic form Trypanosoma brucei. *Mol Biochem Parasitol*, Vol. 82, No. 1, (Nov 12 1996), pp. (67-80).

Florens, L., M. P. Washburn, et al. (2002). A proteomic view of the Plasmodium falciparum life cycle. *Nature*, Vol. 419, No. 6906, (Oct 3 2002), pp. (520-526).

Foley, D. A., J. Kennard, et al. (1978). Plasmodium berghei: infective exoerythrocytic schizonts in primary monolayer cultures of rat liver cells. *Exp Parasitol*, Vol. 46, No. 2, (Dec 1978), pp. (166-178).

Funes, S., E. Davidson, et al. (2002). A green algal apicoplast ancestor. *Science*, Vol. 298, No. 5601, (Dec 13 2002), pp. (2155).

Gardner, M. J., N. Hall, et al. (2002). Genome sequence of the human malaria parasite Plasmodium falciparum. *Nature*, Vol. 419, No. 6906, (Oct 3 2002), pp. (498-511).

Good, M. F. (2001). Towards a blood-stage vaccine for malaria: are we following all the leads? *Nat Rev Immunol*, Vol. 1, No. 2, (Nov 2001), pp. (117-125).

Gowda, D. C., P. Gupta, et al. (1997). Glycosylphosphatidylinositol anchors represent the major carbohydrate modification in proteins of intraerythrocytic stage Plasmodium falciparum. *J Biol Chem*, Vol. 272, No. 10, (Mar 7 1997), pp. (6428-6439).

Granja, A. C., F. Machungo, et al. (1998). Malaria-related maternal mortality in urban Mozambique. *Ann Trop Med Parasitol*, Vol. 92, No. 3, (Apr 1998), pp. (257-263).

Grau, G. E., C. D. Mackenzie, et al. (2003). Platelet accumulation in brain microvessels in fatal pediatric cerebral malaria. *J Infect Dis*, Vol. 187, No. 3, (Feb 1 2003), pp. (461-466).

Guerra, C. A., R. E. Howes, et al. (2010). The international limits and population at risk of Plasmodium vivax transmission in 2009. *PLoS Negl Trop Dis*, Vol. 4, No. 8, 2010), pp. (e774).

Hayes, C. N., D. Diez, et al. (2008). varDB: a pathogen-specific sequence database of protein families involved in antigenic variation. *Bioinformatics*, Vol. 24, No. 21, (Nov 1 2008), pp. (2564-2565).

Hemmer, C. J., F. G. Holst, et al. (2006). Stronger host response per parasitized erythrocyte in Plasmodium vivax or ovale than in Plasmodium falciparum malaria. *Trop Med Int Health*, Vol. 11, No. 6, (Jun 2006), pp. (817-823).

Hjertman, M., J. Wejde, et al. (1997). Evidence for protein dolichylation. *FEBS Lett*, Vol. 416, No. 3, (Oct 27 1997), pp. (235-238).

Hollingdale, M. R. (1992). Is culture of the entire plasmodium cycle, in vitro, now a reality? *Parasitol Today*, Vol. 8, No. 7, (Jul 1992), pp. (223).

Huff, C. G. (1964). Cultivation of the Exoerythrocytic Stages of Malarial Parasites. *Am J Trop Med Hyg*, Vol. 13, No. (Jan 1964), pp. (SUPPL 171-177).

Ibrahim, M., N. Azzouz, et al. (2001). Identification and characterisation of Toxoplasma gondii protein farnesyltransferase. *Int J Parasitol*, Vol. 31, No. 13, (Nov 2001), pp. (1489-1497).

Jordao, F. M., A. Y. Saito, et al. (2011). In vitro and in vivo antiplasmodial activity of risedronate and its interference with protein prenylation in P. falciparum. *Antimicrob Agents Chemother*, Vol. 55, No. 5, (May 2011), pp. (2026-2031).

Kimura, E. A., A. S. Couto, et al. (1996). N-linked glycoproteins are related to schizogony of the intraerythrocytic stage in Plasmodium falciparum. *J Biol Chem*, Vol. 271, No. 24, (Jun 14 1996), pp. (14452-14461).

Lacerda, M. V., M. A. Alexandre, et al. (2004). Idiopathic thrombocytopenic purpura due to vivax malaria in the Brazilian Amazon. *Acta Trop*, Vol. 90, No. 2, (Apr 2004), pp. (187-190).

Lalloo, D. G., A. J. Trevett, et al. (1996). Severe and complicated falciparum malaria in Melanesian adults in Papua New Guinea. *Am J Trop Med Hyg*, Vol. 55, No. 2, (Aug 1996), pp. (119-124).

Langreth, S. G. and W. Trager (1973). Fine structure of the malaria parasite Plasmodium lophurae developing extracellularly in vitro. *J Protozool*, Vol. 20, No. 5, (Nov 1973), pp. (606-613).

Leloir, L. F. (1977). The role of dolichol in protein glycosylation. *Adv Exp Med Biol*, Vol. 83, No. 1977), pp. (9-19).

Lim, L. and G. I. McFadden (2010). The evolution, metabolism and functions of the apicoplast. *Philos Trans R Soc Lond B Biol Sci*, Vol. 365, No. 1541, (Mar 12 2010), pp. (749-763).

Lombardo, M., F. C. Pinto, et al. (2006). Isolation and structural characterization of microcystin-LR and three minor oligopeptides simultaneously produced by

Radiocystis feernandoi (Chroococcales, Cyanobacteriae): a Brazilian toxic cyanobacterium. *Toxicon*, Vol. 47, No. 5, (Apr 2006), pp. (560-566).

Lujan, H. D., M. R. Mowatt, et al. (1995). Isoprenylation of proteins in the protozoan Giardia lamblia. *Mol Biochem Parasitol*, Vol. 72, No. 1-2, (Jun 1995), pp. (121-127).

Luo, S., N. Marchesini, et al. (1999). A plant-like vacuolar H(+)-pyrophosphatase in Plasmodium falciparum. *FEBS Lett*, Vol. 460, No. 2, (Oct 29 1999), pp. (217-220).

Maier, A. G., B. M. Cooke, et al. (2009). Malaria parasite proteins that remodel the host erythrocyte. *Nat Rev Microbiol*, Vol. 7, No. 5, (May 2009), pp. (341-354).

Marsh, K., D. Forster, et al. (1995). Indicators of life-threatening malaria in African children. *N Engl J Med*, Vol. 332, No. 21, (May 25 1995), pp. (1399-1404).

McFadden, G. I., M. E. Reith, et al. (1996). Plastid in human parasites. *Nature*, Vol. 381, No. 6582, (Jun 6 1996), pp. (482).

McTaggart, S. J. (2006). Isoprenylated proteins. *Cell Mol Life Sci*, Vol. 63, No. 3, (Feb 2006), pp. (255-267).

Menendez, C., A. F. Fleming, et al. (2000). Malaria-related anaemia. *Parasitol Today*, Vol. 16, No. 11, (Nov 2000), pp. (469-476).

Miller, L. H., M. F. Good, et al. (1994). Malaria pathogenesis. *Science*, Vol. 264, No. 5167, (Jun 24 1994), pp. (1878-1883).

Millet, P., P. Anderson, et al. (1994). In vitro cultivation of exoerythrocytic stages of the simian malaria parasites Plasmodium fieldi and Plasmodium simiovale in rhesus monkey hepatocytes. *J Parasitol*, Vol. 80, No. 3, (Jun 1994), pp. (384-388).

Millet, P., T. L. Fisk, et al. (1988). Cultivation of exoerythrocytic stages of Plasmodium cynomolgi, P. knowlesi, P. coatneyi, and P. inui in Macaca mulatta hepatocytes. *Am J Trop Med Hyg*, Vol. 39, No. 6, (Dec 1988), pp. (529-534).

Mons, B., C. J. Janse, et al. (1983). In vitro culture of Plasmodium berghei using a new suspension system. *Int J Parasitol*, Vol. 13, No. 2, (Apr 1983), pp. (213-217).

Moore, R. B., M. Obornik, et al. (2008). A photosynthetic alveolate closely related to apicomplexan parasites. *Nature*, Vol. 451, No. 7181, (Feb 21 2008), pp. (959-963).

Moura, I. C., G. Wunderlich, et al. (2001). Limonene arrests parasite development and inhibits isoprenylation of proteins in Plasmodium falciparum. *Antimicrob Agents Chemother*, Vol. 45, No. 9, (Sep 2001), pp. (2553-2558).

O'Donovan, S. M. and J. P. Dalton (1993). An improved medium for Plasmodium chabaudi in vitro erythrocyte invasion assays. *J Eukaryot Microbiol*, Vol. 40, No. 2, (Mar-Apr 1993), pp. (152-154).

Ogura, K. and T. Koyama (1998). Enzymatic Aspects of Isoprenoid Chain Elongation. *Chem Rev*, Vol. 98, No. 4, (Jun 18 1998), pp. (1263-1276).

Ogura, K., T. Koyama, et al. (1997). Polyprenyl diphosphate synthases. Subcellular Biochemistry: Cholesterol: Its function and Metabolism in Biology and Medicine. R. Bottman. New York, Plenum Press. 28: 57-87.

Rodrigues Goulart, H., E. A. Kimura, et al. (2004). Terpenes arrest parasite development and inhibit biosynthesis of isoprenoids in Plasmodium falciparum. *Antimicrob Agents Chemother*, Vol. 48, No. 7, (Jul 2004), pp. (2502-2509).

Use of Radioactive Precursors for Biochemical Characterization the Biosynthesis of Isoprenoids in
Intraerythrocytic Stages of Plasmodium falciparum
113

Rohmer, M., M. Knani, et al. (1993). Isoprenoid biosynthesis in bacteria: a novel pathway for the early steps leading to isopentenyl diphosphate. *Biochem J*, Vol. 295 (Pt 2), No. (Oct 15 1993), pp. (517-524).

Sanghai, S. R. and I. Shah (2010). Plasmodium vivax with acute glomerulonephritis in an 8-year old. *J Vector Borne Dis*, Vol. 47, No. 1, (Mar 2010), pp. (65-66).

Schmidt, R. A., C. J. Schneider, et al. (1984). Evidence for post-translational incorporation of a product of mevalonic acid into Swiss 3T3 cell proteins. *J Biol Chem*, Vol. 259, No. 16, (Aug 25 1984), pp. (10175-10180).

Schneider, C. (2005). Chemistry and biology of vitamin E. *Mol Nutr Food Res*, Vol. 49, No. 1, (Jan 2005), pp. (7-30).

Schwarz, R. T. and R. Datema (1982). The lipid pathway of protein glycosylation and its inhibitors: the biological significance of protein-bound carbohydrates. *Adv Carbohydr Chem Biochem*, Vol. 40, No. 1982), pp. (287-379).

Shen, P. S., J. C. Sanford, et al. (1996). Entamoeba histolytica: isoprenylation of p21ras and p21rap in vitro. *Exp Parasitol*, Vol. 82, No. 1, (Jan 1996), pp. (65-68).

Siqueira, A. M., M. A. Alexandre, et al. (2010). Severe rhabdomyolysis caused by Plasmodium vivax malaria in the Brazilian Amazon. *Am J Trop Med Hyg*, Vol. 83, No. 2, (Aug 2010), pp. (271-273).

Stresing, V., F. Daubine, et al. (2007). Bisphosphonates in cancer therapy. *Cancer Lett*, Vol. 257, No. 1, (Nov 8 2007), pp. (16-35).

Sussmann, R. A. C. (2010). Vitamin E Biosynthesis in intraerythrocytic stages of *Plasmodium falciparum*. Parasitology Department. São Paulo, University of São Paulo. Master Thesis 76.

Tonhosolo, R., F. L. D'Alexandri, et al. (2009). Carotenoid biosynthesis in intraerythrocytic stages of Plasmodium falciparum. *J Biol Chem*, Vol. 284, No. 15, (Apr 10 2009), pp. (9974-9985).

Tonhosolo, R., F. L. D'Alexandri, et al. (2005). Identification, molecular cloning and functional characterization of an octaprenyl pyrophosphate synthase in intra-erythrocytic stages of Plasmodium falciparum. *Biochem J*, Vol. 392, No. Pt 1, (Nov 15 2005), pp. (117-126).

Tonhosolo, R., H. B. Gabriel, et al. (2010). Intraerythrocytic stages of Plasmodium falciparum biosynthesize menaquinone. *FEBS Lett*, Vol. 584, No. 23, (Dec 1 2010), pp. (4761-4768).

Trager, W. and J. B. Jensen (1976). Human malaria parasites in continuous culture. *Science*, Vol. 193, No. 4254, (Aug 20 1976), pp. (673-675).

Trager, W. and J. B. Jensen (1997). Continuous culture of Plasmodium falciparum: its impact on malaria research. *Int J Parasitol*, Vol. 27, No. 9, (Sep 1997), pp. (989-1006).

Valderramos, S. G., D. Scanfeld, et al. (2010). Investigations into the role of the Plasmodium falciparum SERCA (PfATP6) L263E mutation in artemisinin action and resistance. *Antimicrob Agents Chemother*, Vol. 54, No. 9, (Sep 2010), pp. (3842-3852).

Watkins, B. and C. Sibley (2011). Vivax malaria: old drug, new uses? *J Infect Dis*, Vol. 203, No. 2, (Jan 15 2011), pp. (144-145).

WHO (2010a). Guidelines for the treatment of malaria. Switzerland, World Health Organization: 194.

WHO (2010b). World Malaria Report. Switzerland, World Health Organization: 238.

Zhang, F. L. and P. J. Casey (1996). Protein prenylation: molecular mechanisms and functional consequences. *Annu Rev Biochem*, Vol. 65, No. 1996), pp. (241-269).

6

Application of Radioisotopes in Biochemical Analyses: Metal Binding Proteins and Metal Transporters

Miki Kawachi[1,2], Nahoko Nagasaki-Takeuchi[1,3],
Mariko Kato[1] and Masayoshi Maeshima[1]
[1]*Graduate School of Bioagricultural Sciences, Nagoya University, Nagoya*
[2]*Lehrstuhl für Pflanzenphysiologie, Ruhr-Universität Bochum, Bochum,*
[3]*Graduate School of Biosciences, Nara Institute of Science and Technology, Nara*
[1,3]*Japan*
[2]*Germany*

1. Introducition

Radioisotopes (RI) such as ^{3}H, ^{14}C, ^{32}P, and ^{45}Ca are excellent tools in biological research. Most RI are used as tracers in studies of primary and secondary metabolism, drug metabolism, transcription, translation, post-translational modifications such as protein phosphorylation, association of proteins with metals, and transport of metals across biomembranes. Furthermore, some experiments have used neutrons for mutagenesis of microorganisms, animals, and plants. Recent progress in the biological sciences has resulted in novel probes and labeling reagents, which has decreased the need for RI. Experiments with RI require experimental space specialized for RI, careful experimental procedures, and training. Although these are disadvantages, RI are still useful and powerful tools with high resolution compared with non-RI methods. Here, we describe the advantages of RI in biochemical assays, and detailed experimental procedures of metal-binding assays and membrane transport measurements of metal cations, especially calcium and zinc.

2. Advantages of radioisotopes as tracers

Most metabolic pathways that are described in biochemistry textbooks, in various organisms including humans, plants, and microorganisms, could not have been determined without RI such as ^{14}C, ^{35}S, ^{32}P, and ^{3}H. Biochemical experiments with RI provide information on the fates of metabolites, nutrients, and inorganic ions at each periodic stage of living organisms or cells. In the early era of molecular biology, ^{32}P was used as an essential tool in a large number of laboratories to determine DNA sequences and to identify target DNAs or mRNAs. Phosphorylation of serine and/or tyrosine residues is a key covalent modification of proteins. ^{32}P ([γ-^{32}P]ATP) is still used to investigate this biochemical process. ^{35}S ([^{35}S]glutathione) is also used to investigate protein *S*-glutathiolation, which regulates the redox state of cells or detoxifies xenobiotics and natural products.

There are several advantages of RI in biochemical analyses compared with non-RI experimental procedures as follows:

1. High sensitivity: Trace amounts of RI can be detected by using a scintillation counter, X-ray film (autoradiography), or imaging plate. For example, labeling with ^{32}P-nucleotides such as [α-^{32}P]ATP or [γ-^{32}P]ATP is frequently used to label DNAs. In addition to the RI method, labeling of DNA with digoxigenin (DIG) has been used as a non-RI method. DIG labeling can be done by polymerase chain reaction (PCR), and DIG-labeled DNA can be detected by immunochemical methods. Reagents for DIG labeling are available from Roche Diagnostics GmbH (Nonnenwald, Penzberg, Germany). The DIG method can be done without specific equipment and space for RI. However, sensitivity of the RI method is higher than the DIG method. In particular, RI methods have advantageous sensitivity in northern and Southern analyses.

2. High accuracy: A good example is protein phosphorylation in cells. Immunochemical analyses such as immunoblotting are also used to detect phosphorylated proteins. Antibodies specific to phosphoserine, phosphotyrosine, or peptides containing phosphorylated amino acid residues are prepared and used. The accuracy and sensitivity depend on the quality and specificity of the antibodies. In most cases, researchers must pay attention to artifactual signals. In contrast, labeling proteins with RI provides clear and quantitative information on protein phosphorylation. In RI methods, proteins are phosphorylated with [γ-^{32}P]ATP in cell free systems (*in vitro*) or in experiments using cells, tissues, or organisms (*in vivo*), and then proteins are extracted and separated by sodium dodecyl sulfate (SDS)-polyacrylamide gel electrophoresis (PAGE). Proteins in the gel are transferred onto a transfer membrane such as polyvinylidine fluoride (PVDF). The membrane is dried and subjected to autoradiography by contact with an X-ray film at –80°C for a few days. Imaging plates are now generally used for the detection of phosphorylated proteins. Imaging plates have several advantages compared with autoradiography: (i) high sensitivity (quick detection), (ii) good linearity between the content of ^{32}P and the signal, (iii) digital imaging, and (iv) no requirement of a dark room.

3. Reflection of natural conditions: Organic and inorganic compounds containing RI and radioactive elements have the same chemical properties as normal compounds and elements in most cases. Therefore, we can determine and follow the compounds and elements in cells, tissues, and organisms without artificial conditions. This is a critical advantage of RI.

3. Typical radioisotopes used in biochemical and molecular biological studies

3.1 Physicochemical properties of typical isotopes

As mentioned in other chapters, several isotopes are used for biochemical and molecular biological studies. The physical and chemical properties of typical isotopes that are frequently used in the biological sciences are summarized in Table 1. In most laboratories, ^{3}H, ^{14}C, ^{35}S, ^{32}P, and ^{125}I are used for labeling of amino acids, peptides, and proteins. These RI, except for ^{125}I, emit β-rays with different energies (^{3}H, low; ^{14}C, medium; ^{35}S, high). In most cases, ^{35}S is used as L-[^{35}S]-methionine or L-[^{35}S]-cysteine for labeling peptides or proteins in *in vitro* and *in vivo* experiments of protein synthesis. For example, L-[^{35}S]-

methionine is added into *in vitro* translation mixtures that lack cold methionine. In certain experiments, ^{35}S is used for sulfation of peptides or proteins through tyrosine modification by an enzyme tyrosylprotein sulfotransferase (Moore, 2003; Hoffhines et al., 2006). [32]P is used for monitoring protein phosphorylation of serine and/or threonine residues.

RI	Radiation Energy (keV)	Half life	Shielding (1)	Measurement (2)	Monitoring (3)	Application (4)
H-3	β, 18.6	12.3 Y	Not required	ARG, LSC	SM+LSC	As a tracer for metabolites
C-14	β, 156	5730 Y	Not required	ARG, LSC	SM+LSC	As a tracer for metabolites and CO_2
S-35	β, 167	87.5 D	Not required	ARG, LSC	SM+LSC	As a tracer for protein synthesis
P-32	β, 1711	14.3 D	Acryl plate	ARG, LSC	GM	DNA labeling, protein phosphorylation
P-33	β, 249	25.3 D	Acryl plate	ARG, LSC	GM	DNA labeling and macro-array
Ca-45	β, 257	162 D	Acryl plate	ARG, LSC	GM	As a tracer for Ca in organisms and for characterization of Ca-binding proteins and kinetic assays of Ca transport
Zn-65	β, 329 γ, 511 γ, 1116	244 D	Lead blocks	γ counter, ARG, LSC	NaI (T1)	As a tracer for Zn in organisms and membrane transport
I-125	γ, 35.5 X, 27.5	59.4 D	Acryl plate containing lead	ARG, γ counter	NaI (T1) for I-125	Protein labeling
Co-60	β, 310 β, 1480 β, 1173 γ, 1332	5.27 Y	Lead blocks	γ counter	NaI (T1)	Sterilization of medical equipment, radiation source for food irradiation and for mutation of seeds and organisms

(1) Shielding required for safety during experiments. (2) Adequate measurement methods for radioactivity: ARG, autoradiography; LSC, liquid scintillation counter. (3) Detection methods for pollution on the surface of experimental benches with RI: SM, smear method; LSC, liquid scintillation counter; GM, GM survey meter; NaI (T1), NaI (T1) survey meter. (4) Typical application in biochemical research.

Table 1. Physicochemical properties and quantification of typical RI in biological sciences

3.2 Quantification of radioisotopes and measurement of pollution

Several methods have been developed for quantifying RI and measuring pollution with RI during experiments. Here, we introduce two instruments, a Geiger-Müller (GM) survey meter (also known as a GM counter) and a liquid scintillation counter.

A GM survey meter is usually used to monitor β- and γ-rays, but not neutrons, from the surface of substances. Two electrodes are set in a cylinder filled with inert gas, such as neon, helium, or argon. A voltage is applied to the electrodes, i.e., the anode (a central wire or needle) and the cathode (the inside surface of the cylinder). GM survey meters operate under a high voltage of more than several hundred volts. When the ionizing radiation passes through the cylinder, ions and electrons are generated from some of the gas molecules. This reaction generates an electrical current pulse of constant voltage. GM survey meters are usually used for monitoring the surface pollution of RI that radiate β- or γ-rays. The meter cannot count γ-rays efficiently and does not distinguish each isotope generating β-rays.

As an efficient and practical means of quantifying β-ray radiation, liquid scintillation counters are commonly used for biochemical analyses. A liquid scintillation counter measures β-radiation in a solution containing a RI, fluorescent compounds (scintillators), and organic solvents such as xylene, dioxane, or toluene. As a scintillator, 2,5-diphenyloxazole (DPO) and 1,4-bis(5-phenyl-2-oxazolyl)benzene (POPOP) are used. The energy of β-rays (β particles) from RI excites the scintillator, and then the excited fluorescent molecules dissipate the energy by emitting fluorescence. Therefore, radiation of β particles causes a pulse of fluorescent light. Liquid scintillation counters are used for measuring β-ray-emitting RI including ^{3}H, ^{14}C, ^{32}P, ^{45}Ca, and ^{65}Zn because the counting efficiency is high even for nuclides emitting low energy β-rays.

4. Analyses of Ca-binding proteins with ^{45}Ca

There are many types of Ca-binding proteins, such as calmodulin, calreticulin, and annexin (Berridg et al., 2003). Identification and quantitative characterization of Ca-binding proteins provide key information about their biochemical roles in living cells. Here, we briefly introduce biochemical methods to identify and characterize these proteins.

4.1 Identification of Ca-binding proteins

Staining of SDS-PAGE with Stains-all (commercially available from reagent companies such as Sigma Aldrich) is a conventional non-RI method to identify Ca-binding proteins (Campbell et al., 1983). Stains-all is a metachromatic cationic carbocyanine dye that tends to bind acidic proteins. Ca-binding proteins in particular are stained blue relatively stably (Yuasa & Maeshima, 2000). Therefore, this is a useful method to detect candidates for Ca-binding proteins in crude samples prepared from organisms.

Radioisotope ^{45}Ca is necessary to confirm that the protein(s) of interest can bind calcium. The ^{45}Ca overlay assay is one convenient method (Campbell et al., 1983; Yuasa & Maeshima, 2002; Ide et al., 2007; Kato et al., 2010). If a purified preparation is available, aliquots of the purified sample are spotted onto a membrane filter such as a PVDF membrane (ca. 30 × 40 mm). Then the membrane is incubated in a small volume (1 mL) of medium containing ^{45}Ca as $CaCl_2$, 5 mM $MgCl_2$, 60 mM KCl, and 10 mM Mes-KOH, pH 6.5, for 30 min at 25 to 30°C, and then washed with 10 mL of 50% (v/v) ethanol to remove unbound ^{45}Ca (Figure 1) (Nagasaki et al., 2008). The membrane is dried in air at room temperature. $MgCl_2$ and KCl are added into the reaction medium to mimic physiological conditions. An autoradiogram of the $^{45}Ca^{2+}$-labelled proteins on the membrane can be obtained by exposure to an X-ray film for 3 days at $-80°C$.

If the purified protein(s) is not available, proteins separated by SDS-PAGE are transferred onto a transfer membrane such as PVDF as is usually used for immunoblotting (Figure 1). The Ca-binding protein(s) can be detected by the same method as the ^{45}Ca overlay assay mentioned above.

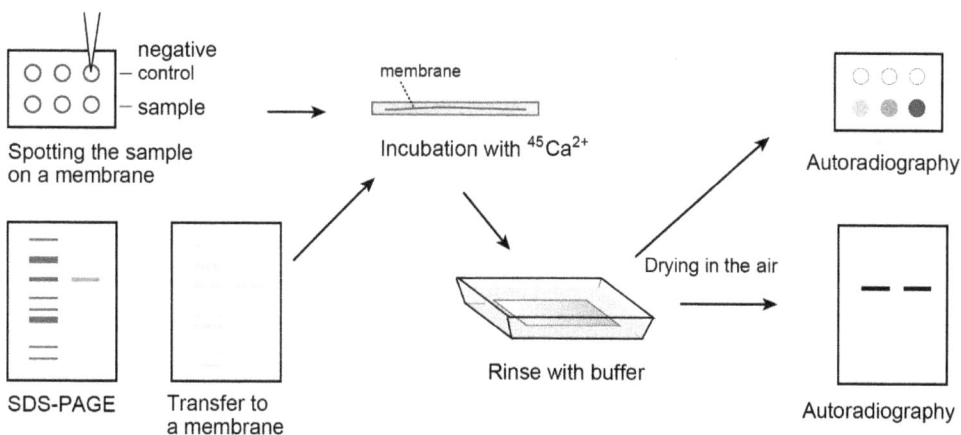

Purified proteins are spotted on a PVDF membrane (left, upper panel). In another method, a protein fraction that contains the Ca-binding protein is subjected to SDS-PAGE and then transferred to a PVDF membrane (left, lower panel). The membrane is incubated with buffer containing $^{45}Ca^{2+}$, rinsed, and then dried in air. By autoradiography, the $^{45}Ca^{2+}$-binding capacity is monitored (right, upper panel). From the membrane blotted after SDS-PAGE, the Ca-binding protein(s) is identified by autoradiography (right, lower panel).

Fig. 1. Detection of Ca-binding protein(s) by ^{45}Ca-overlay assay

4.2 Characterization of kinetic properties of Ca-binding proteins

The dissociation constant (K_d) for Ca^{2+} and the calcium-binding number are important kinetic parameters for understanding these proteins. Several assay methods can be used to measure the Ca-binding kinetics of Ca-binding proteins. For example, there is equilibrium dialysis, flow dialysis, membrane microassay, and spectrophotometry. The former two methods are carried out using $^{45}Ca^{2+}$. Most methods require a relatively large amount of the purified Ca-binding protein. Here, we introduce a special equilibrium dialysis using small dialysis buttons (Figure 2A). A small well of dialysis button is filled with the Ca-binding protein(s) and is sealed with a dialysis membrane. The protein solution in the well is dialyzed against 40 mL of buffer containing $^{45}Ca^{2+}$ at different concentrations. The Ca-binding protein binds the $^{45}Ca^{2+}$ entered into the well. After dialysis for 16 hr at $25°C$, the protein solution in the well of each dialysis button is collected with a needle and syringe. An aliquot of the solution is spotted on a nitrocellulose membrane (13 mm in diameter), and then the membranes are dried in air. Total radioactivity associated with the filter membrane is measured with a liquid scintillation counter. Unbound Ca^{2+} is measured from the radioactivity of the external solution. The amount of Ca^{2+} bound to the Ca-binding protein increases in proportion to the concentration of Ca^{2+} as shown in Figure 2B.

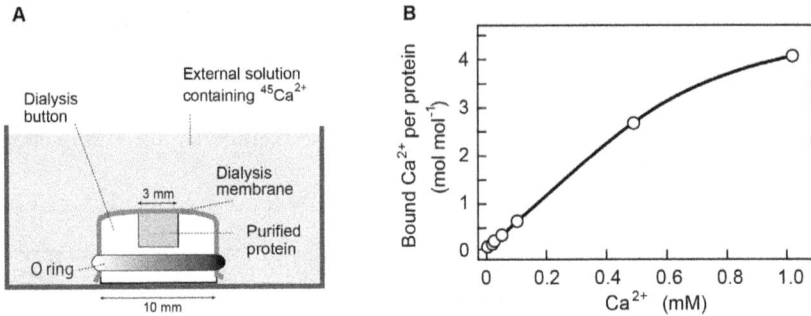

(A) Diagram of the equilibrium dialysis assay with a dialysis button. Twenty microliters of purified protein is put in the well of a dialysis button of 3 mm in diameter, sealed with a dialysis membrane fixed with an O ring, and then dialyzed against 40 mL of 25 mM Mes-KOH, pH 6.0, 150 mM KCl with the indicated concentrations of $^{45}CaCl_2$. The volume of the external solution must be in excess of the sample volume to keep a constant level of Ca^{2+} during the assay. (B) Ratios of Ca^{2+} bound to the Ca-binding protein. The number of Ca^{2+} bound per Ca-binding protein and the K_d value for Ca^{2+} can be calculated from a Scatchard plot (a method of analyzing the binding of a ligand to a macromolecule). The result of radish Ca-binding protein is shown here (Yuasa & Maeshima, 2002).

Fig. 2. Equilibrium dialysis of Ca^{2+}-binding of purified Ca-binding protein

5. Measurement of membrane transport of zinc and calcium

Radioactive elements such as Ca^{2+} and Zn^{2+} are commonly used in ion transport experiments because they provide direct evidence and quantitative information. Here, we introduce a Zn^{2+} transporter and a Ca^{2+} transporter, which work as metal/proton exchangers.

5.1 Determination of kinetic parameters of a Zn transporter across biomembranes

Transporters implicated in Zn transport include members of the metal tolerance protein (MTP), ZRT1/IRT1-like protein (ZIP) (also known as zinc-iron permease), and heavy-metal ATPase (HMA) (P_{1B} subgroup of P-type ATPase) families (Krämer et al., 2007). Here, we introduce an assay procedure for an MTP-type zinc transporter that works as a Zn^{2+}/H^+ exchanger. *Arabidopsis thaliana* MTP1 is localized in the vacuolar membrane, which has two types of proton pumps, vacuolar H^+-ATPase (V-ATPase) and H^+-pyrophosphatase (Enrico et al., 2007). AtMTP1 actively transports excessive zinc in the cytoplasm into vacuoles (Kawachi et al., 2008; Kawachi et al., 2010).

The assay procedure of AtMTP1 expressed in *Saccharomyces cerevisiae* cells is shown in Figure 3. In this case, an *S. cerevisiae* mutant that lacks endogenous zinc transporters COT1 and ZRC1 is used as a host cell for heterologous expression. When ATP is added into the vacuolar membrane vesicle suspension, a pH gradient (ΔpH) is formed across the membrane by yeast endogenous V-ATPase. Then radioactive $^{65}Zn^{2+}$ is added into the reaction mixture as $ZnCl_2$. Under these conditions, AtMTP1 actively incorporates $^{65}Zn^{2+}$ into membrane vesicles using a ΔpH (Figure 3A). The reaction medium contains 300 mM sorbitol, 5 mM MES-Tris pH 6.9, 25 mM KCl, 1 mM dithiothreitol, 5 mM $MgCl_2$, 0.2 mM NaN_3, 0.1 mM Na_3VO_4, and 3 mM ATP-Tris. The uptake reaction is started by adding 5 μM $^{65}ZnCl_2$. Vacuolar membranes from plants and yeast contain metal-translocating ATPases that have the ability to transfer Zn^{2+} into vacuoles. Therefore, the activities of these ATPases must be inhibited. Sodium azide and vanadate are potent inhibitors of the F-type and P-type ATPases, respectively. The K_m value for Zn^{2+} has been reported to be 0.30 µM for AtMTP1 (Kawachi et al., 2008). The value is comparable to the *S. cerevisiae* endogenous zinc transporter ZRC1 (0.16 µM) (MacDiarmid et al., 2002), *Escherichia coli* ZitB (1.4 µM) (Anton et al., 2004), and human hZIP4 (2.5 µM) (Mao et al., 2007). This method is applicable to assay the zinc transport activity of vacuolar membrane vesicles from plant tissues. Vacuolar membrane vesicles can be prepared from plant tissues such as mung bean hypocotyls by conventional differential centrifugation (Maeshima and Yoshida, 1989).

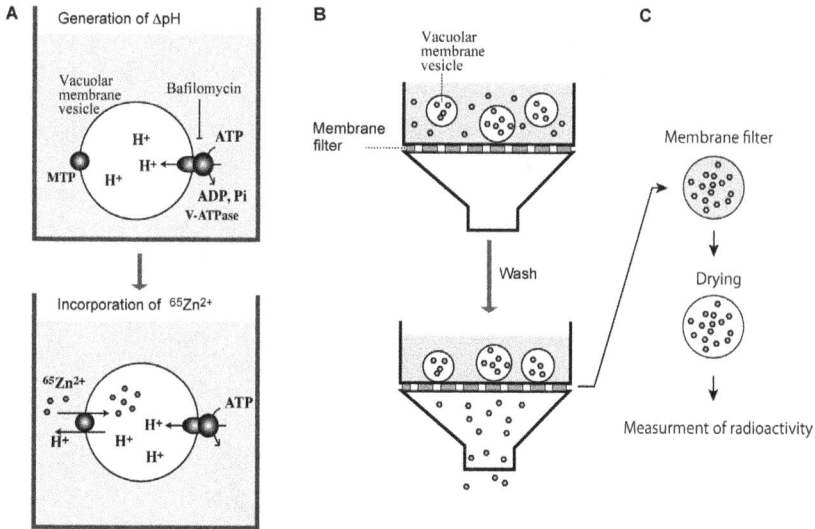

(A) Vacuolar membrane vesicles prepared from yeast cells or plant tissues are activated by adding ATP into the suspension. Vacuolar H^+-ATPase (V-ATPase) acidifies membrane vesicles and generates a pH gradient across the membrane.Bafilomycin A_1 is a potent inhibitor of V-ATPase and used to assess the V-ATPase-dependent (ΔpH-dependent) activity of theZn^{2+}/H^+ exchanger. When radioactive $^{65}Zn^{2+}$ is added, membrane vesicles actively uptake $^{65}Zn^{2+}$ using a pH gradient in a Zn^{2+}/H^+ exchanger-dependent manner. (B) Membrane vesicles are filtrated and washed with the buffer. (C) The radioactivity of $^{65}Zn^{2+}$ membrane vesicles trapped on the membrane filter is determined by a scintillation counter.

Fig. 3. Assay of Zn^{2+} transport into membrane vesicles through a Zn^{2+}/H^+ exchanger

The catalytic domain of V-ATPase (V_1 sector) is exposed to the cytoplasm. Therefore, only right-side-out membrane vesicles, in which the V_1 sector faces to the reaction mixture, can be energized by V-ATPase. Approximately half of the membrane vesicles are right-side-out when plant tissues and yeast cells are homogenized. In the remaining half, the V_1 sector faces the vesicle lumen and cannot utilize ATP. The inside-out vesicles have no ability to uptake $^{65}Zn^{2+}$ or to export zinc under this assay condition. Therefore, this experiment determines the Zn^{2+} uptake activity of the right-side-out membrane vesicles. The inside-out membrane vesicles do not interfere with the Zn^{2+} transport of right-side-out vesicles.

After an adequate period of uptake reaction, incorporated $^{65}Zn^{2+}$ must be separated from the un-incorporated ions. Aliquots (for example, 100 µL) of the membrane vesicle suspension are transferred to funnels with nitrocellulose membrane filters that are presoaked with the buffer at appropriate intervals. Filter units with a 0.45-μm nitrocellulose membrane of 13 mm in diameter are easy to use for assays of multiple samples. The filter units are washed with 1.5 mL of cold wash buffer without $^{65}Zn^{2+}$. The wash buffer contains 300 mM sorbitol, 5 mM MES-Tris pH 6.9, 25 mM KCl, and 0.1 mM $ZnCl_2$. The addition of cold $ZnCl_2$ is essential to remove $^{65}Zn^{2+}$ from the surface of the membrane vesicles thoroughly. Finally, the radioactivity of $^{65}Zn^{2+}$ is determined by a γ scintillation counter.

When measuring the zinc transport activity of plant vacuolar membranes, vacuolar H^+-pyrophosphatase (V-PPase) also works as a useful proton pump (Maeshima, 2001). V-PPase hydrolyzes pyrophosphate (diphosphate) instead of ATP as a substrate. Therefore, metal-translocating ATPases do not work in the assay medium when assayed with V-PPase.

To demonstrate the active translocation of Zn^{2+} by exogenous zinc transporters, the membrane sample from the yeast mutant with a vacant vector is assayed as a control. Zinc ionophore pyrithione is usually used to collapse the concentration gradient of Zn^{2+} across the membrane at a concentration of 5 μM (MacDiarmid et al., 2002). If Zn^{2+} is actively incorporated into the membrane vesicles, the addition of pyrithione releases Zn^{2+} from membrane vesicles as shown in Figure 4A. Also, uptake experiments without ATP or with 0.2 μM bafilomycin A_1 is done by the same protocol. The bafilomycin A_1-sensitive zinc uptake activities are plotted as V-ATPase-dependent zinc uptake in time-course or substrate saturation analysis.

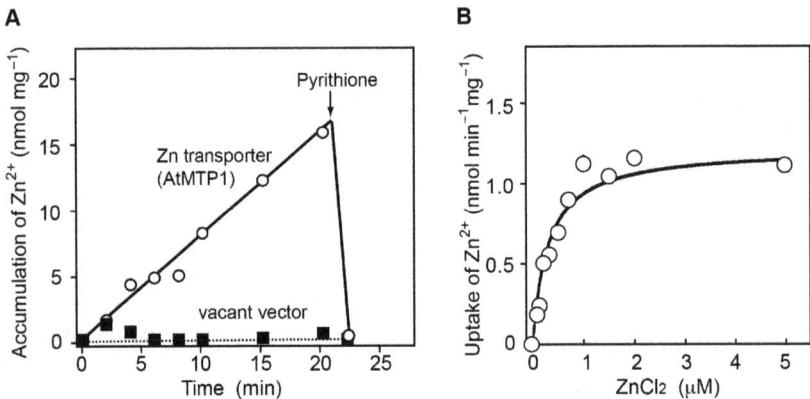

The vacuolar type Zn^{2+}/H^+ exchanger of *Arabidopsis thaliana* (AtMTP1) was heterologously expressed in a yeast (*Saccharomyces cerevisiae*) mutant that lacks endogenous vacuolar membrane zinc transporters (*zrc1 cot1* mutant). The vacuolar membrane-enriched fraction was prepared from yeast cells expressing AtMTP1 (circles) or vacant vector (closed squares) and assayed for zinc uptake activity. (A) Membrane vesicles were pre-incubated in uptake medium (1.0 mL) containing 3 mM ATP for 10 min at 25°C to generate a pH gradient across the membrane as shown in Figure 3A. The same reaction media supplemented with 0.2 mM bafilomycin A_1 were also prepared and assayed to measure bafilomycin A_1-sensitive zinc activity. The reaction was started by the addition of 5 μM $^{65}ZnCl_2$ at time 0 and continued for the indicated period. Aliquots (100 μL) of the reaction suspensions were filtered though a nitrocellulose membrane and washed with 1.5 mL of cold wash buffer. The radioactivity of $^{65}Zn^{2+}$ in the membrane vesicles was determined. The bafilomycin A_1-sensitive zinc uptake activities are plotted as V-ATPase-dependent zinc uptake. A zinc ionophore pyrithione is added into the reaction medium to make a final concentration of 5 μM to confirm the active transport of zinc. (B) Zinc uptake activity is measured at the indicated concentration of $^{65}ZnCl_2$ and shown as a substrate-saturation curve.

Fig. 4. Assay of zinc uptake by a zinc transporter

5.2 Determination of kinetic parameters of Ca transporters across biomembranes

The Ca^{2+}-ATPase (calcium pump) belongs to the P-type ATPase family that includes the Na^+,K^+-ATPase (Morth et al., 2011), and actively translocates Ca^{2+} across the membrane coupled with ATP hydrolysis. Ca^{2+}-ATPases in eukaryotes are divided into to the ER-type and calmodulin-activated plasma-membrane-type Ca^{2+}-ATPases. The Ca^{2+}-ATPase is localized in the plasma membrane, ER, Golgi apparatus, and vacuole, and maintains calcium homeostasis in the cytoplasm and lumen spaces.

The Ca^{2+}/H^+ exchanger is the other type of active Ca^{2+} transporter (Ueoka-Nakanishi et al., 2000; Kamiya & Maeshima, 2004). As an energy source, the exchangers use a pH gradient across the membrane that is generated by proton pumps. Plant and fungal cells have the P-type H^+-ATPase in their plasma membranes, and H^+-ATPase in their vacuolar membranes as primary proton pumps. In plants, an additional proton pump, H^+-pyrophosphatase (V-PPase), functions as an efficient proton pump (Martinoia et al., 2007). Thus, the Ca^{2+}/H^+ exchanger is categorized as a secondary active Ca^{2+} transporter and exports excess Ca^{2+} from the cytoplasm to the extracellular space or vacuoles.

The physiological roles of Ca^{2+}-ATPase cannot be understood without information of their kinetic parameters. Quantitative analysis of calcium pumps and other active calcium transporters is usually performed by using $^{45}Ca^{2+}$. Assay procedures for plant vacuolar Ca^{2+}-ATPase and Ca^{2+}/H^+ exchanger are similar to the Zn^{2+}/H^+ exchanger. Aliquots (10 μg of protein) of vacuolar membranes prepared from plant tissues are suspended in 100 μL of assay medium consisting of 0.25 M sorbitol, 5 mM Mes-Tris pH 7.2, 50 mM KCl, 0.5 mM dithiothreitol, 3 mM $MgCl_2$, 1 mM Tris-ATP (ATP solution neutralized with Tris), and 100 μM $CaCl_2$ ([^{45}Ca], 37–220 kBq mL^{-1}) for the Ca^{2+}-ATPase assay. The uptake reaction is started by adding $CaCl_2$. After an adequate reaction period, the mixture is filtered through a presoaked 0.45-μm nitrocellulose filter (13 mm in diameter) as described for the Zn^{2+}/H^+ exchanger. The filter is washed twice with 200 μL of 0.25 M sorbitol, 5 mM Mes-Tris pH 7.2, 50 mM KCl, 0.5 mM dithiothreitol, 0.25 mM $MgCl_2$, and 1 mM ethylene glycol tetraacetic acid (EGTA). EGTA has a higher affinity for Ca^{2+} than for Mg^{2+} and is added to remove the unabsorbed $^{45}Ca^{2+}$ in the medium. Radioactive Ca^{2+} incorporated into membrane vesicles is trapped on the nitrocellulose membrane. The radioactivity associated with the filter membrane is measured with a liquid scintillation counter. The reaction medium for the V-PPase-dependent Ca^{2+}/H^+ exchanger contains 1 mM sodium pyrophosphate (Na_2PPi) instead of Tris-ATP. The membrane vesicles are pre-incubated with Na_2PPi for 3 min, and then the Ca^{2+} transport reaction is started by adding $CaCl_2$.

Background values resulting from incubations without ATP or Na_2PPi are subtracted from the corresponding values in the presence of ATP or Na_2PPi. Bafilomycin A_1 and carbonyl cyanide m-chlorophenylhydrazone (CCCP) dissolved in dimethyl sulfoxide (DMSO) are used to inhibit V-ATPase and collapse the pH gradient, respectively. The DMSO concentration in the assay medium should be less than 1% (by volume) to avoid artificial effects of the solvent. Calcium ionophore A23187 is also dissolved in DMSO and used to confirm the active transport of Ca^{2+} through Ca^{2+}-ATPase or Ca^{2+}/H^+ exchanger. A23187 is a mobile Ca^{2+} carrier produced by *Streptomyces chartreusensis* as an antibiotic.

Figure 5 shows typical substrate-saturation curves of the Ca^{2+}-ATPase and Ca^{2+}/H^+ exchanger of vacuolar membranes prepared from mung bean hypocotyls (Ueoka-Nakanishi et al., 1999). Experiments with radioisotope ^{45}Ca provide quantitative information of their transport kinetics. Ca^{2+}-ATPase is recognized as a high-affinity, low-capacity transporter, while the Ca^{2+}/H^+ exchanger is low-affinity, high capacity. These two active transporters maintain calcium homeostasis in the cytoplasm through their characteristic properties.

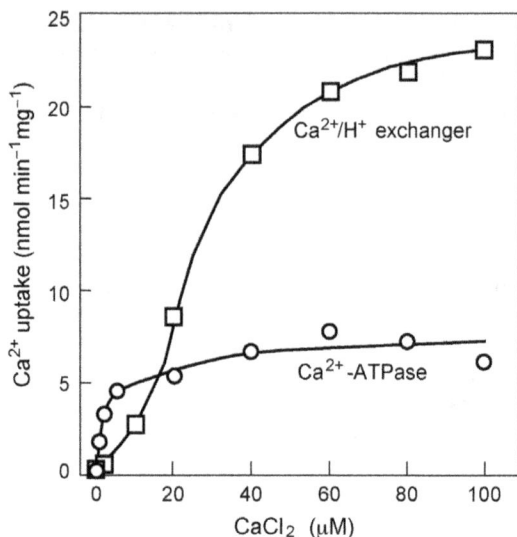

Activity of Ca^{2+}-ATPase (circles) in the vacuolar membrane was determined in the presence of 1 mM ATP, 0.1 mM bafilomycin A_1, and indicated concentrations of $^{45}CaCl_2$. Ca^{2+}/H^+ exchanger activity (squares) was determined after pre-incubation with 1 mM NaPPi for 3 min. V_{max} values of Ca^{2+}-ATPase and the Ca^{2+}/H^+ exchanger were 6.9 and 21 nmol min^{-1} mg^{-1} of protein, respectively. Apparent K_m values of Ca^{2+}-ATPase and the Ca^{2+}/H^+ exchanger were 2.6 and 25 μM, respectively.

Fig. 5. Calcium transport activity of Ca^{2+}-ATPase and Ca^{2+}/H^+ exchanger in vacuolar membranes.

6. Conclusion

High sensitivity and accuracy in quantitative assay are the actual merits of RI. Information concerning kinetic parameters of ion transporters cannot be obtained without RI, such as ^{65}Zn and ^{45}Ca, as described for Zn^{2+}/H^+ exchanger and Ca^{2+}/H^+ exchanger, and Ca^{2+}-ATPase. The obtained values of K_m and V_{max} are fundamental to evaluate physiological importance of individual transporters quantitatively. The data presented here are typical examples, which show advantages of RI in biochemical analyses. Although the use of RI is regulated by the laws established in each country, university and research institute, these rules keep the safety for the users and people. RI in biochemistry is one of the peaceful use of atomic energy and will be utilized as an essential tool to develop our scientific knowledge.

7. References

Anton, A., Weltrowski, A., Haney, C. J., Franke, S., Grass, G., Rensing, C. & Nies, D. H. (2004) Characteristics of zinc transport by two bacterial cation diffusion facilitators from *Ralstonia metallidurans* CH34 and *Escherichia coli*. *Journal of Bacteriology*, Vol.186, No.22, pp. 7499–7507, ISSN 0021-9193.

Berridge, M.J., Bootman, M.D. & Roderick, H.L. (2003) Calcium signalling: dynamics homeostasis and remodelling. *Nature Reviews Molecular Cell Biololgy,* Vol.4, No.7, pp. 517-529. ISSN 1471-0072.

Campbell, K.P., MacLennan, D.H. & Jorgensen, A.O. (1983). Staining of the Ca^{2+}-binding proteins, calsequestrin, calmodulin, troponin C, and S-100, with the cationic carbocyanine dye "Stains-all." *Journal of Biological Chemistry,* Vol.258, No.18, pp. 11267–11273, ISSN 0021-9258.

Hoffhines, A.J., Damoc, E., Bridges, K.G., Leary, J.A., Moore, K.L. (2006). Detection and purification of tyrosine-sulfated proteins using a novel anti-sulfotyrosine monoclonal antibody. *Journal of Biological Chemistry,* Vol.281 No.49, pp. 37877–87. ISSN 0021-9258.

Ide, Y., Nagasaki, N., Tomioka, R., Suito, M., Kamiya, T. & Maeshima, M. (2007). Molecular properties of novel, hydrophilic cation-binding proteins associated with the plasma membrane. *Journal of Experimental Botany,* Vol.58, No.5, 1173–1183. ISSN 0022-0957.

Kamiya, T. & Maeshima, M. (2004) Residues in internal repeats of the rice cation/H^+ exchanger are involved in the transport and selection of cations. *Journal of Biological Chemistry,* Vol.279, No.1, pp. 812–819. ISSN 0021-9258.

Kato, M., Nagasaki-Takeuchi, N., Ide, Y. & Maeshima, M. (2010) An *Arabidopsis* hydrophilic Ca^{2+}-binding protein with a PEVK-rich domain, PCaP2, is associated with the plasma membrane and interacts with calmodulin and phosphatidylinositol phosphates. *Plant and Cell Physiology,* Vol.51, No.3, 366–379. ISSN 0032-0781.

Kawachi, M., Kobae, Y., Mimura, T. & Maeshima, M. (2008). Deletion of a histidine-rich loop of AtMTP1, a vacuolar Zn^{2+}/H^+ antiporter of *Arabidopsis thaliana*, stimulates the transport activity. *Journal of Biological Chemistry,* Vo.283, No.13, pp. 8374–8383. ISSN 0021-9258.

MacDiarmid, C. W., Milanick, M. A. & Eide, D. J. (2002) Biochemical properties of vacuolar zinc transport systems of *Saccharomyces cerevisiae*. *Journal of Biological Chemistry,* Vol.277, No.42, pp. 39187–39194. ISSN 0021-9258.

Maeshima, M. (2001) Tonoplast transporters: organization and function. *Annual Review of Plant Physiology and Plant Molecular Biology,* Vol.52, pp. 469–497. ISSN 1040-2519.

Mao, X., Kim, B.-E., Wang, F., Eide, D. J. & Petris, M. J. (2007) A histidine-rich cluster mediates the ubiquitination and degradation of the human zinc transporter, hZIP4, and protects against zinc cytotoxicity. *Journal of Biological Chemistry,* Vol.282, No.10, pp. 6992–7000. ISSN 0021-9258.

Moore, K.L. (2003) The biology and enzymology of protein tyrosine O-sulfation. Vol.278, No.27, pp. 24243–24246. ISSN 0022-0957.

Morth, J. P., Pedersen, B.P., Buch-Pedersen, M.J., Andersen, J.P., Vilsen, B., Palmgren, M.G. & Nissen, P. (2011) A structural overview of the plasma membrane Na^+,K^+-ATPase and H^+-ATPase ion pumps. *Nature Review Molecular and Cell Biology,* Vol.12, No.1, pp. 60-70. ISSN 1471-0072.

Nagasaki, N., Tomioka, R. & Maeshima, M. (2008). A hydrophilic cation-binding protein of *Arabidopsis thaliana*, AtPCaP1, is localized to plasma membrane via N-myristoylation and interacts with calmodulin and the phosphatidylinositol phosphates, $PtdIns(3,4,5)P_3$ and $PtdIns(3,5)P_2$. *FEBS Journal,* Vol.275, No.9, pp. 2267–2282. ISSN 0021-9258.

Ueoka-Nakanishi, H, Nakanishi, Y., Tanaka, Y. & Maeshima, M. (1999) Properties and molecular cloning of Ca^{2+}/H^+ antiporter in the vacuolar membrane of mung bean. *European Journal of Biochemistry*, Vol.262, No.2, pp. 417–425. ISSN 0014-2956.

Ueoka-Nakanishi, H., Tsuchiya, T., Sasaki, M., Nakanishi, Y., Cunningham, K.W. & Maeshima, M. (2000) Functional expression of mung bean Ca^{2+}/H^+ antiporter in yeast and its intracellular localization in the hypocotyl and tobacco cells. *European Journal of Biochemistry*, Vol.267, No.10, pp. 3090–3098. ISSN 0014-2956.

Yuasa, K. & Maeshima, M. (2000). Purification, properties and molecular cloning of a novel Ca^{2+}-binding protein in radish vacuoles. *Plant Physiology*, Vol.124, No.3, 1069–1078. ISSN 0032-0889.

Yuasa, K. & Maeshima, M. (2002). Equilibrium dialysis measurements of the Ca^{2+}-binding properties of recombinant radish vacuolar Ca^{2+}-binding protein expressed in *Escherichia coli Biosciences, Biotechnology and Biochemistry*, Vol.66, No.11, pp. 2382–2387. ISSN 0916-8451

7

The Use of Radioisotopes to Characterise the Abnormal Permeability of Red Blood Cells from Sickle Cell Patients

Anke Hannemann[1], Urszula Cytlak[1], Robert J. Wilkins[2],
J. Clive Ellory[2], David C. Rees[3] and John S. Gibson[1]
[1]Department of Veterinary Medicine, University of Cambridge,
[2]Department of Physiology, Anatomy and Genetics, University of Oxford,
[3]Department of Molecular Haematology, King's College Hospital, London,
UK

1. Introduction

1.1 Membrane transport and radioisotopes

In order to function, cells of necessity must transport a variety of substances across their plasma membranes. Aside from simple non-polar entities, most of these will require the presence of special transport proteins. Radioisotopes, used as tracers for these substrates, have provided an invaluable tool for understanding the mechanism and regulation of such transport pathways.

In this chapter, we will cover some theoretical and practical issues concerning the measurement of transport with radioactive tracers. As an illustration, we will focus on membrane transport in red blood cells, a tissue much used in the study of membrane permeability. In particular, we will look at the abnormal cation permeability of red blood cells from sickle cell patients to show how radioactive tracer methodologies can be used to investigate the pathophysiology of membrane permeability.

Chemical analysis of transport is feasible, but can often be tedious and slow. That by radioactive tracers has the advantage of being relatively immediate whilst retaining comparative simplicity. The first artificial radioisotopes were produced by Curie and Juliot in 1934 when they synthesised phosphorus-30 by exposure of aluminium-27 to α particles. Over the next 50 years or so, numerous different radioisotopes became available and were widely used to follow transport across biological membranes. Seminal examples include the use of $^{24}Na^+$ to investigate active Na^+ transport across the giant axon of cuttlefish in the mid 1950s (Hodgkin & Keynes, 1955a, 1955b; as elegantly retold by Boyd, 2011, following Keynes' death last year). The perceptive analysis of these studies underpinned the hypothesis and identification of the Na^+/K^+ pump by Skou, Post, Jolly and colleagues (Skou, 1957; Post & Jolly, 1957; Glynn, 2002; Skou, 2003), whilst in the late 1960s onwards Glynn and colleagues also used $^{24}Na^+$ and $^{42}K^+$ to study the pump (Garrahan & Glynn, 1967) and especially the existence of occluded ions (Glynn et al., 1984; Glynn, 2002).

With more recent methodogical advances and the more restrictive practices surrounding use of radioactive compounds, other non-radioactive methods (such as ion-sensitive microelectrodes or fluorophores, patch clamp methodology and molecular methods) have overtaken the use of radioactive tracers in popularity. Although the use of radioactive isotopes for transport studies has declined, however, they remain a valuable tool. This chapter provides a timely opportunity to illustrate their continuing value.

2. Flux analysis

2.1 The theory of transport measurement

The flux of a substance (J, typically given as moles per unit time) can be taken as its rate of movement from one defined region to another. Biologically relevant fluxes, of course, usually take place across a selective permeability barrier, which is often the cellular plasma membrane, or that of an organelle, or perhaps across a whole tissue. (For simplicity, the following account will usually refer only to cells but implicit in this term is the whole gamut of these other compartments).

In biology, often the two compartments which lose and accept the substance are the outside and inside of cells. As such, rather than measuring flux in units of moles per unit time, it is more usual to measure it in terms of cell number or cell volume, with units of moles.(cell volume.time)$^{-1}$ or moles.(cell number.time)$^{-1}$.

From Fick's first law of diffusion, flux will depend on time (t), the concentration difference across the barrier (ΔC), the permeability of the barrier (P) to the substrate in question and its area (A). One form of Fick's first law of diffusion therefore is:

$$J = - P. A. \Delta C \tag{1}$$

The permeability coefficient, P, with units of distance per unit time (often cm.s^{-1}) is simply derived from Einstein's diffusion coefficient (D). The units of D (often cm^2.s^{-1}) are derived from the average distance, x, over which a substance diffuses in one dimension per unit time, where:

$$x^2 = 2 D. T \tag{2}$$

If we want to know the diffusion co-efficient within the barrier separating the two compartments, as well as measuring flux we need to know the concentration of substrate within the membrane on each side (and not simply the concentration in aqueous fluid bathing the barrier). This requires knowledge of both the partition coefficient (K) and the functional thickness of the barrier (d). These are related such that:

$$P = D. K / d \tag{3}$$

Whilst it is relatively simple to measure the permeability co-efficient, ignorance of the partition coefficient and the exact thickness of the major barrier to movement make its separation into its three component parts more problematic.

A further complication arises with highly permeant substrates which move rapidly between compartments. Here exact measurement of the time that substrate and barrier are in contact, and also the presence of unstirred layers, become important issues. In these cases, establising permeabilities accurately will require sophisticated devices for rapid mixing and separation of cells and substrate.

From seminal work in the earlier part of the twentieth century investigating the permeability coefficients of substrates using large cells such as the plant algae *Chara ceratophylla* (Collander, 1949), the membrane of cells was originally considered to be a hydrophobic sieve, where transport reflected the lipid solubility and size of the substrate. We now know, of course, that many substances use specific protein-mediated pathways for their transport. Flux analysis is therefore often used to tell us how rapidly substances cross from one side to the other using these more-or-less specific transport systems. By utilising different conditions, one can measure the affinity of the protein for its substrate, potential inhibitors and their mechanism of action, and factors involved in regulating the activity of these protein transport system.

2.2 Uptake assays

Uptake studies involve measurement of the rate of entry of a substance into a compartment. This may involve a relatively simple procedure to incubate cells and radioactive test substance for the requisite amount of time. The ratio of the volume of cells to that of the extracellular medium should be low. This serves to limit any change in extracellular composition, especially concentration of the test substance during incubation. Afterwards, it is then necessary to remove that part of the substance remaining extracellularly.

An important consideration is prevention of surface binding and any loss, degradation or metabolism of the test substance which will artefactually alter the measured value. After separating out the cells at the end of incubation, it is also necessary to avoid or account for any residual trapped extracellular medium. If thorough separation of target compartment and extracellular medium is impractical, impermeable space markers can be added to measure accurately the volume of trapped extracellular medium and thereby calculate the extent of unincorporated test substance.

A key property of space markers is to penetrate all the remaining extracellular volume whilst being excluded from the compartment under study. Typical ones include inulin, mannitol, sucrose, polyethylene glycol (PEG), sorbitol (which can all be labelled with ^{3}H or ^{14}C). Alternative commonly used extracellular markers are represented by ^{51}Cr, ^{57}Co, ^{60}Co-labelled EDTA, ^{125}I-labelled albumin and various other ions such as ^{22}Na, ^{24}Na, ^{42}K, ^{86}Rb, ^{35}SO$_4$ and ethane sulphonate. For organic molecules, it is sometimes useful to employ non-transported stereoisomers (eg L-glucose as a space marker for D-glucose; D-alanine for L-alanine).

If both space marker and substrate are radioactive, their activity must be distinguished during counting. There are several ways to expedite this. For example, the space marker chosen may have a much shorter half-life than the test substance (ie count sample twice, before and after decay of the space marker) or a different type of radioactive emission (ie allowing the different energy of emitted particles to be distinguished). A further consideration is that the volume of trapped extracellular medium is likely to be small relative to target compartment, thus accurate determination of its volume will require a high specific activity of space marker relative to test substance.

2.3 Separation methods

Following incubation with the radioactive test substance, it is necessary to separate the target compartment from unincorporated medium. For larger tissues (tissue slices, whole muscles or nerves, etc) this is achieved simply by exchange of radioactive extracellular medium with an excess of a non-radioactive stop solution. Smaller target compartments require other methods. These are usually represented by washing techniques, oil separation methods and filtration techniques.

Washing methods are very versatile. High speed (15,000g) bench centrifuges can be used to rapidly pellet samples within a few seconds. Supernatant is then aspirated leaving the cell pellet. This is a common method for isolated cells such as red cells, lymphocytes, platelets, isolated hepatocytes or epithelial cells. For incubations of short duration or for more labile substrates, incubations can be terminated by addition of ice-cold wash solutions (to reduce or prevent transporter activity) prior to centrifugation. Estimates of the volume of target compartment (eg of red cells from the haematocrit) and trapped extracellular medium (usually 5-20% that of the cells) can be made to determine the number of washes required to dilute adequately unincorporated test substance. For example, for red cells at 2-5% haematocrit in 1ml reaction volume, we find that 4 washes (5 spins) with 1ml wash solution are adequate. In this case, volume of cells will be 20-50µl, that of trapped extracellular medium about 2-10µl, diluted 100-fold with each wash, ie 10^8-fold with 4 washes. From the activity of the starting reaction medium and the expected rate of incorporation of test substance, it is simple to determine whether this will be adequate to interfere with the signal.

Washing can also be refined by addition of specific inhibitory strategies to limit further uptake or loss – ice-cold media, specific chelators (eg EGTA for Ca^{2+}, EDTA for Mg^{2+}), transporter inhibitors (eg Ba^{2+} for K^+ channels, phloretin or $HgCl_2$ for glucose transporters, nitrobenzylthioinosine for nucleosides) or with "chase" strategies adding excess unlabelled solutes (though care must be exercised should exchange-diffusion, ie trans stimulation, be a possibility as this will artefactually reduce incorporated solute).

Oil separation methods may be used for more rapid fluxes where the more protracted washing processes may result in unacceptable loss of radioactive tracer. Flux durations as short as about 5s are practical with these (cf about 30s for washing methods). The strategy is to choose an oil with density greater than that of the incubation medium but less than that of the cells. On centrifugation, cells spin out through the oil leaving saline suspended above. Diphthalate esters have proved particularly useful. Diethyl, dibutyl and di-iso-octyl diphthalates have densities of 1.118, 1.047 and 0.981 g.ml^{-1}, respectively. They can be mixed to achieve an appropriate density relative to that of the cell under investigation. An appropriate stop solution (often ice-cold) is placed on top of the oil layer, the sample is added and tubes rapidly centrifuged to pellet cells through the oil layer. In addition, a third layer at the bottom (such as formic acid) may be used to lyse cells quickly and prevent intracellular metabolism should this be an issue.

Using oils, separation of cells and medium is usually less complete than for washing techniques. In the case of red blood cells, for example, residual trapped medium is about 2% cell volume. Depending on this ratio of cell volume to trapped medium, it may be necessary again to add appropriate space markers. It then remains to aspirate the saline and oil and carefully clean the inside of the centrifuge tube (with tissue or cotton bud) to remove residual tracer adherent to the sides.

Filtration methods may also be used to separate cells from unincorporated radioactive tracer. This involves choosing filters of appropriate pore size to retain the target tissues. A range of pore sizes and filter materials are available. Common ones comprise nylon, PVC, PTFE, cellulose acetate, methylcellulose, cellulose acetate/nitrate mixtures, polycarbonate and glass fibre. For example, nitrocellulose filters of 0.2-0.45µm pore size are suitable for many isolated cells. The filter usually rests on a rigid support and unincorporated media and a suitable wash solution is sucked through using negative pressure. Potential problems involve binding of tracer to the filter (this can be reduced through pre-washing with excess cold substrate) or poor trapping of the target tissue (eg small vesicles). In addition, cell lysis during the filtration or subsequent washing can be substantial. Refinements of this technique include the use of small chromatography columns which bind free substrate whilst allowing the target tissue to pass through.

2.4 Efflux

Efflux studies measure the appearance of tracer into the extracellular medium. They have certain advantages, but also disadvantages, over uptake methods. Thus in principle, it is simpler to obtain cell-free extracellular media, by centrifugation or filtration than it is to separate out an uncontaminated cell pellet. The limitations involve initial loading where high intracellular specific activity is often required. Other issues include intracellular compartmentalisation of tracer, metabolism, surface binding and also kinetic analysis of the results. In addition, unstable cells may lyse during incubation or the separation process and, containing high specific activity of label, will obviously pose considerable threat of heavy extracellular contamination.

2.5 Kinetic considerations

Other reviews (such as Schultz, 1980; Stein, 1986) should be consulted for full kinetic analysis and only a brief overview is presented here. In general, there are three main parameters associated with flux measurements: uptake or loss over a fixed period of time (sometimes referred to as an "influx" or "efflux"), the rate constant (which provides information on the initial rate of tracer flux) and the affinity of a substrate with its transporter (equivalent to the Michaelis constant, K_m, of enzyme systems).

Influx over a defined duration is the simplest to determine but it is important to establish that the flux is linear over this time and does not exhibit saturation.

If uptake or loss is followed more carefully as a function of time, the initial linear rate will decrease following an exponential relationship until intracellular and extracellular concentrations become equal, at which point the rate of tracer influx and efflux are the same. For uptake:

$$C_t = C_\infty \cdot \left(1 - e^{-kt}\right) \qquad (4)$$

where C gives the radioactivity in the cells at time (C_t) and after equilibration (C_∞) and k is the rate constant.
For efflux:

$$C_t = C_0 \cdot e^{-kt} \qquad (5)$$

where C_t is defined as previously and C_0 is the internal radioactivity at time zero. Thus plotting ln ((C_∞ - C_t) / C_∞ against t for uptake and ln (C_t / C_0) for efflux should give a straight line relationship whose slope represents the rate constant, k.

This analysis assumes zero-trans conditions apply initially, ie that the internal concentration of tracer for uptake and the extracellular concentration of tracer for efflux is very low compared to that on the opposite side of the membrane. In the presence of non-radioactive trans substrate, the linear phase is prolonged as the radioactive tracer is diluted. Complications arise when systems carry out exchange of substrates when trans-stimulation can occur.

The rate constant, k, has dimensions of time^{-1}. It can be converted to a flux by multiplying by the concentration of the substrate (units of moles per unit volume per unit time). In addition, the rate constant can be used to derive the apparent permeability, P, by multiplying with the ratio of volume to surface area (V / A), with units (as before) usually given as cm.s^{-1}. For instance, with human red blood cells typical values for V / A are 3.4×10^{-5}.

Transport involving carrier-mediated systems will typically show saturation. A plot of flux against concentration gives a rectangular hyperbola which can be analysed using simple Michaelis-Menten kinetics to find an apparent affinity of substrate with transporter. In this case:

$$J = J_{max}. s \, / \, (s + K_T)$$

(6)

where J is flux, J_{max} the maximal flux, s is concentration and K_T the apparent dissociation constant (the equivalent of the Michaelis constant, K_M).

Graphical analysis of this relationship is used to linearise the plot. The Lineweaver-Burk plot of 1 / J vs 1 / s is well known but is less accurate because it gives large weight to fluxes at small concentrations. Better are the Eadie-Hofstee or Woolf plots of (J / s) vs (s) or (J) vs (J / s).

Often there may be several mechanisms by which substrates cross membranes which complicates analysis, as J vs s then represents the sum of several components. The simplest case would be a saturable transporter in parallel with a diffusion non-saturable component, in which case:

$$J = (J_{max}. s \, / (s + K_T)) + D. s$$

(7)

It is possible, however, to have much more complicated arrangements – such as the multiple systems involved in amino acid transport across red cell membranes (Barker & Ellory, 1990).

2.6 Some problems with use of radioisotopes

One needs to be sure that any radioactive counts actually originate from the substrate under consideration and that it is located in the correct site. There are a number of issues here. Purity of the radioactive tracer may be a problem, though commercial sources are usually reliable. Loss of label from the substrate can occur, however, especially with organic molecules containing ^3H or ^{14}C or ^{32}P. These groups (especially ^3H) may be labile and exchange with various targets, notably that of water. That labels remain in the desired compounds can be checked using chromatography – especially worthwhile with older stocks of radioactive tracers. Ions are usually more stable but can change valency (eg Cr^{2+}, Cr^{3+} and CrO_4^{2-}) so that again radioactive label is no longer in the requisite substrate. Metabolic substances may also be rapidly metabolised by cells and tracer may then be lost as by-products (such as ^{14}C as $^{14}CO_2$, ^3H as ^3H_2O). On the other hand, incorporation of substance into other compounds or into subcompartments may give a misleading overestimate of intracellular accumulation. Thus amino acids may be synthesised rapidly into proteins or glutathione; Ca^{2+} may be pumped avidly into intracellular stores. An obvious strategy is to use non-metabolisable analogues like 3-O-methylglucose for glucose or ATP-γ-S for ATP or to prevent secondary movement of tracer with appropriate inhibitors (like thapsigargin for the Ca^{2+} pump of certain intracellular organelles or vanadate for that of the plasma membrane). When counting is carried out, it is also important to account for the possibility of quenching. This may occur through contamination of the samples with coloured material or soluble macromolecules and has the effect of shifting the energy spectrum to lower values. Channels ratio analysis, or the use of external standards, are common ways of dealing with this problem. Finally, it should be realised that flux measurement with radioisotopes provides a unidirectional value and does not give any information on net flux movements. This will be an important caveat in assessing the significance of any results.

3. Red blood cells and membrane transport

3.1 Red blood cells as a transport paradigm
Red blood cells are simple cells, readily obtainable and fairly homogeneous, and also lacking intracellular organelles which may complicate flux studies in other tissues. As such, they are highly amenable to transport studies. It is not surprising that they have provided an invaluable model system for analysis of the physiology of membrane permeability.

Seminal studies include those of Overton in the early 1900s (Al-Awquati, 1999), and of Ponder and Davson around 1940 (Davson & Danielli, 1943; Davson & Ponder, 1938). Later, in the 1960s red cells were used by Hoffman and Tosteson to develop the pump-leak model of volume regulation (Tosteson & Hoffman, 1960). The stoichiometry of the Na^+/K^+ pump was also established using red blood cells (Post & Jolly, 1957; Garrahan & Glynn, 1967b). Latterly, much has been learnt about the kinetics of facilitated transporters and co-/counter-transporters – for example those of the glucose transporter and the amino acid transporters (Barker & Ellory, 1990; Naftalin, 2010).

In addition to these physiological studies, the permeability characteristics of the red blood cell membrane can be important pathologically, as seen for example in haemolytic anaemias and hereditary stomatocytoses (Ellory et al., 1998; Stewart, 2003). Here we focus on the abnormal permeability exhibited by red cells from patients with sickle cell disease (SCD).

3.2 Sickle cell disease
SCD is one of the commonest severe inherited disorders (Bunn & Forget, 1986; Steinberg et al., 2001). Its first impact on Western medicine dates from the early 1900s when Herrick and colleagues working in Chicago described a low cell count and the presence of abnormally shaped, elongated red blood cells in a blood smear from a Grenadan dental student, Walter Noel (Herrick, 1910). Noel died aged 32 from acute chest syndrome (Serjeant, 2001). Patients, however, may present with very heterologous signs and levels of severity. The numerous complications divide into two main groups – a chronic regenerative anaemia and acute ischaemic episodes. The latter may manifest as pain, organ dysfunction (stroke, retinopathy, osteonecrosis, acute chest syndrome, etc) and ultimately early mortality (Steinberg, 1999; Nagel & Platt, 2001). Some patients are severely and frequently affected by these complications, others much less so, so that in some cases the individual may even be unaware that they have the condition.

The aetiology of SCD, which was established by Pauling over 50 years ago (Pauling et al., 1949), is well known – the presence of the abnormal haemoglobin HbS in patients' red blood cells, instead of the usual HbA. HbS results from a single amino acid substitution at residue 6 of the β chain of haemoglobin with valine replacing glutamic acid (Ingram, 1957). The resulting loss of a negative charge enabled identification of HbS by electrophoresis (Pauling et al., 1949). The amino acid change usually reflects a single mutation in codon 6 (an adenine to thymine alteration, so GAG becomes GTG) (Marotta et al., 1977). About 2/3rd patients are homozygous for HbS (HbSS) but 1/3rd are heterozygotes of HbS and HbC (HbSC) (Nagel & Steinberg, 2001; Nagel et al., 2003). In addition, there are a few rarer genotypes (such as HbS-β thalassaemia or HbS-HbO-Arab) (Steinberg, 2001).

Although aetiology is clear, how the presence of HbS results in the clinical complications of SCD, however, is not. A key feature is the propensity of neighbouring HbS molecules to polymerise upon deoxygenation, encouraged through the loss of glutamate's negative charge. Long rods of HbS aggregates distort the red cell shape into the eponymous sickles and other bizarre shapes (like holly leaves). The deleterious effect on red blood cell rheology (Hebbel et al., 1980; Eaton & Hofrichter, 1987), together with altered nitric oxide scavenging by haemoglobin following intravascular haemolysis (Hebbel, 1991; Reiter et al., 2002; Gladwin et al., 2004; Kato et al., 2006) and also release of inflammatory cytokines (Setty et al., 2008; Setty & Betal, 2008), are thought to contribute to SCD complications. Our interest lies in the involvement in pathogenesis of the abnormal permeability of the red blood cell membrane in SCD patients (Joiner, 1993; Ellory et al., 1998; Gibson & Ellory, 2002; Lew & Bookchin, 2005).

3.3 Membrane transport in sickle cell disease

Human red blood cells have a particularly high Cl$^-$ permeability mediated via the anion exchanger (AE1) and also as yet unidentified Cl$^-$ channels. As such, their membrane potential usually reflects the Nernst equilibrium potential for Cl$^-$ (rather than that of K$^+$ as in many other cell types). By contrast, their permeability to cations is relatively modest. The presence of a low capacity Na$^+$/K$^+$ pump is sufficient to carry out volume regulation, repelling ions attracted by the fixed intracellular negative charges of (mainly) haemoglobin whilst keeping intracellular K$^+$ levels high and Na$^+$ levels low (Tosteson & Hoffman, 1960; Ellory & Lew, 1976). A high capacity plasma membrane Ca^{2+} pump, together with a low passive Ca^{2+} permeability, also serves to keep intracellular Ca^{2+} at very low values, some 30nM (Lew et al., 1982).

The commoner red cell cation transport pathways and their inhibitors are summarised in Table 1, listed in a potential order of significance to ion homeostasis in red blood cells from SCD patients. All can be distinguished using radioactive isotopes and requisite inhibitors or other manoeuvres (such as ion substitutions). There is an important caveat, however, in that few inhibitors (aside from the Na$^+$/K$^+$ pump inhibitor, ouabain) are totally specific. In addition, channels, as conductive transport systems, are perhaps better suited to electrophysiological analysis such as, patch clamp studies.

HbS-containing red cells (HbS cells) have an abnormally high cation permeability. This was first described in the 1950s, in seminal work by Tosteson and colleagues (Tosteson et al., 1955) who showed that red blood cells from SCD patients lost potassium chloride and osmotically obliged water, thereby causing them to shrink. The effect was exacerbated upon deoxygenation. Over the subsequent 50 years, the pathways responsible have been well investigated. We now know that three systems are particularly implicated (Joiner, 1993; Gibson & Ellory, 2002; Lew & Bookchin, 2005): the KCl cotransporter (KCC, likely a mix of isoforms KCC1 and KCC3; Crable et al., 2005), the Gardos channel (KCNN4, IK1; Ishii et al., 1997; Hoffman et al., 2003) and a deoxygenation-induced "cation" conductance (sometimes termed P_{sickle}; Joiner et al., 1988; Joiner et al., 1993; Lew et al., 1997). Despite its central role in dehydration, the molecular identity of the latter, unlike the other two, remains unknown.

Name	Predominant substrate(s)	Inhibitors (IC$_{50}$)
K$^+$-Cl$^-$ cotransporter (KCC)	K$^+$, Cl$^-$	Bumetanide (180µM(KCC1)-900µM(KCC4))[1] Furosemide (50µM-sheep RBC)[2] DIOA (10µM)[3]
Ca^{2+}-activated K$^+$ channel (Gardos channel)	K$^+$	Clotrimazole (0.05-0.1µM)[4,5] - Senicapoc (ICA-17043) (0.011µM)[6] Charybdotoxin (0.8nM)[7] Tioconazole (0.3µM)[4] Miconazole (1.5µM)[4] Econazole (1.8µM)[4] Nifedipine[8] Quinine (5µM)[9] Nitrendipine[10]
Deoxygenation-induced cation channel (P$_{sickle}$)	Na$^+$, K$^+$, Ca^{2+}, Mg^{2+}, Li$^+$, Rb$^+$, Cs$^+$	Stilbene derivatives (partially)[11,12,13] - DIDS - SITS Dipyridamole[14]
Plasma membrane Ca^{2+}-ATPase (PMCA, Ca^{2+}-pump)	Ca^{2+}	Vanadate (non-specific) (3µM)[15]
Na$^+$/K$^+$-ATPase (Na$^+$/K$^+$-pump)	Na$^+$, K$^+$	Ouabain (1µM)[16] Hydroxyxanthones -MB7 (1.6µM)[16] Oligomycin[17] Vanadate (non-specific)[18]
Na$^+$-K$^+$-2Cl$^-$ cotransporter (NKCC)	Na$^+$, K$^+$, Cl$^-$	Bumetanide (~0.3-0.6µM)[19] Furosemide (500µM)[20]
Na$^+$/H$^+$ exchanger (NHE)	Na$^+$, H$^+$	Amiloride - EIPA (0.8µM)[21] - DMA (0.023µM-rat RBC)[22,23]
K$^+$(Na$^+$)/H$^+$ exchanger	K$^+$, Na$^+$, H$^+$	Amiloride - EIPA[24] Chloroquine (66µM)[24] Quinacrine (81µM)[24]

Name	Predominant substrate(s)	Inhibitors (IC_{50})
Non-selective cation channel (NSCC):	Monovalent cations: Na^+, H^+, Rb^+, NH_4^+	
Voltage-dependent (NSVDC)	Divalent cations: Ca^{2+}, Mg^{2+}, Ba^{2+}	Stilbene derivatives (partially) -DIDS[25] Ruthenium red (3.7µM)[26] IAA (480µM)[26]
Volume-dependent		Amiloride -EIPA (0.6µM)[27]
Oxidation-dependent		Amiloride (partially) -EIPA[28] Gd^{3+} [28]
Magnesium transport:	Na^+, Mg^{2+}	
Na^+-dependent: 2Na^+/Mg^{2+} exchanger		Quinidine (50µM)[29] Mn^{2+} (0.5-1.0mM)[29] Imipramine (25µM)[30]
Na^+-independent: Mg^{2+}/H^+ exchanger		DIDS (40µM)[31] SITS (30µM)[31] Amiloride (0.4mM)[31]
Mg^{2+}/anion cotransporter		

Table 1 list of abbreviations: DIDS: 4,4'-diisothiocyanatostilbene-2,2'-disulfonate; DIOA: (dihydroindoenyl)oxy]alkanoic acid; DMA: 5-(N,N-dimethyl)amiloride; EIPA: 5-(N-ethyl-N-isopropyl)amiloride; IAA: iodoacetamide; MB7: 3,4,5,6-tetrahydroxyxanthone; SITS: 4-acetamido-4'-isothiocyanatostilbene-2,2'-disulfonate.

Table 1 list of references: [1]Mercado et al., 2000; [2]Lauf, 1984; [3]Garay et al., 1988; [4]Alvarez et al., 1992; [5]Brugnara et al., 1993; [6]Stocker et al., 2003; [7]Wolff et al., 1988; [8]Kaji, 1990; [9]Reichstein & Rothstein, 1981; [10]Ellory et al., 1992; [11]Joiner, 1990; [12]Ellory et al., 2007; [13]Clark & Rossi, 1990; [14]Joiner et al., 2001; [15]Tiffert & Lew, 2001; [16]Zhang et al., 2010; [17]Sachs, 1980; [18]Cantley et al., 1978; [19]Russell, 2000; [20]Bernhardt et al., 1987; [21]Pedersen et al., 2007; [22]Counillon et al., 1993; [23]Masereel et al., 2003; [24]Weiss et al., 2004; [25]Halperin et al., 1989; [26]Bennekou et al., 2004; [27]Lang et al., 2003; [28]Duranton et al., 2002; [29]Feray & Garay, 1986; [30]Feray & Garay, 1988; [31]Gunther & Vormann, 1990.

Table 1. Principal cation permeability pathways in human red blood cells and their inhibitors

In HbS cells, KCC is highly active, over-expressed and also shows abnormal regulation (Gibson et al., 1998; Su et al., 1999). These features are distinct from the situation in red blood cells from normal HbAA individuals (HbA cells). In HbA cells, KCC activity is usually modest even in the youngest cell population, whilst it becomes quiescent as cells age and is inactive at low O_2 tensions (Hall & Ellory, 1986; Gibson et al., 1998). In HbS cells, KCC activity is high and maintained, with unusual features, showing continued activity even in the absence of O_2 (Brugnara et al., 1986; Hall & Ellory, 1986; Ellory et al., 1991; Gibson et al., 1998). The main stimulus is probably reduction in pH from the normal plasma value of pH 7.4 to 7 (Brugnara et al., 1986; Ellory et al., 1989). When active, KCC mediates coupled KCl efflux. Regulation involves a cascade of conjugate pairs of protein kinase and phosphatase enzymes (Gibson & Ellory, 2003).

The Gardos channel (Gardos, 1958) is a Ca^{2+}-activated K^+ channel, the activity of which reflects abnormal Ca^{2+} homeostasis in red blood cells from sickle cell patients. These cells show a high Ca^{2+} leak and lower active Ca^{2+} removal. As such, and particularly under deoxygenated conditions, the Gardos channel is activated and mediates rapid conductive K^+ efflux with Cl^- following electrically (Etzion et al., 1993; Lew et al., 1997). Activity begins at around 100nM, becoming maximal at a few µM (Bennekou & Christophersen, 2003).

The third transport system, P_{sickle}, is not seen in normal red blood cells. However, since its molecular identity has not yet been established it remains to be established whether it is in fact present. Its apparent absence in HbA cells may in fact result from lack of appropriate stimuli, such as HbS polymerisation. P_{sickle} in HbS cells is activated by deoxygenation and HbS polymerisation (Mohandas et al., 1986; Gibson et al., 2001). It has characteristics of a cation channel and, importantly, is permeable to Ca^{2+}. Entry of Ca^{2+} through P_{sickle} leads to activation of the Gardos channel which mediates rapid K^+ efflux with Cl^- following electrically through separate channels. P_{sickle} is especially responsible for the high Ca^{2+} influx and altered Ca^{2+} homeostasis (Rhoda et al., 1990; Joiner et al., 1995; Lew et al., 1997).

The three pathways thereby combine to cause KCl loss with oxygen tension playing an important role in regulating membrane permeability. There is also a degree of positive feedback. For example, KCl loss will lower intracellular [Cl^-] resulting in Cl^- entry and HCO_3^- loss via the anion exchanger AE1, thereby acidifying the cell and further stimulating KCC (Lew et al., 1991).

The co-operative function of these three transport systems mediates solute loss, with water following osmotically, resulting in red blood cell shrinkage. This can be extensive and rapid (Bookchin et al., 1991), or alternatively it may result from modest but repeated episodes as the red blood cells circulate. Red blood cells generally lack the ability to regain these lost solutes and eventually shrinkage will be significant. The consequent elevation in [HbS] markedly encourages polymerisation and sickling, as the lag to polymerisation following deoxygenation is inversely proportional to a very high power of [HbS] (Eaton & Hofrichter, 1987). Hence the abnormal cation permeability of HbS cells is a significant feature of disease, of importance to pathogenesis.

All these systems are particularly amenable to study with radioactive tracer techniques making these a powerful way of analysing red blood cell behaviour. In the following, we show how radioisotopes can be used to follow the activity of these systems and thereby investigate the abnormal cation permeability of HbS cells. *In vivo* conditions are mimicked as much as possible. In particular, we use tonometry and gas mixing to replicate the oxygen tensions experienced by the red blood cells as they traverse the circulation. Plasma levels of important ions, notably K^+, Na^+, Ca^{2+} and Mg^{2+}, also mimic *in vivo* parameters.

4. Experimental considerations

4.1 Methodology

Blood samples: Routine discarded blood samples were acquired from SCD patients of both main genotypes (HbSS and HbSC) using EDTA as anticoagulant. Samples are best left refrigerated as whole blood (with plasma) if not wanted immediately. For longer storage, it may be appropriate to incorporate additives for sterility. As required, whole blood was washed in saline to remove plasma and buffy coat, and red blood cells stored on ice until required. Notwithstanding washing, it should be appreciated that contamination of the red blood cell fraction with white cells or platelets can occur and may affect results – a particular problem when assessing enzymatic activity.

Salines and inhibitors: Nitrate-containing MOPS-buffered saline (N-MBS) comprised (in mM): $NaNO_3$ 145, $CaCl_2$ 1.1, MOPS 10, glucose 5, pH 7.4 at 37°C. Cl^--containing MBS (Cl-MBS) had similar composition but with NaCl replacing $NaNO_3$. Wash solution (W-MBS) was isotonic $MgCl_2$ solution: $MgCl_2$ 107, MOPS 10, pH 7.4 at 0°C. The buffer chosen, here MOPS, usually reflects the temperature of incubation. It should be noted that most buffers have a significant temperature dependence (with pH usually declining as temperature increases). Phosphate buffer has a particularly low temperature coefficient and is useful if studies are being carried out over a temperature range. Here we used nitrate to replace Cl^-. Other possibilities include methylsulphate, although there is probably no completely inert replacer for Cl^- (Payne et al., 1990).   Except for the experiment shown in Figure 1, ouabain (100μM) and bumetanide (10 μM) were present during all fluxes to inhibit K^+ uptake via the Na^+/K^+ pump and Na^+-K^+-$2Cl^-$ cotransporter (NKCC), respectively. Where required, clotrimazole (10μM) was added to inhibit the Gardos channel. To analyse red blood cell shape, aliquots of cells were placed in saline containing 0.3% glutaraldehyde before examination under light microscopy.

Tonometry: Several of the transport systems found in red blood cells are O_2-sensitive (Gibson et al., 2000; Gibson & Ellory, 2002). It is therefore important to regulate O_2 tension during incubation. We used Eschweiler tonometers, coupled to a Wösthoff gas mixing pump to set the O_2 tension at the requisite level from 150mmHg to 0 (by mixing pre-warmed and humified air and N_2). Typically, cells were placed in the tonometers at 10-fold the haematocrit needed for transport assay and gently equilibrated at the requisite O_2 tension. They were then diluted 10-fold into test tubes, also pre-equilibrated at the required O_2 level. Tubes were also gassed during incubation, but not bubbled as this lyses cells. Humidified gas is necessary to prevent dehydration of the samples. In addition, to prevent condensation, all glassware and tubing should be submerged and kept at the same temperature which will usually be 37°C for mammalian red blood cells.

Radioisotope and measurement of fluxes: We used uptake to measure transporter activity. $^{86}Rb^+$ was used as a K^+ congener, chosen in preference to $^{42}K^+$ as its half life is longer (about 18 days cf 12.4 hours). After dilution of the red cell samples into the test tubes, the flux was started by addition of $^{86}Rb^+$ (final activity about 0.05MBq.ml^{-1}) to warm (37°C) cell suspensions. We used $^{86}Rb^+$ in a solution of 150mM KNO_3 added at a 1 in 20 dilution to give a final extracellular $[K^+]$ of 7.5mM. (Adding K^+ with $^{86}Rb^+$ is particularly useful if fluxes are carried out at different $[K^+]$s, for example to study the affinity of a transport system to K^+.) If cold-start is preferred, time taken for suspensions to reach the incubation temperature should be taken into account (about 1min for a 1.5ml eppendorf tube). Symmetrical cold-stop / cold-start protocols are used to allow for this with tubes kept on ice before placing them at 37°C, then being returned to ice for 1min before washing. The duration of uptake here was 10min - control experiments have established that uptake is linear over this time period – and determinations were usually carried out in triplicates. Uptake was stopped by diluting aliquots of the cell suspension into ice-cold W-MBS. Unincorporated $^{86}Rb^+$ was removed by centrifugation (10s at 15,000g), aspiration of supernatant and addition of further wash solution (4 washes and 5 spins in total). After each centrifugation step, cells were resuspended by gentle vortexing, though it was noticeable that HbS cells become stickier at lower O_2 tensions. This can be important for trapping extracellular medium. Following the final wash, the cell pellet was lysed with Triton X-100 (0.1%) and protein (mainly haemoglobin in the case of red blood cells) precipitated with trichloroacetic acid (TCA, 5%). A final centrifugation step was used to separate off the clear, colourless supernatant before counting. Activity was measured as Čerenkov radiation by liquid scintillation (Packard Tri-carb 2100TR).

Incubation conditions: The test tubes contained transport inhibitors in Cl-MBS or N-MBS as required. Na^+/K^+ pump activity is then given as the ouabain-sensitive K^+ uptake, KCC as the Cl$^-$-dependent K^+ uptake, NKCC as the bumetanide-sensitive K^+ uptake and P_{sickle} as the K^+ uptake in the N-MBS in the presence of all 3 inhibitors.

Flux calculation: There are several ways of doing this. The general approach is:

$$\text{Flux} = \{(\text{Sample counts - background})/\text{specific activity}\} \times$$
$$\times (60/\text{flux time in min}) \times (10^3/\text{final haematocrit}) \tag{8}$$

where, specific activity is total counts of $^{86}Rb^+$ per tube/total mmol K^+ per tube. The first term converts the sample counts to mmol K^+; the second converts the time into flux per hour; the third converts volume into flux per litre of packed cells. Therefore, flux units become mmol K^+(l cells.h)$^{-1}$, i.e. if 50µl $^{86}Rb^+$ in 150 mM KCl are added to final volume of 1 ml, and standard counts are the counts in every 10µl of this $^{86}Rb^+$ stock, for a 10min flux: Flux = (sample counts – background counts)/(standard counts – background counts) x (10x10^{-6} x 150) x (60/flux time) x 10^3 x (1/haematocrit as a fraction). Background counts are provided by the same volumes of Triton X-100 and TCA used for sample preparation but lacking cells. In practice, the standard counts will be very high relative to samples and background and it is not necessary to subtract the background from them.

4.2 Results
4.2.1 The main cation transport in red blood cells from sickle cell patients

In the first series of experiments, the activities of the five main K^+ transport systems present in the membrane of red cells from homozygous (HbSS) SCD patients were established in fully oxygenated (100mmHg O_2) and fully deoxygenated (0 O_2) conditions (Figure 1).

Na^+/K^+ pump activity was present at a similar level to that seen in red blood cells from normal individuals (Garrahan & Glynn, 1967) and also many other mammals (Shaw, 1955). Its activity was unaffected by O_2 tension.

Similarly, NKCC activity was low and showed little change on deoxygenation, notwithstanding that it is usually stimulated by low O_2 in red blood cells from other species (Gibson et al., 2000). Apparent lack of oxygen sensitivity here may be because activity was minimal. Residual K^+ movement, in nitrate medium in the presence of ouabain, bumetanide and clotrimazole, was minimal, indicative of small movement through other transport pathways.

Two other systems, KCC and Gardos channel, by contrast differ markedly from their behaviour in normal red blood cells. KCC is usually quiescent in all but the youngest red blood cells from normal individuals. Here it can be seen that influx was high. The high K^+

content of red blood cells means that this system will mediate KCl loss at about 10-fold greater rates (Dunham & Ellory, 1981). On deoxygenation, KCC activity remained at substantial levels. Again, this differs with the situation in normal red blood cells.

The fourth transport system, P_{sickle}, assayed as a Cl^- independent K^+ flux, is not found in normal red blood cells. Figure 1 shows that on dexoygenation, P_{sickle} activity in patient red blood cells became activated.

One of the main actions of P_{sickle} is to allow entry of Ca^{2+}, thereby perturbing the normal pump-leak balance and enabling intracellular Ca^{2+} to accumulate sufficiently to activate the Gardos channel. This, too, was apparent. Gardos channel activity was minimal in oxygenated cells – the small component probably coming from irreversibly sickled cells. On deoxygenation, the Gardos channel was also activated.

Similar findings were seen in red blood cells from heterozygous (HbSC) patients. Here KCC activity was high, cells sickled on deoxygenation and also showed activation of P_{sickle} and the Gardos channel. There was one exception, however. Thus although KCC activity was high in oxygenated cells, on deoxygenation it became inhibited – by a mean of $86\pm4\%$, n=11, $p<0.05$, cf levels in fully oxygenated cells - as seen in red blood cells from normal HbAA individuals.

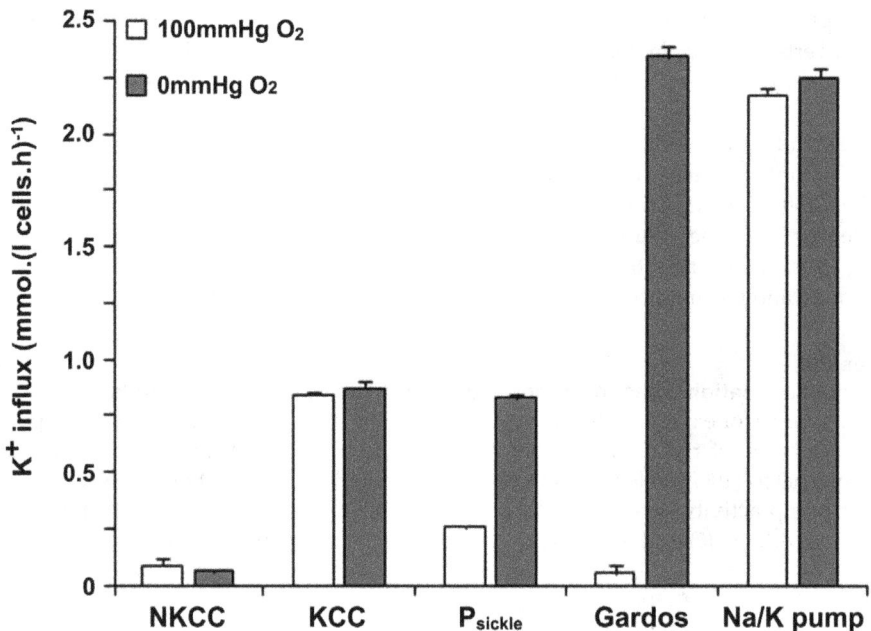

Fig. 1. The activity of the five main cation transport systems of human red blood cells were measured under fully oxygenated (100mmHg O_2; open histograms) or fully deoxygenated (0mmHg O_2; filled histograms) conditions in samples from patients with sickle cell disease. Histograms represent means$\pm$S.E.M., n=4, with K^+ influx given as mmol.(l cells.h)$^{-1}$. NKCC = Na^+-K^+-$2Cl^-$ cotransporter; KCC = K^+-Cl^- cotransporter; P_{sickle} = the deoxygenation-induced cation pathway; Gardos = Ca^{2+}-activated K^+ channel (Gardos channel); Na/K pump = Na^+,K^+-ATPase.

4.2.2 O_2 dependence of cation transport in red blood cells from sickle cell patients

In the second series of experiments, we followed the activity of the three main transport systems (KCC, Gardos channel and P_{sickle}) across the physiological range of O_2 tensions as red blood cells were deoxygenated from arterial O_2 tensions down to levels pertaining to metabolically active tissues. Results are shown in Figure 2.

As the red blood cells were deoxygenated, they began to show the sickling shape change. Under fully deoxygenated conditions, over 80% of cells showed evidence of sickling. KCC activity was high at arterial O_2 tensions, started to decrease as cells were deoxygenated but then became activated again at the lowest O_2 tensions. Both P_{sickle} and Gardos channel activity were reduced at the higher O_2 tensions and showed maximal activity when cells were fully deoxygenated. Changes to these four parameters all became marked at about the P_{50} of O_2 saturation of haemoglobin. Activity of the transport systems correlated with the sickling shape change and hence HbS polymerisation.

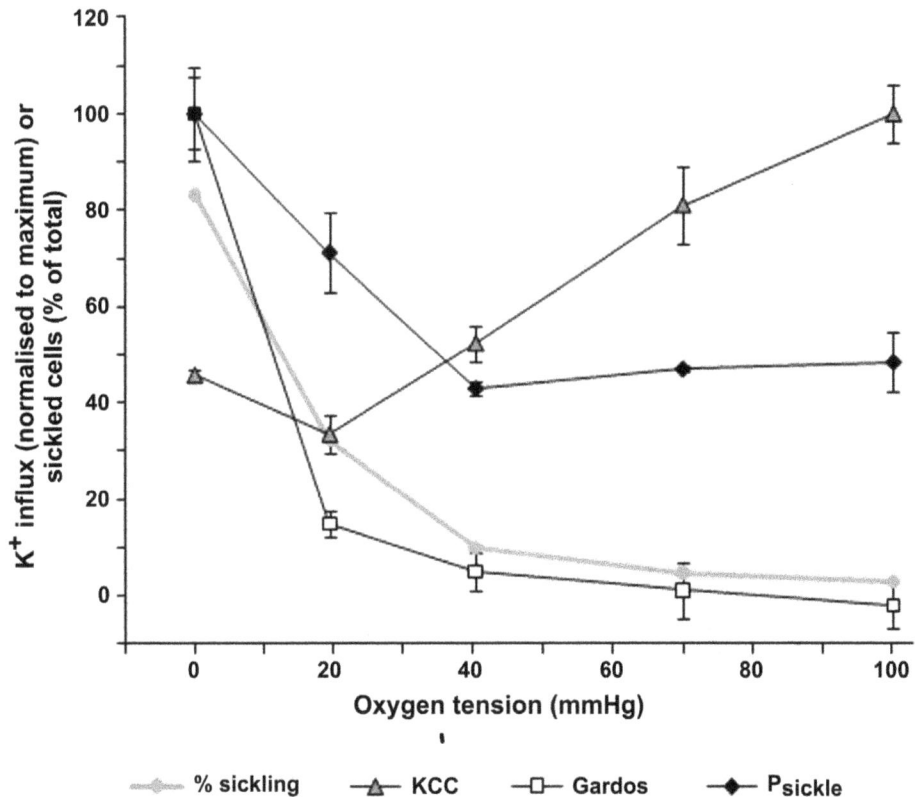

Fig. 2. Effect of oxygen tension on sickling and the activity of KCC, Gardos and P_{sickle} (defined as in Figure 1) in red blood cells from sickle cell patients. All are given as normalised percentages (%) of maximal transport activity or of complete sickling. Symbols represent means±S.E.M., n=3.

5. Conclusions

This chapter illustrates the value of using radioactive isotopes for measurement of fluxes across biological membranes. It serves to highlight the relative simplicity of these methods. This, coupled with the ability to acquire useful information concerning a variety of transport systems through use of ion substitutions or inhibitors, make radioisotopes of continuing value. We show the abnormal permeability of red blood cells from SCD patients and how it is affected by O_2 tension. Historically, fluxes across the red blood cell membrane used methods such as haemolysis, such that more permeable species result in faster cell lysis. More recent methods have included nuclear magnetic resonance, fluorophores and electrophysiological approaches. As shown here, radioisotopes nevertheless remain a valuable tool for investigating this field.

6. Acknowledgements

We thank the Medical Research Council, BBSRC and the British Heart Foundation for financial support.

7. References

Al-Awquati, Q. (1999). One hundred years of membrane permeability: Does Overton still rule? *Nature Cell Biology* 1, E210-E202.

Barker, G. A. & Ellory, J. C. (1990). The identification of neutral amino acid transport systems. *Experimental Physiology* 75, 3-26.

Bennekou, P., Barksmann, T. L., Kristensen, B. I., Jensen, L. R. & Christophersen, P. (2004). Pharmacology of the human red cell voltage-dependent cation channel. Part II: Inactivation and blocking. *Blood Cells, Molecules and Disease* 33, 356-361.

Bennekou, P. & Christophersen, P. (2003). Ion channels, in *Red Cell Membrane in Health and Disease*, Eds. I. Bernhardt & J. C. Ellory, pp. 139-152, Springer, 3-540-44227-8, Berlin.

Bernhardt, I., Erdmann, A, Vogel, R. & Glaser, R. (1987). Factors involved in the increase of K^+ efflux of erythrocytes in low chloride media. *Biochimica et Biophysica Acta* 46, S36-S40.

Bookchin, R. M., Ortiz, O. E. & Lew, V. L. (1991). Evidence for a direct reticulocyte origin of dense red cells in sickle cell anemia. *Journal of Clinical Investigation* 87, 113-124.

Boyd, R., (2011). R. D. Keynes (1913-2010): The legacy from two papers. *Physiology News* 82, 24-25.

Brugnara, C., Bunn, H. F. & Tosteson, D. C. (1986). Regulation of erythrocyte cation and water content in sickle cell anemia. *Science* 232, 388-390.

Brugnara, C., de Franceschi, L. & Alper, S. L. (1993). Inhibition of Ca^{++}-dependent K^+ transport and cell dehydration in sickle erythrocytes by clotrimazole and other imidazole derivatives. *Journal of Clinical Investigation* 92, 520-526.

Bunn, H. F. & Forget, B. G. (1986). *Hemoglobin: Molecular, Genetic and Clinical Aspects*, Saunders, Philadelphia.

Cantley, L. C. Jr., Cantley, L. G. & Josephon, L. (1978). A characterization of vanadate interactions with the (Na,K)-ATPase. Mechanistic and regulatory implications. *Journal of Biological Chemistry* 253, 7361-7368.

Clark, M. R. & Rossi, M. E. (1990). Permeability characteristics of deoxygenated sickle cells. *Blood* 76, 2139-2145.

Collander, R. (1949). The permeability of plant protoplasts to small molecules. *Physiologia Plantarum* 2, 300-311.

Counillon, L., Scholz, W. Lang, H. J. & Pouyssegur, J. (1993). Pharmacological characterization of stably transfected Na$^+$/H$^+$ antiporter isoforms using amiloride analogues and a new inhibitor exhibiting anti-ischemic properties. *Molecular Pharmacology* 44, 1041-1045.

Crable, S. C., Hammond, S. M., Papes, R., Rettig, R. K., Zhou, G.-P., Gallagher, P. G., Joiner, C. H. & Anderson, K. P. (2005). Multiple isoforms of the KCl cotransporter are expressed in sickle and normal erythroid cells. *Experimental Hematology* 33, 624-631.

Davson, H. & Danielli, J. F. (1943). *The Permeability of Natural Membranes*, Cambridge University Press, Cambridge.

Davson, H. & Ponder, E. (1938). Studies on the permeability of erythrocytes: The permeability of "ghosts" to cations. *Biochemical Journal* 32, 756-762.

Dunham, P. B. & Ellory, J. C. (1981). Passive potassium transport in low potassium sheep red cells: Dependence upon cell volume and chloride. *Journal of Physiology* 318, 511-530.

Duranton, S. M., Huber, S. M. & Lang, F. (2002). Oxidation induces a Cl-dependent cation conductance in human red blood cells. *Journal of Physiology* 539, 847-855.

Eaton, W. A. & Hofrichter, J. (1987). Hemoglobin S gelation and sickle cell disease. *Blood* 70, 1245-1266.

Ellory, J. C., Gibson, J. S. & Stewart, G. W. (1998). Pathophysiology of abnormal cell volume in human red cells. *Contributions to Nephrology* 123, 220-239.

Ellory, J. C., Hall, A. C. & Ody, S. A. (1989). Is acid a more potent activator of KCl co-transport than hypotonicity in human red cells? *Journal of Physiology* 420, 149P.

Ellory, J. C., Kirk, K., Culliford, S. J., Nash, G. B. & Stuart, J. (1992). Nitrendipine is a potent inhibitor of the Ca^{2+}-activated K$^+$ channel of human erythrocytes. *FEBS letters* 296, 219-221.

Ellory, J. C. & Lew, V. L., Eds. (1976). *Membrane Transport in Red Cells*, Academic Press, London.

Ellory, J. C., Robinson, H. C., Browning, J. A., Stewart, G. W., Gehl, K. A. & Gibson, J. S. (2007). Abnormal permeability pathways in human red blood cells. *Blood Cells, Molecules and Disease* 39, 1-6.

Ellory, J.C., Hall, A. C., Ody, S. A., de Figueiredos, C. E., Chalder, S. & Stuart, J. (1991). KCl cotransport in HbAA and HbSS red cells: Activation by intracellular acidity and disappearance during maturation, in *Red Blood Cell Ageing*, Eds. M. Mangani & A. DeFlora, Plenum Press, New York.

Etzion, Z., Tiffert, T., Bookchin, R. M. & Lew, V. L. (1993). Effects of deoxygenation on active and passive Ca^{2+} transport and on the cytoplasmic Ca^{2+} levels of sickle cell anemia red cells. *Journal of Clinical Investigation* 92, 2489-2498.

Feray, J-C. & Garay, R. (1986). An Na$^+$ -stimulated Mg^{2+} -transport system in human red blood cells. *Biochimica et Biophysica acta* 856, 76-84.

Feray, J. C. & Garay, R. (1988). Demonstration of a Na$^+$:Mg^{2+} exchange system in human red cells by its sensitivity to tricyclic antidepressant drugs. *Naunyn Schmiedebergs Archives of Pharmacology* 338, 332-337.

Garay, R. P., Nazaret, C. Hannaert, P. A. & Cragoe, E. J. (1988). Demonstration of a [K+,Cl-]-cotransport system in human red cells by its sensitivity to [(dihydroindenyl)oxy] alkanoic acids: Regulation of cell swelling and distinction from the bumetanide-sensitive [Na+,K+,Cl-]-cotransport system. *Molecular Pharmacology* 33, 696-701.

Gardos, G. (1958). The function of calcium and potassium permeability of human erythrocytes. *Biochimica Biophysica Acta* 30, 653-654.

Garrahan, P. J. & Glynn, I. M. (1967a). The behaviour of the sodium pump in red cells in the absence of external potassium. *Journal of Physiology* 192, 159-174.

Garrahan, P. J. & Glynn, I. M. (1967b). The stoichiometry of the sodium pump. *Journal of Physiology* 192, 217-235.

Gibson, J. S., Cossins, A. R. & Ellory, J. C. (2000). Oxygen-sensitive membrane transporters in vertebrate red cells. *Journal of Experimental Biology* 203, 1395-1407.

Gibson, J. S. & Ellory, J. C. (2002). Membrane transport in sickle cell disease. *Blood Cells, Molecules and Disease* 28, 1-12.

Gibson, J. S. & Ellory, J. C. (2003). K+-Cl- cotransport in vertebrate red cells, in *Red Cell Membrane in Health and Disease*, Eds. I. Bernhardt & J. C. Ellory, pp. 197-220, Springer, 3-540-44227-8, Berlin.

Gibson, J. S., Khan, A., Speake, P. F. & Ellory, J. C. (2001). O_2 dependence of K+ transport in sickle cells: The effects of different cell populations and the substituted benzaldehyde, 12C79. *FASEB Journal* 15, 823-832.

Gibson, J. S., Speake, P. F. & Ellory, J. C. (1998). Differential oxygen sensitivity of the K+-Cl- cotransporter in normal and sickle human red blood cells. *Journal of Physiology* 511, 225-234.

Gladwin, M. T., Crawford, J. H. & Patel, R. P. (2004). The biochemistry of nitric oxide, nitrite and hemoglobin: Role in blood flow regulation. *Free Radical Biology and Medicine* 15, 707-717.

Glynn, I. M. (2002). A hundred years of sodium pumping. *Annual Review of Physiology* 64, 1-18.

Glynn, I. M., Hara, Y. & Richards, D. E. (1984). The occlusion of sodium ions within the mammalian sodium-potassium pump. *Journal of Physiology* 351, 531-547.

Gunther, T. & Vormann, J. (1990). Characterization of Na+-independent Mg^{2+} efflux from erythrocytes. *FEBS letters* 271, 149-151.

Hall, A. C. & Ellory, J. C. (1986). Evidence for the presence of volume-sensitive KCl transport in 'young' human red cells. *Biochimica Biophysica Acta* 858, 317-320.

Halperin, J. A., Brugnara, C., Tosteson, M. T., Van Ha, T. & Tosteson, D. C. (1989). Voltage-activated cation channels in human erythrocytes. *American Journal of Physiology* 257, C986-C996.

Hebbel, R. P. (1991). Beyond hemoglobin polymerization: The red blood cell membrane and sickle cell disease pathophysiology. *Blood* 77, 214-237.

Hebbel, R. P., Boogaerts, M. A., Eaton, J. W. & Steinberg, M. H. (1980). Erythrocyte adherence to endothelium in sickle cell anaemia: A possible determinant of disease severity. *New England Journal of Medicine* 302, 992-995.

Herrick, J. B. (1910). Peculiar elongated and sickle-shaped red blood corpuscles in a case of severe anaemia. *Archives of Internal Medicine* 6, 517.

Hodgkin, A. L. & Keynes, R. D. (1955a). Active transport of cations in giant axons from sepia and loligo. *Journal of Physiology* 128, 28-60.

Hodgkin, A. L. & Keynes, R. D. (1955b). The potassium permeability of a giant nerve fibre. *Journal of Physiology* 128, 61-88.

Hoffman, J. F., Joiner, W., Nehrke, K., Potapova, O., Foye, K. & Wickrema, A. (2003). The hSK4 (KCNN4) isoform is the Ca^{2+}-activated K^+ channel (Gardos channel) in human red blood cells. *Proceedings National Academy of Sciences USA* 100, 7366-7371.

Ingram, V.M. (1957). Gene mutations in human haemoglobin, the chemical difference between normal and sickle cell haemoglobin. *Nature* 180, 326-328.

Ishii, T. M., Silvia, C., Hirschberg, B., Bond, C. T., Adelman, J. P. & Maylie, J. (1997). A human intermediate conductance calcium-activated potassium channel. *Proceedings of the National Academy of Sciences* 94, 11651-11656.

Joiner, C. H. (1990). Deoxygenation-induced cation fluxes in sickle cells: II. Inhibition by stilbene disulfonates. *Blood* 76, 212-220.

Joiner, C. H. (1993). Cation transport and volume regulation in sickle red blood cells. *American Journal of Physiology* 264, C251-C270.

Joiner, C. H., Dew, A. & Ge, D. L. (1988). Deoxygenation-induced fluxes in sickle cells. I. Relationship between net potassium efflux and net sodium influx. *Blood Cells* 13, 339-348.

Joiner, C. H., Jiang, M., Claussen, W. J., Roszell, N. J., Yasin, Z. & Franco, R. S. (2001). Dipyridamole inhibits sickling-induced cation fluxes in sickle red blood cells. *Blood* 97, 3976-3983.

Joiner, C. H., Jiang, M. & Franco, R. S. (1995). Deoxygenation-induced cation fluxes in sickle cells IV. Modulation by external calcium. *American Journal of Physiology* 269, C403-C409.

Joiner, C. H., Morris, C. L. & Cooper, E. S. (1993). Deoxygenation-induced cation fluxes in sickle cells. III. Cation selectivity and response to ph and membrane potential. *American Journal of Physiology* 264, C734-C744.

Kaji, D. M. (1990). Nifedipine inhibits calcium -activated k transport in human erythrocytes. *American Journal of Physiology* 259, C332-C339.

Kato, G. J., McGowan, V., Machado, R. F., Little, J. A., Taylor, J., Morris, C. R., Nichols, J. S., Wang, X., Poljakovic, M., Morris, S. M. & Gladwin, M. T. (2006). Lactate dehydrogenase as a biomarker of hemolysis-associated nitric oxide resistance, priapism, leg ulceration, pulmonary hypertension, and death in patients wtih sickle cell disease. *Blood* 107, 2279-2285.

Lang, K. S., Myssina, S., Tanneur, V., Wieder, T., Huber, S. M., Lang, F. & Duranton, C. (2003). Inhibition of erythrocyte cation channels and apoptosis by ethylisopropylamiloride. *Naunyn Schmiedebergs Archives of Pharmacology* 367, 391-396.

Lauf, P. K. (1984). Thiol-dependent passive K/Cl transport in sheep red cells: IV. Furosemide inhibition as a function of external Rb^+, Na^+ and Cl^-. *Journal of Membrane Biology* 77, 57-62.

Lew, V. L. & Bookchin, R. M. (2005). Ion transport pathology in the mechanism of sickle cell dehydration. *Physiological Reviews* 85, 179-200.

Lew, V. L., Freeman, C. J., Ortiz, O. E. & Bookchin, R. M. (1991). A mathematical model of the volume, pH and ion content regulation in reticulocytes. *Journal of Clinical Investigation* 87, 100-112.

Lew, V. L., Ortiz, O. E. & Bookchin, R. M. (1997). Stochastic nature and red cell population distribution of the sickling-induced Ca^{2+} permeability. *Journal of Clinical Investigation* 99, 2727-2735.

Lew, V. L., Tsien, R. Y., Miner, C. & Boockchin, R. M. (1982). Physiological $[Ca^{2+}]_i$ level and pump-leak turnover in intact red cells measured using an incorporated Ca chelator. *Nature* 298, 478-481.

Marotta, C. A., Wilson, J. T., Forget, B. J. & Weissman, S. M. (1977). Human beta-globin messenger RNA III. Nucleotide sequences derived from complementary DNA. *Journal of Biological Chemistry* 252, 5040-5051.

Masereel, B., Pochet, L. & Laeckmann, D. (2003). An overview of inhibitors of Na^+/H^+ exchanger. *European Journal of Medicinal Chemistry* 38, 547-554.

Mercado, A., Song, L., Vazquez, N., Mount, D. B. & Gamba, G. (2000). Functional comparison of the K^+-Cl^- cotransporters KCC1 and KCC4. *Journal of Biological Chemistry* 275, 30326-30334.

Mohandas, N., Rossi, M. E. & Clark, M. R. (1986). Association between morphologic distortion of sickle cells and deoxygenation-induced cation permeability increases. *Blood* 68, 450-454.

Naftalin, R. J. (2010). Reassessment of models of facilitated transport and cotransport. *Journal of Membrane Biology* 234, 75-112.

Nagel, R. L., Fabry, M. E. & Steinberg, M. H. (2003). The paradox of hemoglobin SC disease. *Blood Reviews* 17, 167-178.

Nagel, R. L. & Platt, O. S. (2001). General pathophysiology of sickle cell anemia, in *Disorders of Hemoglobin: Genetics Pathophysiology, and Clinical Management*, Eds. M. H. Steinberg, B. G. Forget, D. R. Higgs & R. L. Nagel, pp. 494-526, Cambridge University Press, 0-521-632-668, Cambridge.

Nagel, R. L. & Steinberg, M. H. (2001). Hemoglobin SC disease and HbC disorders, in *Disorders of Hemoglobin: Genetics Pathophysiology, and Clinical Management*, Eds. M. H. Steinberg, B. G. Forget, D. R. Higgs & R. L. Nagel, pp. 756-785, Cambridge University Press, 0-521-632-668, Cambridge.

Pauling, L., Itano, H. A., Singer, S. J. & Wells, I. C. (1949). Sickle cell anaemia, a molecular disease. *Science* 110, 1141-1152.

Payne, J. A., Lytle, C. & McManus, T. J. (1990). Foreign anion substitution for chloride in human red blood cells: Effect on ionic and osmotic equilibria. *American Journal of Physiology* 259, C819-C827.

Pedersen, S. F., King, S. A., Nygaard, E. B., Rigor, R. R. & Cala, P. M. (2007). NHE1 inhibition by amiloride- and benzoylguanidine-type compounds. Inhibitor binding loci deduced from chimeras of NHE1 homologues with endogenous differences in inhibitor sensitivity. *Journal of Biological Chemistry* 282, 19716-19727.

Post, R. L. & Jolly, P. C. (1957). The linkage of sodium, potassium and ammonium active transport across the human erythrocyte. *Biochimica et Biophysica Acta* 25, 118-128.

Reichstein, E. & Rothstein, A. (1981). Effects of quinine on Ca^{2+}-induced K^+ efflux from human red blood cells. *Journal of Membrane Biology* 59, 57-63.

Reiter, C. D., Wang, X., Tanus-Santos, J. E., Hogg, N., Cannon, R. O. III., Schechter, A. N. & Gladwin, M. T. (2002). Cell-free hemoglobin limits nitric oxide bioavailability in sickle-cell disease. *Nature Medicine* 8, 1383-1389.

Rhoda, M. D., Apovo, M., Beuzard, Y. & Giraud, F. (1990). Ca^{2+} permeability in deoxygenated sickle cells. *Blood* 75, 2453-2458.

Russell, J. M. (2000). Sodium-potassium-chloride cotransport. *Physiological Reviews* 80, 211-276.

Sachs, J. R. (1980). The order of release of sodium and addition of potassium in the sodium - potassium pump reaction mechanism. *Journal of Physiology* 302, 219-240.

Schultz, S. G. (1980). *Basic Principles of Membrane Transport*. Cambridge University Press, Cambridge.

Serjeant, G. R. (2001). The emerging understanding of sickle cell disease. *British Journal of Haematology* 112, 3-18.

Setty, B. N. Y. & Betal, S. G. (2008). Microvascular endothelial cells express a phosphatidyl serine receptor: A functionally active receptor for phosphatidyl serine-positive erythrocytes. *Blood* 111, 905-914.

Setty, B. N., Betal, S. G., Zhang, J. & Stuart, M. J. (2008). Heme induces endothelial tissue factor expression: Potential role in hemostatic activation in patients with hemolytic anemia. *Journal of Thrombosis and Haemostasis* 6, 2202-2209.

Shaw, T. L. (1955). Potassium movements in washed erythrocytes. *Journal of Physiology* 129, 464-475.

Skou, J. C. (1957). The influence of some cations on an adenosine triphosphatase from peripheral nerves. *Biochimica et Biophysica Acta* 23, 394-401.

Skou, J. C. (2003). The identification of the sodium potassium pump, in *Nobel Lectures, Chemistry 1996-2000*, Ed. I. Grenthe, Scientific Publishing Co., Singapore.

Stein, W. D. (1986). *Transport and Diffusion across Cell Membranes*, Academic Press, Orlando.

Steinberg, M. H. (1999). Management of sickle cell disease. *The New England Journal of Medicine* 340, 1021-1030.

Steinberg, M. H., Forget, B. G., Higgs, D. R. & Nagel, R. L., Eds. (2001). *Disorders of Hemoglobin: Genetics Pathophysiology, and Clinical Management*, Cambridge University Press, 0-521-632-668, Cambridge.

Steinberg, M. H. (2001). Compound heterozygous and other sickle hemoglobinopathies, in *Disorders of Hemoglobin: Genetics Pathophysiology, and Clinical Management*, Eds. M. H. Steinberg, B. G. Forget, D. R. Higgs & R. L. Nagel, pp. 786-810, Cambridge University Press, 0-521-632-668, Cambridge.

Stewart, G. W. (2003). The hereditary stomatocytosis and allied conditions: Inherited disorders of Na^+ and K^+ transport, in *Red Cell Membrane in Health and Disease*, Eds. I Bernhardt & J. C. Ellory, pp. 511-523, Springer, 3-540-44227-8, Berlin.

Stocker, J. W., De Franceschi, L. D., McNaughton-Smith, G. A., Corrocher, R., Beuzard, Y. & Brugnara, C. (2003). ICA-17043, a novel gardos channel blocker, prevents sickled red blood cell dehydration in vitro and in vivo in sad mice. *Blood* 101, 2412-2418.

Su, W., Shmukler, B. E., Chernova, M. A., Stuart-Tilley, A. K., De Franceschi, L., Brugnara, C. & Alper, S. L. (1999). Mouse K-Cl cotransporter KCC1:   Cloning, mapping, pathological expression, and functional regulation. *American Journal of Physiology* 277, C899-C912.

Tiffert, T. & Lew, V. L. (2001). Kinetics of inhibition of the plasma membrane calcium pump by vanadate in intact human red cells. *Cell Calcium* 30, 337-342.

Tosteson, D. C., Carlsen, E. & Dunham, E. T. (1955). The effects of sickling on ion transport. I. Effect of sickling on potassium transport. *Journal of General Physiology* 39, 31-53.

Tosteson, D. C. & Hoffman, J. F. (1960). Regulation of cell volume by active cation transport in high and low potassium sheep red cells. *Journal of General Physiology* 44, 169-194.

Weiss, E., Lang, H. J. & Bernhardt, I. (2004). Inhibitors of the $K^+(Na^+)/H^+$ exchanger of human red blood cells. *Bioelectrochemistry* 62, 135-140.

Wolff, D., Cecchi, X., Spalvins, A. & Canessa, M. (1988). Charybdotoxin blocks with high affinity the Ca-activated K^+ channel of hba and hbs red cells: Individual differences in the number of channels. *Journal of Membrane Biology* 106, 243-252.

Zhang, Z., Li, Z., Tian, J., Jiang, W., Wang, Y., Zhang, X., Li, Z., You, Q., Shapiro, J. I., Si, S. & Xie, Z. (2010). Identification of hydroxyanthones as Na/K-ATPase ligands. *Molecular Pharmacology* 77, 961-967.

8

Undesirable Radioisotopes Induced by Therapeutic Beams from Medical Linear Accelerators

Adam Konefał

Institute of Physics, Department of Nuclear Physics and
Its Application, University of Silesia in Katowice
Poland

1. Introduction

Contemporary linear accelerators called often linacs, used in radiation medicine generate electrons and X-rays with energies up to over 20 MeV. Such energies are enough to induce nuclear reactions in which neutrons and radioisotopes are produced. These neutrons and radioisotopes are undesirable in therapy, because they are source of an additional dose to patients and to staff operating the medical accelerators. The therapeutic electrons and X-rays can induce electronuclear (e,e′n) and photonuclear (γ,n) reactions, respectively. These reactions take place inside the therapeutic beam in massive components of an accelerator head, mainly and in air. In the case of the X-rays the main neutron sources are the collimators of the beam, flattening filter giving the appropriate profile of the beam and the target in which electrons are converted into X-ray radiation. In the case of the electron beams the majority of neutrons are produced in the collimator system and in the scattered foils. The neutrons originated in both mentioned type of reactions have the broad energy spectrum with the high-energy end of more than ten MeV. Majority of the neutrons reach the concrete walls, ceiling and floor of the radiotherapy facility. Concrete is a good moderator. In this medium the neutrons undergo elastic collisions with nuclei of hydrogen, mainly. The slowed down neutrons may get out of concrete and return to air, contributing to the specific distribution of neutron energy inside the radiotherapy facility. Kinetic energies of the slowed down neutrons are distributed according to the Maxwell-Boltzmann distribution law. The neutrons can easy induce the simple capture (n,γ) reactions in the thermal and resonance energy range and radioisotopes are produced. The neutron field is almost uniform in whole accelerator room. Thus the radioisotopes originating from the neutron reactions can be produced in the accelerator components and accessories as well as in the wall, ceiling and floor of the radiotherapy facility. Moreover, the neutrons can induce simple capture reactions in the entrance door of the radiotherapy facility. The penetrative gammas are emitted as a result of these neutron reactions. Therefore, the gamma radiation can appear close to the radiotherapy facility door in the operator room during emission the high-energy therapeutic beams.

In the paper the radioisotopes originating in the accelerator components and in the accessories as well as in the walls, ceiling, floor and door of the radiotherapy facility and in air are identified for the typical commercial linacs used in European oncological centres. Knowledge of the radioisotopes originating during radiotherapy by the high-energy electron and X-ray therapeutic beams is very significant for the radiological protection of staff operating accelerators as well as for constructors of medical linacs.

2. Characteristic of photonuclear, electronuclear and simple capture reactions

In this paragraph a short characteristic of photonuclear, electronuclear and simple capture reactions will be presented. The dependences between the cross sections and energies of gammas, electrons and neutrons are presented. All cross sections presented in this chapter are taken from web retrieval system of National Nuclear Data Center in Brookhaven National Laboratory, basing on various data base.

2.1 Photonuclear reactions - (γ,n)
Gammas with energy of more than ten MeV are produced mainly in accelerators. Such generated gamma radiation is often called X-rays (bremsstrahlung) and it comes into existence when accelerated electrons are slow down in the force field of the atomic nuclei in a target. The gamma radiation produced in an accelerator is characterized by a spectrum with a continuous distribution of energies. The maximum energy E_{max} of the spectrum corresponds to maximum energy of electrons hitting a target. The high-energy gammas generated by accelerators can induce photonuclear reactions – (γ,n), (γ,2n) and (γ,p). The gamma gives its energy to a nucleus in this interaction. The delivered energy must be at least as great as the binding energy of a neutron or proton in order to eject them from a nucleus in a target. However, separation of a neutron is more probably than separation of a proton because it has no charge contrary to a proton that additionally has to pass through a coulomb barrier of a nucleus. Thus, in a photonuclear reaction a nucleus changes into the other with less a mass number. For heavy nuclei the production of two neutrons in the photonuclear reaction - (γ,2n) is also possible. However, separation of two neutrons needs more photon energy. Energy threshold of the two neutron production is usually near the high-energy end of the therapeutic X-ray beam spectrum. The photonuclear cross section has a resonance character. The maximum value of the cross section depends on the atomic number. It is in the range from several milibarns for light nuclei to several hundred milibarns for heavy nuclei. Moreover, the maximum of the photonuclear cross section corresponds to gamma energy of about 22 MeV for light nuclei and it shifts to about 12 MeV as the atomic number increases. The energy threshold of the photonuclear reactions is about 8 MeV to 10 MeV for most isotopes. In Figure 2.1 the energy spectrum of gamma radiation produced in a medical linear accelerator working in a mode with nominal potential of 20 MV is compared with the photonuclear cross sections for chosen materials.

2.2 Electronuclear reactions - (e,e'n)
Electrons with energies of several MeV or greater can induce electronuclear reactions – (e,e'n). This interaction is an inelastic scattering of an electron in the force field of the atomic nuclei. An electron gives a part of its energy to a nucleus. This energy is partially used to separate a neutron from a nucleus and it is partially change into a kinetic energy of the separated neutron. The cross sections of electronuclear reactions increase with an increasing

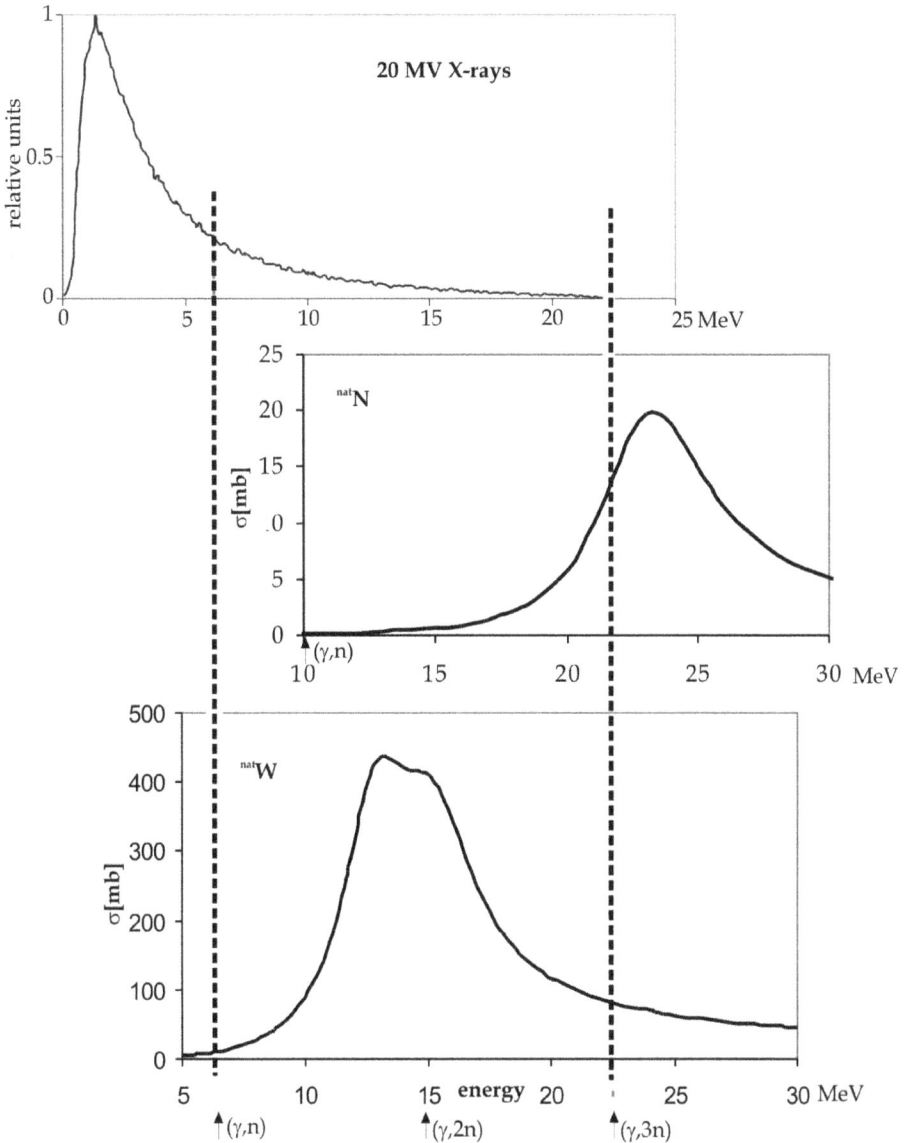

Fig. 2.1. Comparison between the energy range of spectrum of therapeutic 20 MV X-rays and the energy ranges of the photonuclear cross section [12] for natural nitrogen and tungsten. The presented 20 MV X-ray spectrum was derived by the use of Monte Carlo calculation based on the GEANT4 code.

energy of electrons. The neutron production yield in the (e,e'n) reactions is much less than in the case of the (γ,n) reactions. The cross sections of electronuclear reactions are about 3 orders of magnitude less than those for photonuclear reactions in the energy range of the therapeutic beams generated by medical accelerators.

2.3 Simple capture reactions - (n,γ)

Thermal and resonance neutrons can be captured by an atomic nucleus during interaction between a neutron and a nucleus. A photon is emitted by a nucleus as a result of this interaction. Energy of the photon Eγ is equal to $\Delta m \cdot c^2$, where Δm (= $M_{neut} + M_{nuc} - M'_{nuc}$) is the difference between the sum of the masses of the neutron M_{neut} and the nucleus M_{nuc} before the neutron capture and the mass of the nucleus M'_{nuc} after the neutron capture, c is the velocity of light. Therefore Eγ is called binding energy of a neutron. The (n,γ) reaction can be induced by thermal neutrons for nearly all isotopes. It can also occur in the resonance neutron energy range where the cross sections of the neutron capture reactions have the high peaks (resonance peaks) for a number of isotopes. The cross sections of simple capture reactions as a function of neutron energy for the chosen isotopes are presented in Figure 2.2.

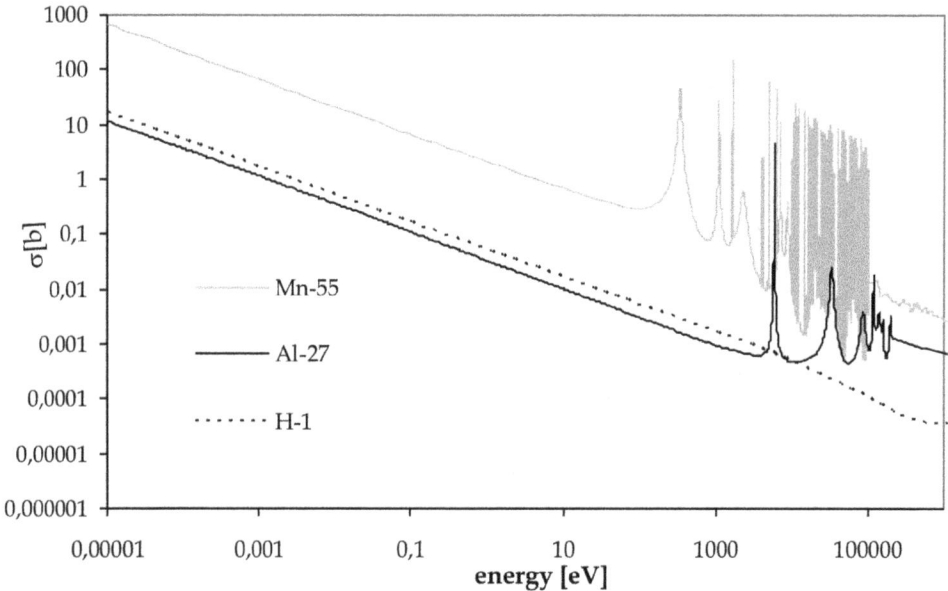

Fig. 2.2. The exemplary cross sections of simple capture reactions [12].

3. Method of identifications of radioisotopes

The base method of identification of the radioisotopes induced by therapeutic beams is a spectroscopy of gammas emitted by the originated unstable atomic nuclei.

In this paragraph the details of the method using the high-purity germanium detector is described (energy calibration of the detection system, the use of calibration sources and verification of the calibration during measurements of the spectra of gammas emitted by the radioisotopes etc.).

The complementary method of identification of the radioisotopes can be also based on the Monte Carlo computer simulation of physical processes and particle transportation inside the radiotherapy facility and in the walls, ceiling and floor of the room with a linac. The computation method based on the GEANT4 code will be described. The exemplary energy spectrum derived by the MC computer simulation will be presented and analysed.

3.1 The high-purity germanium detector

One of the base detectors that can be applied to identify radioisotopes originated inside the radiotherapy facility is the high-purity germanium detector applied for a field spectrometry. The detector is connected to a multichannel analyzer installed in a PC computer (usually in a laptop). The work of the detector is operated by the special software. All detection system is relatively small and it can be easy displaced. The high-purity germanium detectors are characterized by a good efficiency and resolution. It makes it possible to measure the photon energies from several dozen keV to several MeV. The low-energy limit depends on the thickness of the germanium crystal shield whereas the high-energy limit is connected with the size of the germanium crystal. The energy calibration (i.e. determination of energy for each channel of the multichannel analyzer) is usually performed with the use of set of commercial calibration sources. The view of the ORTEC high-purity germanium detector and its exemplary energy calibration curve is presented in Figure 3.1.

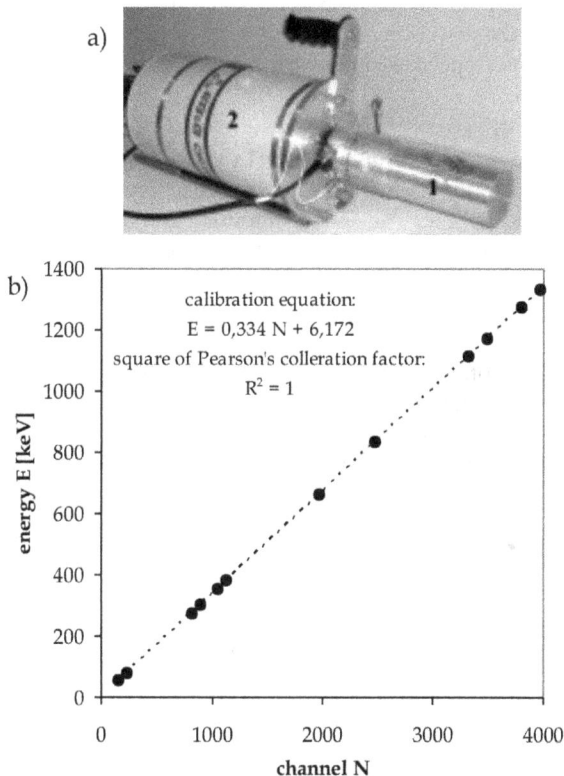

Fig. 3.1. The view of the high-purity germanium detector by ORTEC (1 – the aluminium shield of the germanium crystal, 2 - container with liquid nitrogen) a) and its energy calibration curve b). The following calibration sources were used to derive the calibration equation: ^{22}Na, ^{54}Mn, ^{60}Co, ^{74}Se, ^{133}Ba, ^{137}Cs, ^{241}Am. These seven sources give twelve points in the energy calibration curve (from 12 photons with various energies). The linear function is fitted to get the calibration equation. The square of the Pearson's correlation factor value - $R^2 = 1$ for the fit shows that the determined function is a good description of the obtained energy calibration curve.

The energy calibration can be checked in the measuring place. For this purpose, the block of the ^{115}In isotope is located inside the radiotherapy facility (for example on the therapeutic couch) in the field of neutrons from (γ,n) reactions occurring during the emission of the high-energy therapeutic X-ray beam. Indium is activated by thermal and resonance neutrons. The following neutron capture reaction takes place:

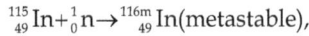

$$^{115}_{49}\text{In} + ^{1}_{0}\text{n} \rightarrow ^{116m}_{49}\text{In (metastable)},$$

followed by radioactive decay:

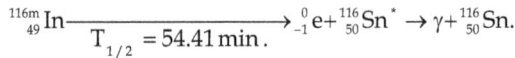

$$^{116m}_{49}\text{In} \xrightarrow[\text{T}_{1/2} = 54.41\,\text{min}.]{} {}^{0}_{-1}\text{e} + ^{116}_{50}\text{Sn}^{*} \rightarrow \gamma + ^{116}_{50}\text{Sn}.$$

Six strong photopeaks from the deexcitation of the ^{116}Sn* state are used for the verification of the energy calibration of the detection system.

The details of the gamma spectroscopy were presented in many publications [see for example 1, 2].

3.2 Monte Carlo computer simulation by GEANT4

The spectra of gammas from the (e,e'n), (γ,n) and (n,γ) reactions can be derived by the Monte Carlo calculations realized with the use of the computer simulations.

The Monte Carlo calculation method seems to be the most comprehensive and potentially the most accurate method to derive such energy spectra. At present one of the dominant software that makes it possible to simulate the above mentioned reactions and others[1] occurring during emission of the high-energy therapeutic beams is the GEANT4 (GEometry ANd Tracking) code. This software has a form of C++ libraries prepared in CERN. GEANT4 is a good tool for modeling objects with complicated shape like a medical linac because it has plentiful predefined solid geometry types that can be combined via Boolean operations. This feature of the GEANT4 permits to make an accurate copy of the real object, which is the main condition to obtain the sensitive results. In Figure 3.2 the visualization of a fragment of a radiotherapy facility with a linac and a patient is presented. More information on GEANT4 is on the web side of the GEANT4 project [3] and in many publications [for example 4-6].

4. Radioisotopes originating from photonuclear and electronuclear reactions

The therapeutic gammas and electrons with appropriately high energy induce the nuclear reactions (γ,n) and (e,e'n) in which radioisotopes come into existence (see the explanation in Introduction). The therapeutic beam is collimated therefore the area of occurrence of these reactions is well determined except for the scattered gammas / electrons leaving the beam. However, the fluence of the scattered radiation outside the therapeutic beam is several

---

[1] The processes occurring during emission of the high-energy therapeutic electrons and X-rays: bremsstrahlung production, ionization, multiple scattering for electrons and positrons, positron annihilation, and additionally, photoelectric effect, Compton interaction, gamma conversion and Rayleigh scattering for photons, neutron capture, elastic and inelastic neutron scattering, decay process and some others of the lower significance, for example, the Auger effect ect.

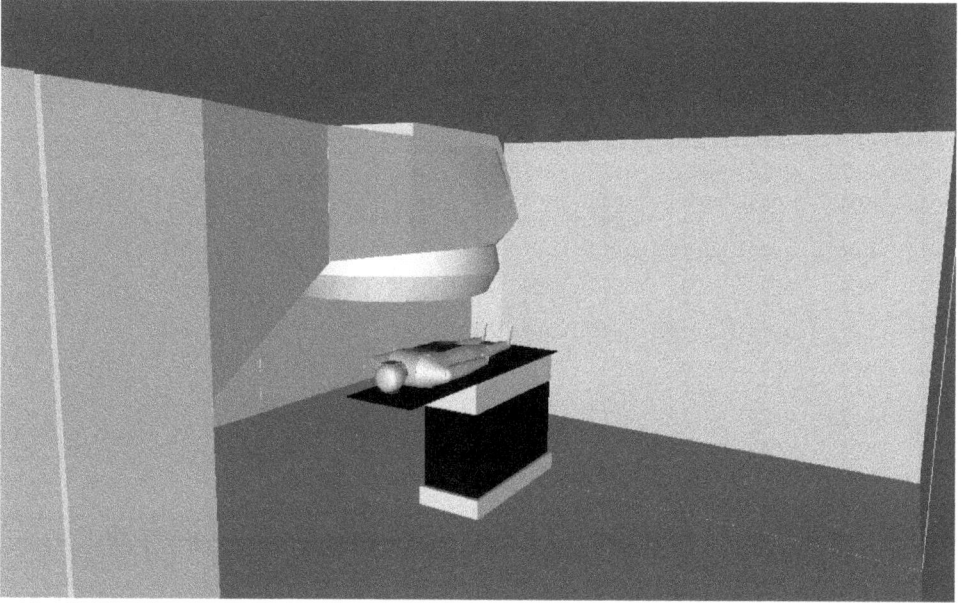

Fig. 3.2. The visualization of a fragment of the virtual model applied for the computer simulation of the high-energy X-ray irradiation of a patient. The simulation program was written with the use of the GEANT4 code. The picture was generated by the simulation program in a graphic mode of the VRML script.

orders of magnitude less than inside the beam. Thus, the photonuclear and electronuclear reactions occur mainly inside the therapeutic beam i.e. in the accelerator head and in air. Interactions between the therapeutic gammas and nuclei of ^{14}N and ^{16}O cause the photonuclear reactions giving two radioisotopes ^{13}N and ^{15}O:

$$\gamma + {}^{14}_{7}N(99.6\%) \rightarrow {}^{1}_{0}n + {}^{13}_{7}N, \qquad \gamma + {}^{16}_{8}O(99.8\%) \rightarrow {}^{1}_{0}n + {}^{15}_{8}O.$$

Abundance is given after the symbol of the nucleus in parenthesis. Such way of a reaction notation is used in this chapter. The nuclei: ^{13}N and ^{15}O disintegrate by β+ decay and electron capture (EC) into stable nuclei. The decay schemes of these two unstable nuclei are presented in Figure 4.1. The total cross section of the photonuclear reactions with ^{16}O is shown in Figure 4.2. The activation of the air was considered by many authors (see, for example [7]).

The contemporary commercial medical linacs have usually collimator system made of tungsten. However, target and flattening filter are made of various materials, dependently on the kind and energy of the therapeutic beam and on a manufacturer of an accelerator. In Figure 4.3 a spectrum measured under the head of one of the commercial medical accelerator with a gold target is presented. The peaks from gammas emitted by radioisotopes originating in the (γ,n), (e,e'n) and (n,γ)[2] reactions are marked.

---

[2] Radioisotopes originating from simple capture reactions are discussed in the next paragraph.

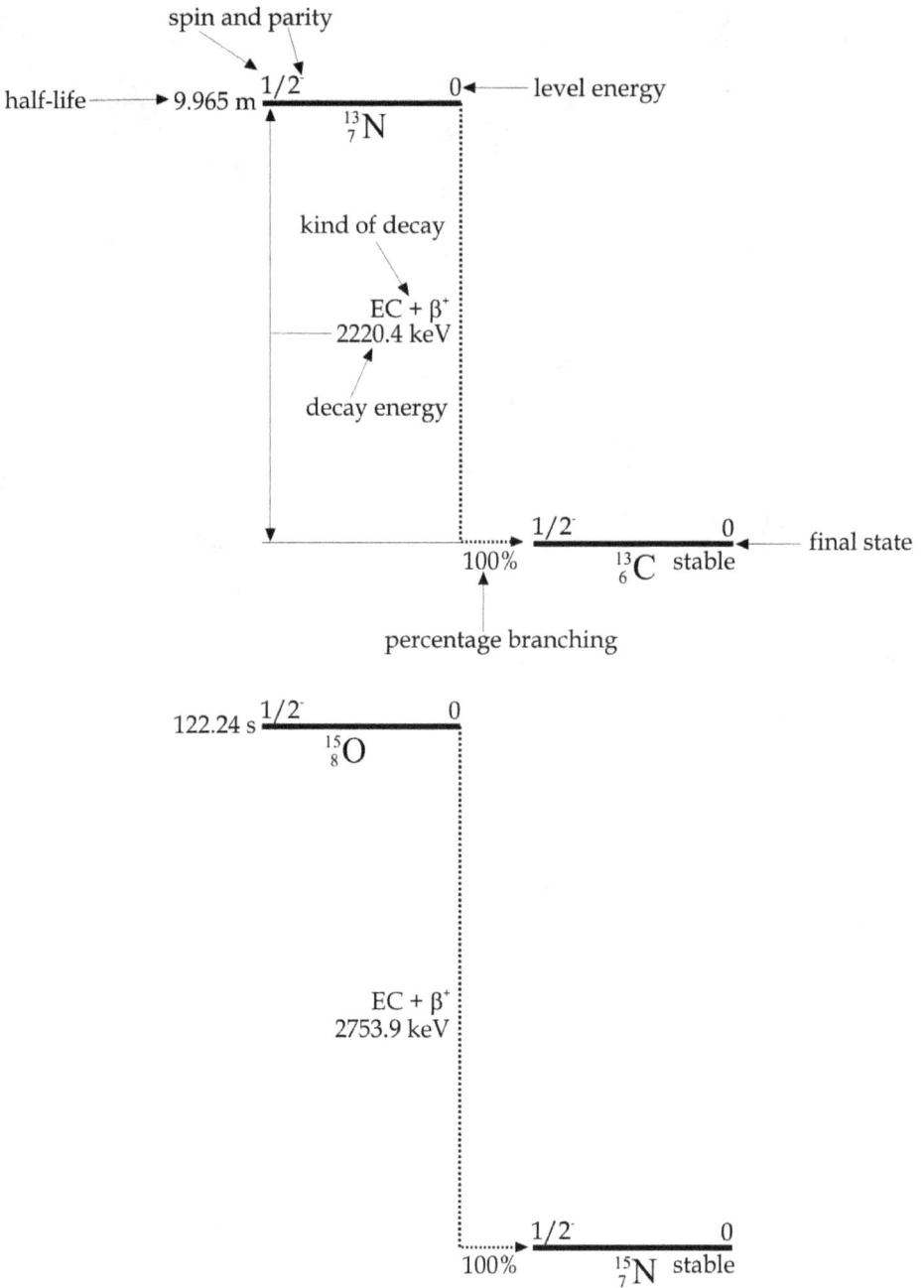

Fig. 4.1. Decay schemes of ^{13}N and ^{15}O – nuclei originating from photonuclear reactions induced by gammas of therapeutic X-ray beams. Explanation of the description of the formalism of all decay schemes presented in this chapter is given using the decay of ^{13}N.

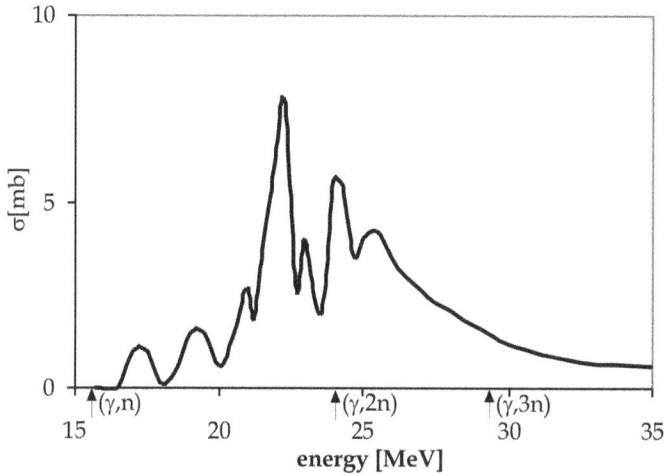

Fig. 4.2. The cross sections of the photonuclear reaction $^{16}O(\gamma,n)^{15}O$. The cross section for the analogous reaction with ^{14}N is presented in Figure 2.1. [12]

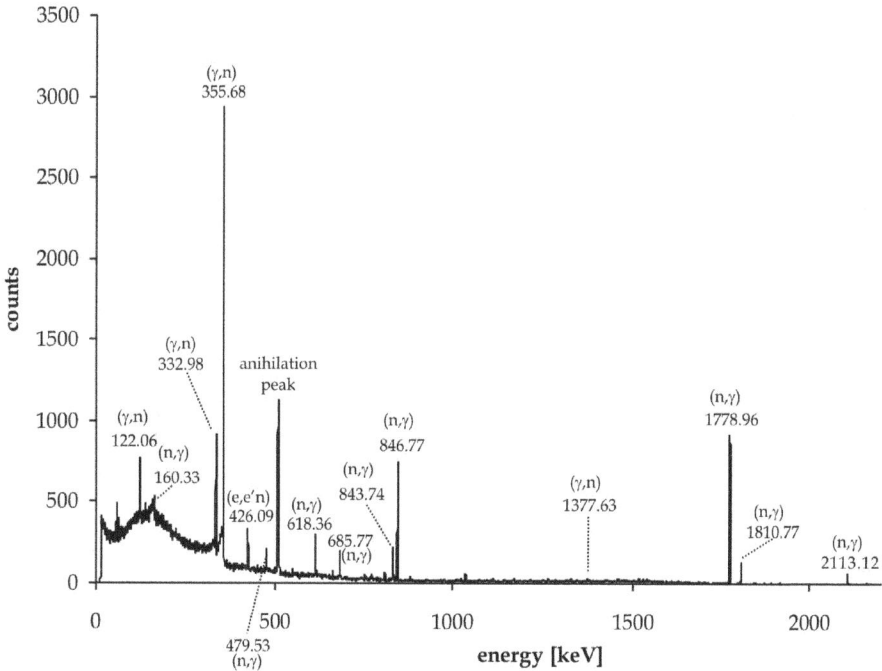

Fig. 4.3. The energy spectrum of gamma radiation measured under the head of a commercial medical linac with the gold target. The peaks come from decays of radioisotopes originated from photonuclear, electronuclear and simple capture reactions (see notation near the peaks).

The annihilation peak[3] visible in Figure 4.3 comes from β^+-emitters. Moreover, gamma with energies over 1022 keV can create a pair of electron-positron in the nucleus field. The created positron reaching thermal energy can undergo annihilation. Two radioisotopes: ^{196}Au and ^{57}Ni came into being as a result of the photonuclear and electronuclear reactions. The radioisotope ^{196}Au can originate from the following reactions:

$$\gamma + {}^{197}_{79}\text{Au}(100\%) \rightarrow {}^{1}_{0}\text{n} + {}^{196}_{79}\text{Au}, \qquad {}^{0}_{-1}\text{e} + {}^{197}_{79}\text{Au} \rightarrow {}^{0}_{-1}\text{e}' + {}^{1}_{0}\text{n} + {}^{196}_{79}\text{Au}.$$

The maximum value of the cross section of the reaction ^{197}Au$(\gamma,n)^{196}$Au is about 580 mb and it corresponds to neutron energy of about 14 MeV [11]. Gold is characterized by the relatively high cross section for the neutron capture reaction - ^{197}Au$(n,\gamma)^{198}$Au (i.e. the total thermal neutron capture cross section of 96 barns and the resonance activation integral of 1558 barns [8, 9]). However, the gammas (411.8 keV) from the decays of nuclei of ^{198}Au are absent in the presented spectrum measured under the accelerator head since the fast neutrons originate in the target and they leave it before slowing down to the thermal and resonance energies. The production of another radioisotope ^{195}Au (EC + β^+, $T_{1/2}$ = 186.1 d, decay energy of 226.8 keV, energy of emitted gammas of 98.9 keV [13]) is possible in the $(\gamma,2n)$ reaction. However, the energy threshold of this reaction is 15 MeV [11]. The photonuclear reaction from which the radioisotope - ^{57}Ni originates, can be expressed as

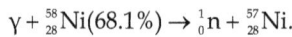

$$\gamma + {}^{58}_{28}\text{Ni}(68.1\%) \rightarrow {}^{1}_{0}\text{n} + {}^{57}_{28}\text{Ni}.$$

The maximum value of the cross section of this reaction is about 30 mb and it corresponds to gamma energy of about 17 MeV [11]. Nickel is the main component of stainless steel used in a construction of medical accelerators. The decay schemes of nuclei of ^{196}Au and ^{57}Ni are presented in Figures 4.4 and 4.5, respectively.

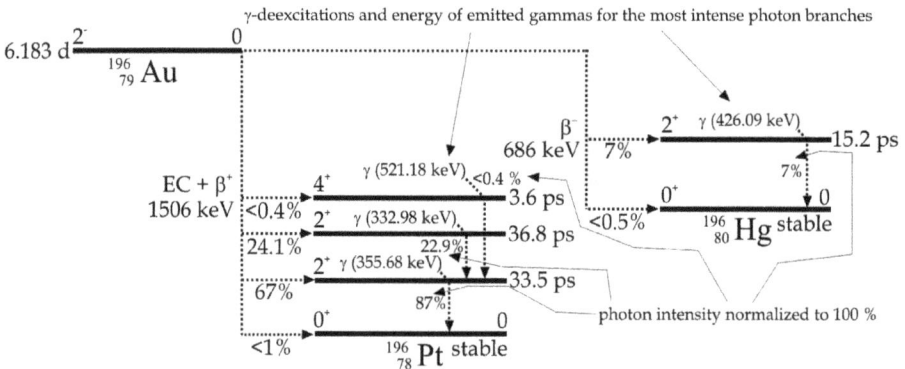

Fig. 4.4. Decay scheme of the ^{196}Au nucleus originating from photonuclear reaction induced by gammas of therapeutic X-ray beam in the target of medical linac. The γ-deexcitation notation used for the excited states of ^{196}Pt and ^{196}Hg is valid in this entire chapter. The peaks from gammas with energies of 355.68 keV, 332.98 keV and 426.09 keV are visible in the spectrum measured under the accelerator head (see Figure 4.3).

---

[3] The annihilation peak is created by gammas with energy of 511 keV corresponding to the rest mass of an electron or a positron.

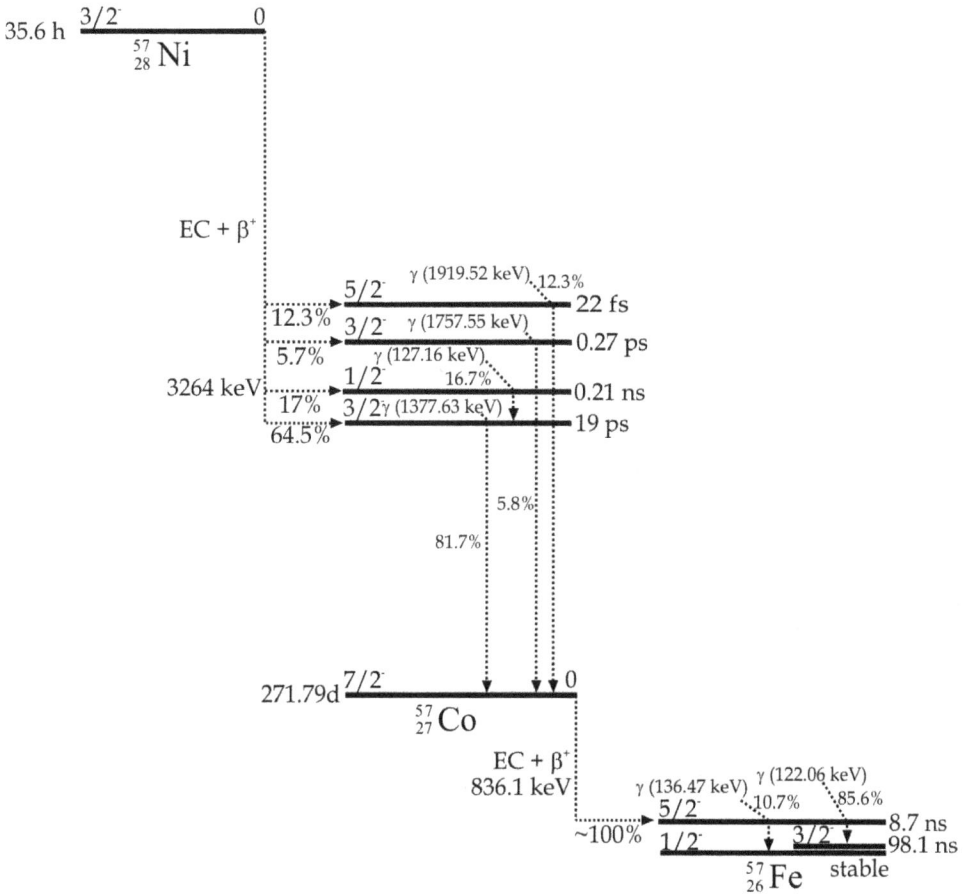

Fig. 4.5. Decay schemes of the [57]Ni nucleus and its daughter nucleus - [57]Co. The peaks at 1377.55 keV and at 122.06 keV from gammas of two most intense photon branch are visible in the spectrum measured under the accelerator head (see Figure 4.3).

Massive accessories of medical linacs, like wedges, electron applicators etc., are often located inside a high-energy therapeutic beam and they are activated in the (γ,n) and (e,e'n) reactions. The linac accessories can also be activated outside the therapeutic beam, by neutrons from the photonuclear and electronuclear reactions. Massive accessory of linacs are designed to collimate the beam or to change the beam profile etc. They are often made of lead, tungsten and stainless steel. The exemplary energy spectrum of gammas emitted by radioisotopes induced in the typical medical accelerator accessory is presented in Figure 4.6.

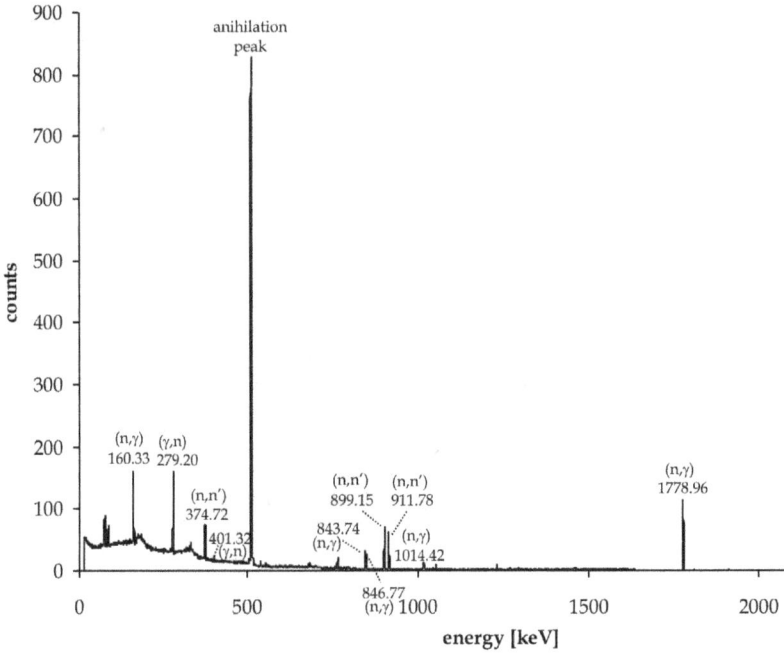

Fig. 4.6. The typical energy spectrum of gammas emitted by a medical linac accessory. The notation near the peaks is as in Figure 4.3.

The peaks at 279.20 keV and at 401.32 keV are a result of the following photonuclear reaction:

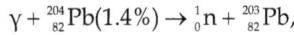

$$\gamma + {}^{204}_{82}Pb(1.4\%) \rightarrow {}^{1}_{0}n + {}^{203}_{82}Pb,$$

followed by electron capture and β^+ decay giving the excited state of ${}^{203}Tl^*$ (Figure 4.7).

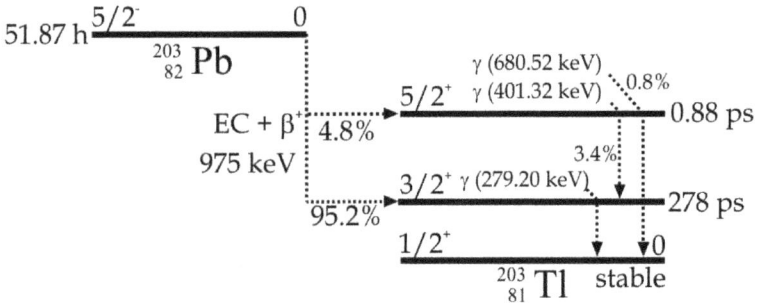

Fig. 4.7. Decay scheme of the ${}^{203}Pb$ nucleus. The gammas with energies of 279.20 keV and of 401.32 keV, from the deexcitation of ${}^{203}Tl^*$ give the peaks visible in the spectrum of gammas emitted by a medical linac accessory (see Figure 4.6).

There are four natural stable isotopes of lead: ${}^{204}Pb$ (abundance of 1.5 %), ${}^{206}Pb$ (24.1 %), ${}^{207}Pb$ (22.1 %) and ${}^{208}Pb$ (52.3 %) which can give three radioisotopes: ${}^{203}Pb$ (mentioned above), ${}^{205}Pb$ and ${}^{209}Pb$ (Figure 4.8). First two of them are a result of photonuclear reactions whereas third originates from neutron capture reaction (see next paragraph). Decays of ^{205}Pb and ^{209}Pb nuclei occur without emission of gammas. Thus, identification of these radioisotopes cannot be based on the gamma spectroscopy. However, appearance of the radioisotope ^{203}Pb decaying with gamma emission, originating from the isotope ^{204}Pb with abundance smallest of all natural lead isotopes indicates to appearance of the radioisotopes: ^{205}Pb and ^{209}Pb, because the (γ,n) cross sections are of the similar value for all isotopes of lead (the total photoneutron cross section for natural lead ~3047 mb [11]), and the total thermal neutron cross section for ^{208}Pb(n,γ)^{209}Pb is relatively high to be 0.49 b[12]. Additionally, ^{202}Pb can originate as a result of (γ,2n) reaction for the beam with a nominal potential of 18 MV or higher (photon energy threshold of this reaction ~ 15 MeV).

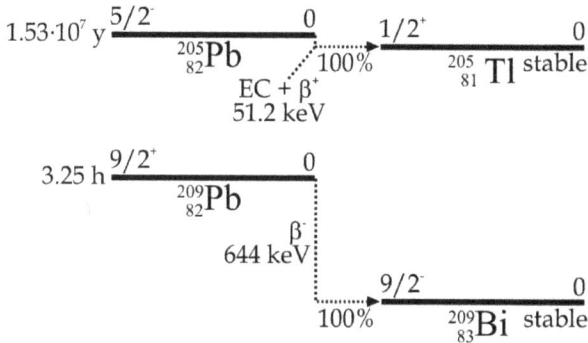

Fig. 4.8. Decay schemes of the ^{205}Pb and ^{209}Pb nuclei.

In general, the analogical considerations can be conducted for other radioisotopes. In the presented way the appearance of the radioisotope: ^{59}Ni and ^{63}Ni can be evidenced. The first radioisotope comes into being from ^{60}Ni (26.2 %) in the photonuclear reaction and the second from ^{62}Ni (3.6 %) in the neutron capture reaction. The radioisotope ^{63}Ni can also originate from the photonuclear reaction with ^{64}Ni. The neutron capture reaction with ^{64}Ni can give radioisotope ^{65}Ni but it is not considered because of a small abundance of ^{64}Ni (less than 1 %). Decay schemes of ^{59}Ni and ^{63}Ni nuclei are presented in Figure 4.9.

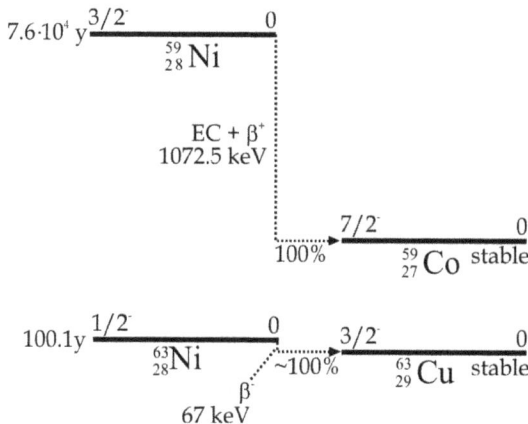

Fig. 4.9. Decay schemes of the ^{59}Ni and ^{63}Ni nuclei.

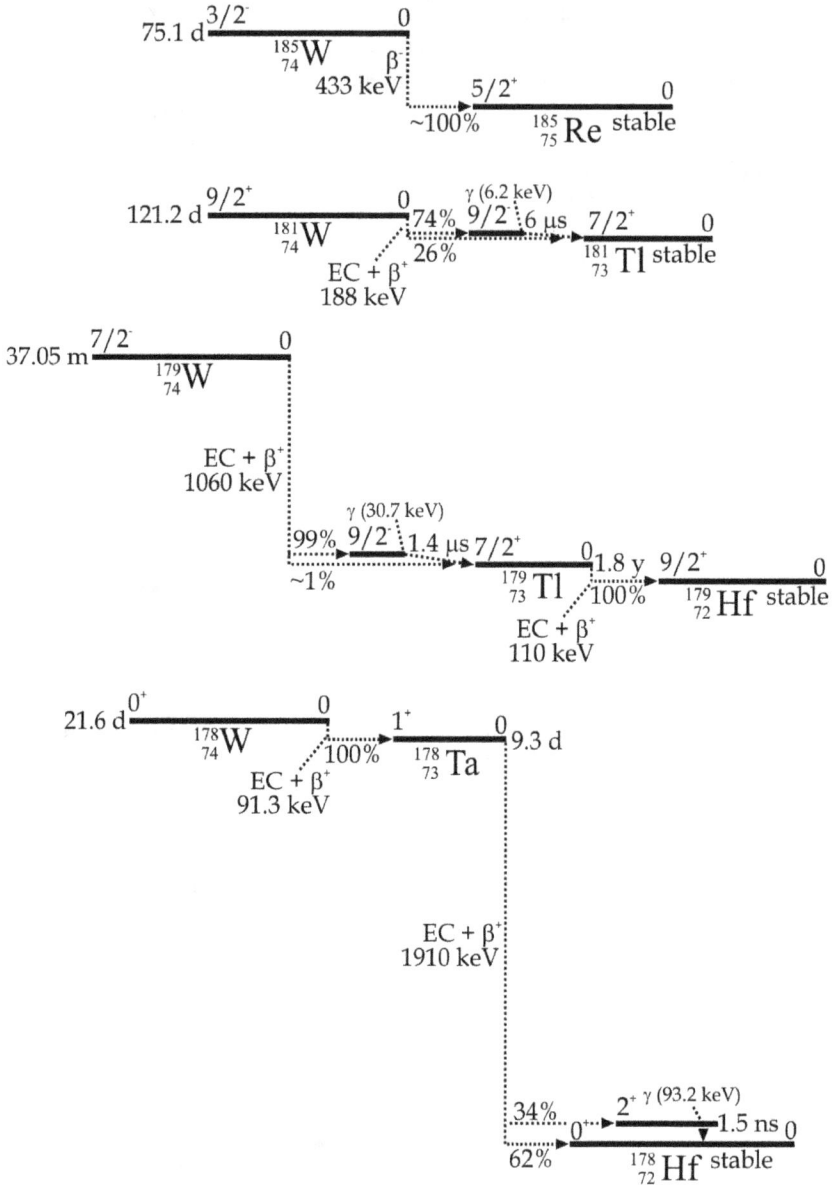

Fig. 4.10. Decay schemes of the 178W, 179W, 181W and 185W nuclei originating from photonuclear or electronuclear reactions. The gammas from the decays of these tungsten radioisotopes are difficult to identify by the gamma spectroscopy because of relatively low energy. The peaks from these gammas are not visible in the spectrum measured under the accelerator head contrary to the gammas from the decay of 187W (see Figure 4.3). Low-energy peaks from decays of nuclei can be concealed among peaks from scattered radiation.

As it was mentioned above, the collimator system of the contemporary medical linacs is usually made of tungsten. Thus, constructions of the linac heads are characterized by a great amount of tungsten. There are five natural stable isotopes of tungsten: ^{180}W (0.1 %), ^{182}W (26.3 %), ^{183}W (14.3 %), ^{184}W (30.7 %) and ^{186}W (28.6 %). Three of them (i.e. ^{180}W, ^{182}W and ^{186}W) can be changed into radioisotopes in photonuclear and electronuclear reactions. Two of them (i.e. ^{183}W and ^{184}W) change into stable nuclei in the mentioned reactions. The radioisotopes: ^{179}W, ^{181}W and ^{185}W originate from the nuclei of ^{180}W, ^{182}W and ^{186}W, respectively, after separation of one neutron. The radionuclide ^{178}W comes into existence after separation of two neutrons from the nucleus of ^{180}W. The energy threshold for the reaction – ^{180}W(γ,2n)^{178}W is about 15 MeV. The total photoneutron cross section for natural tungsten is 2854 mb [11]. Decay schemes of the tungsten radioisotopes originating from photonuclear or electronuclear reactions are presented in Figure 4.10.

5. Radioisotopes originating from simple capture reactions

The neutrons from the photonuclear and electronuclear reactions induced by high-energy electron and X-ray therapeutic beams can interact with nuclei of atoms of matters inside the radiotherapy facility. The neutrons interact mainly in simple capture reactions. The (n,γ) reaction cross sections are particularly high in the range of thermal and resonance energies (see paragraph 2.3). Inelastic scattering of neutrons is less possible. Radioisotopes decaying by emission of β^- radiation mainly come into existence as a result of the simple capture reactions whereas radioisotopes originating from photonuclear and electronuclear reactions disintegrate by electron capture or β^+ decay. It is connected by the fact that simple capture reactions cause the increase of a number of neutrons in a nucleus. A nucleus becomes neutron-rich and it attains more stable configuration by a change of a neutron into a proton. Analogically, photonuclear and electronuclear reactions cause the decrease of a number of neutrons. A nucleus becomes proton-rich and it attains "curve of stability" by a change of a proton into a neutron.

The simple capture reactions can occur in the whole area of the treatment room, because the field of thermal and resonance neutrons is approximately uniform around the working linac [10]. Thus, radioisotopes can originate in all objects inside the treatment room, regardless of their distance from the therapeutic beam.

The radioisotopes: ^{181}W and ^{185}W mentioned in the previous paragraph can also originate from simple capture reactions even outside the therapeutic beam. The radioisotope ^{187}W comes into existence only by the capture of neutron:

$$_0^1 n + {}_{74}^{186}W(28.6\%) \rightarrow \gamma + {}_{74}^{187}W.$$

This reaction is characterized by the neutron thermal cross section of 37.9 b [12] greatest of all natural tungsten isotopes and by the relatively large resonance activation integral of 355 barns [14]. Decay of the originated radioisotope - ^{187}W is connected with emission of gammas (Figure 5.1). Thus, this radioisotope can be identified by the gamma spectroscopy (see Figure 4.3). The maximum range of electrons from the β^- decay of ^{187}W is $R_{m,air}$ = 429 cm in air and $R_{m,bt}$ = 0.6 cm in a biological tissue[4].

---

[4] The maximum ranges of electrons from the β^- decay in air (denoted as $R_{m,air}$) and in a biological tissue (denoted as $R_{m,bt}$) were calculated on the base of equivalent values of exposure constant given by Gostkowska in [15]. The air density of 0.0013 g/cm^3 and the biological tissue density of 1 g/cm^3 were taken for the calculations of the electron ranges.

$23.72\ h\ \dfrac{3/2^-}{^{187}_{74}W}\qquad 0$

β^-
1311.2 keV

54.9% $5/2^-$
4.7% $3/2^+$
$9/2^-$
$5/2^+$
29.8%

γ (685.77 keV)
γ (479.53 keV)   27.3%
21.8%
γ (618.36 keV)
6.3%

6.1 ps
9.7 ps

555.3 ns
0   $4.35 \cdot 10^{10}$ y

$^{187}_{75}Re$

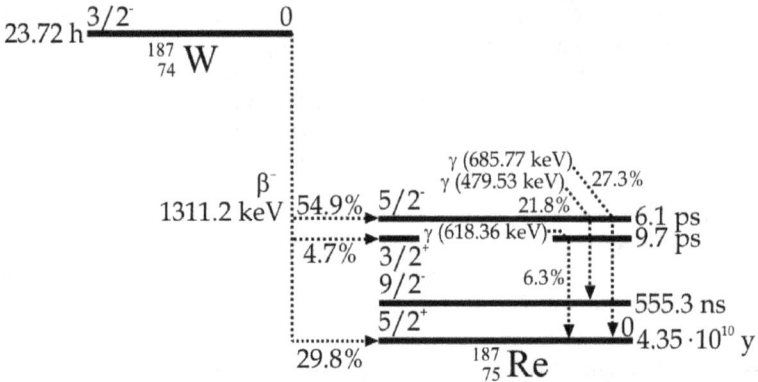

Fig. 5.1. Decay scheme of the ^{187}W nucleus. The gammas with energies of 479.53 keV, 618.36 keV and of 685.77 keV, from the deexcitation of the state ^{203}Re* give the peaks visible in the spectrum measured under the accelerator head (see Figure 4.3). This scheme includes the most photon branches of the decay.

In the spectrum measured under the medical linac head (Figure 4.3) as well as in the spectrum of gammas emitted by a medical linac accessory (Figure 4.6) the peaks from the decay of ^{56}Mn (Figure 5.2) are visible. Manganese - ^{55}Mn in the stable state (the only natural isotope of manganese), like Nickel, is one of the fundamental components of stainless steel used in a construction of medical accelerators. This isotope is easily activated by neutrons because it has a large cross section for the neutron capture i.e. the thermal neutron capture cross section is 13.2 b, and the resonance activation integral is 15.7 b (the high resonance peak occurs at 337 eV) [12]. The simple capture reaction giving the ^{56}Mn radioisotope can be expressed as follow:

$$^{1}_{0}n + ^{55}_{25}Mn(100\%) \rightarrow \gamma + ^{56}_{25}Mn.$$

The ^{56}Mn radioisotope disintegrates by emission of β^- radiation. The maximum range of emitted electrons is $R_{m,air}$ = 1256 cm in air and $R_{m,bt}$ = 1.6 cm in a biological tissue.
The peak at 1778.96 keV, visible in the spectra presented in Figures 4.3 and 4.6, comes from the decay of the ^{28}Al radioisotope originating from the following reaction:

$$^{1}_{0}n + ^{27}_{13}Al(100\%) \rightarrow \gamma + ^{28}_{13}Al.$$

The cross section for this reaction is 0.231 b at the thermal energy and it decreases for higher energy and one greater resonance of 4.685 b occurs at 5904 keV [12]. The only natural isotope of aluminum – ^{27}Al is a material applied in the radiation absorbers located inside the primary collimator under the target and also in the construction of other components of a linac head, for example in dose chambers and in wedge trays. The decay scheme of the ^{28}Al radioisotope is presented in Figure 5.3. Decay of the ^{28}Al nucleus is connected not only with emission of gammas but also with emission of β^- radiation. The maximum range of electron from this decay is close to that from the decay of the ^{56}Mn nuclei i.e. $R_{m,air}$ = 1265 cm, $R_{m,bt}$ = 1.6 cm.

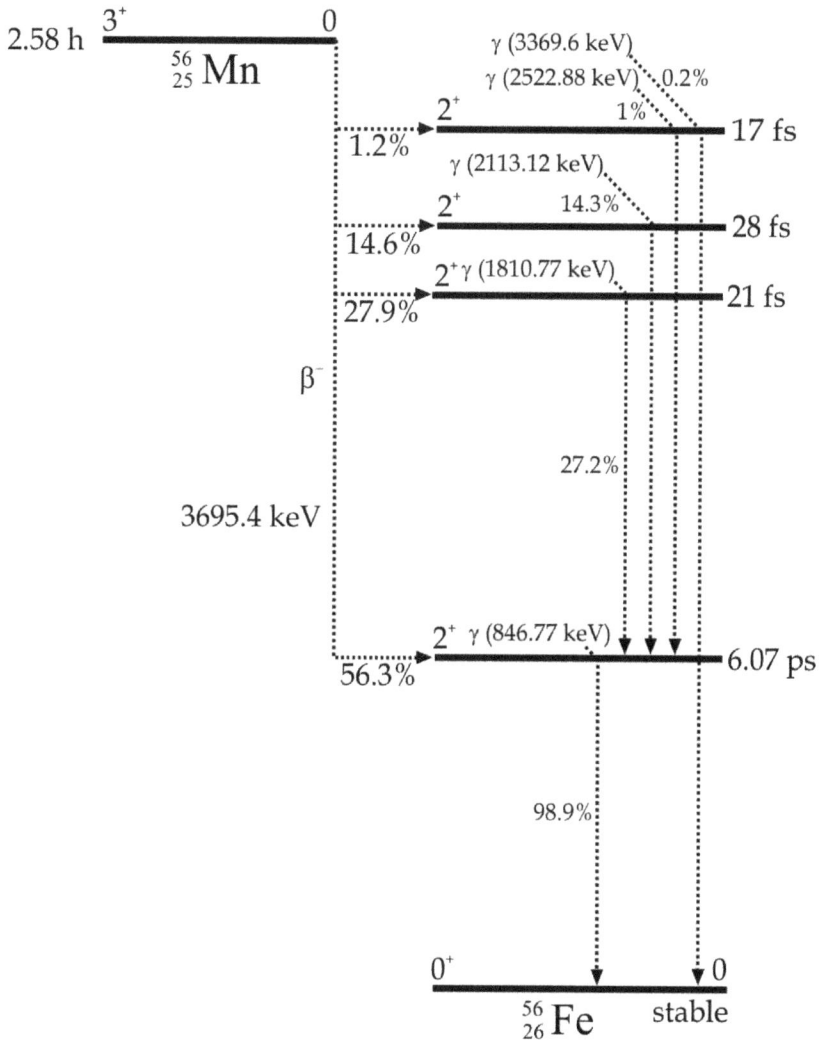

Fig. 5.2. Decay scheme of the ^{56}Mn nucleus.

2.24 m 3^+ 0

$^{28}_{13}\mathrm{Al}$

β^-

4642.15 keV

2^+ γ (1778.96 keV)   475 fs

100%

100%

0^+ 0

$^{28}_{14}\mathrm{Si}$ stable

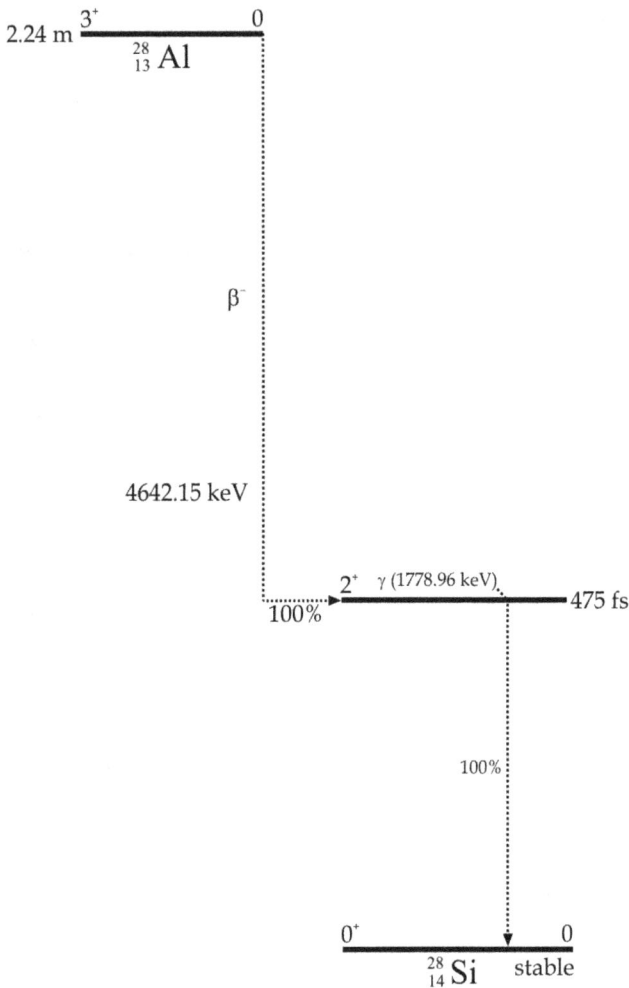

Fig. 5.3. Decay scheme of the ^{28}Al nucleus.

One of the strongest peaks in the spectrum of gammas emitted by the accessory of a medical linac (Figure 4.6) is that at 160.33 keV. It comes from the disintegration of the ^{123}Sn nuclei originated from the following reaction:

$$^{1}_{0}\mathrm{n} + {}^{122}_{50}\mathrm{Sn}(4.7\%) \rightarrow \gamma + {}^{123}_{50}\mathrm{Sn}.$$

The decay scheme of the ^{123}Sn nucleus is presented in Figure 5.4. Electrons from the β^- decay of this radioisotope have $R_{m,air} = 400$ cm and $R_{m,bt} = 0.5$ cm.

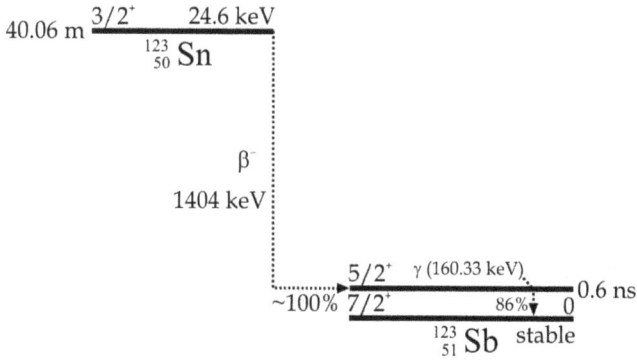

Fig. 5.4. Decay scheme of the [123]Sn nucleus.

The neutrons from (γ,n) reactions taking place in the components of the medical accelerator head have a broad energy spectrum inside the radiotherapy facility (see [17, 18]). Fast neutrons undergo elastic and inelastic scatterings. The neutron inelastic interaction provides nucleons of a nucleus with energy. The radioisotopes and nuclei in metastable states can be induced as a result of such interactions. In the spectrum of gammas emitted by the linac accessory the [204m]Pb metastable state was identified. It comes from inelastic interactions between neutrons and nuclei of [204]Pb:

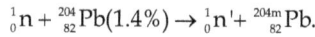

$$_0^1n + \,_{82}^{204}Pb(1.4\%) \rightarrow \,_0^1n' + \,_{82}^{204m}Pb.$$

The scheme of disintegration of the metastable [204m]Pb state is presented in Figure 5.5.

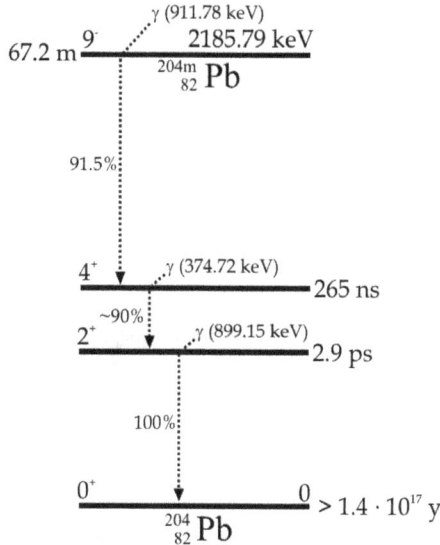

Fig. 5.5. Decay scheme of the [124m]Pb state. The peaks at 374.72 keV, 899.15 keV and of 911.78 keV are visible in the spectrum of gammas emitted by the accessory of a medical linac (Figure 4.6).

In the previous paragraph the ^{209}Pb radioisotope was mentioned to be a result of the simple capture reaction. The ^{209}Pb nuclei disintegrate by β- decay (Figure 4.8). The maximum range of electrons from this β- decay is $R_{m,air}$ = 173 cm in air and $R_{m,bt}$ = 0.2 cm in a biological tissue. Contribution of this radioisotope to a β- radiation level around the linac head can be significant because it originates from ^{208}Pb - the isotope with the abundance greatest of all natural lead isotopes. Moreover, lead is often used for construction of big components of the accelerator head like radiation shields, and also for individual shields[5] for patients.

Gold is often used in the construction of medical linacs, for example, connectors of wires are often made of Au (see also considerations for Figure 4.4). Gold has a relatively high cross section of the slowed down neutron simple capture reaction - ^{197}Au(n,γ)^{198}Au. Therefore the radioisotope ^{198}Au can easy come into existence in whole accelerator room. The energy spectrum of the gamma radiation from disintegration of the ^{198}Au nuclei, measured after a neutron activation of gold is presented in Figure 5.6. Decay scheme of the ^{198}Au nucleus is shown in Figure 5.7.

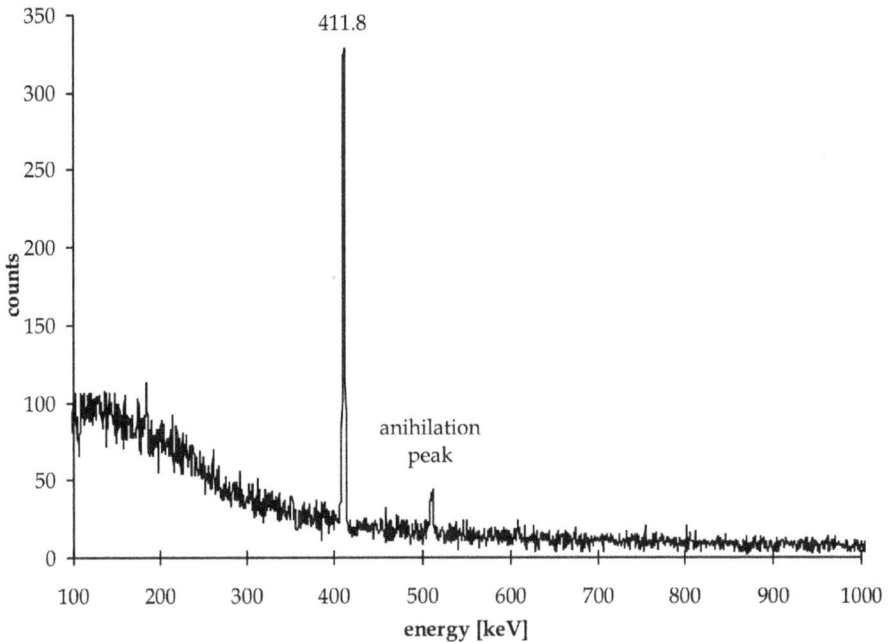

Fig. 5.6. The spectrum of gammas emitted by the ^{198}Au radioisotope. The spectrum was measured after the end of the neutron activation of gold.

---

[5] The individual shields for patients are often made of wood's melt which includes over 60 % of lead.

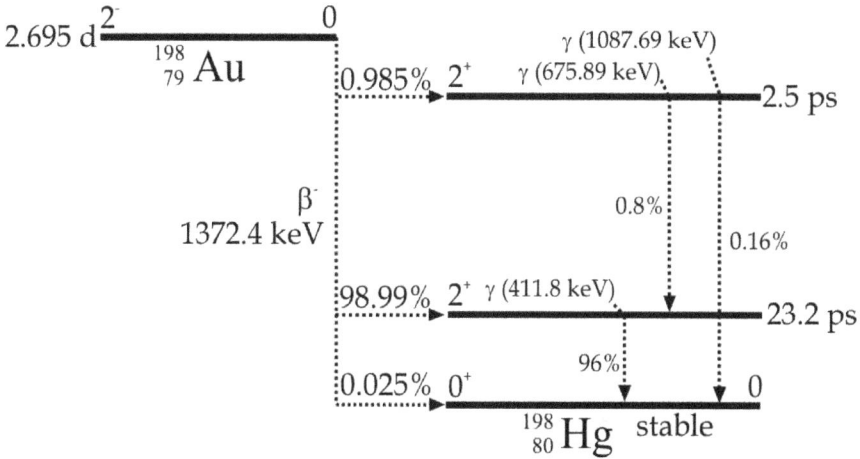

Fig. 5.7. Decay scheme of the ^{198}Au nucleus.

The simple capture reactions can be identified by binding energy of a neutron. Photons from these reactions are emitted by a nucleus immediately after capture of a neutron. However, in this case there appears the experimental difficulty connected with the fact that the gamma energy spectrum has to be measured during the occurrence of simple capture reactions. Thus, measuring apparatus has to be in the neutron field during such measurement and the neutrons interact with a detector and electronics of the detection system. The radioisotopes can be induced in materials of the detection system as a result of theses interactions. It can give an additional significant contribution to the measured energy spectrum and it can even disturb a measurement in many cases. The exemplary energy spectrum of gammas from simple capture reactions: ^{206}Pb(n,γ)^{207}Pb and ^{207}Pb(n,γ)^{208}Pb, induced mainly by thermal and resonance neutrons[6] is presented in Figure 5.8. To avoid the above described problems, the spectrum was obtained with the use of the Monte Carlo calculations.

---

[6] The reaction ^{206}Pb(n, γ)^{207}Pb has the high resonance of 309 b at 3.36 keV whereas the reaction ^{207}Pb(n, γ)^{208}Pb has the high resonance of 267 b at 3.06 keV [14].

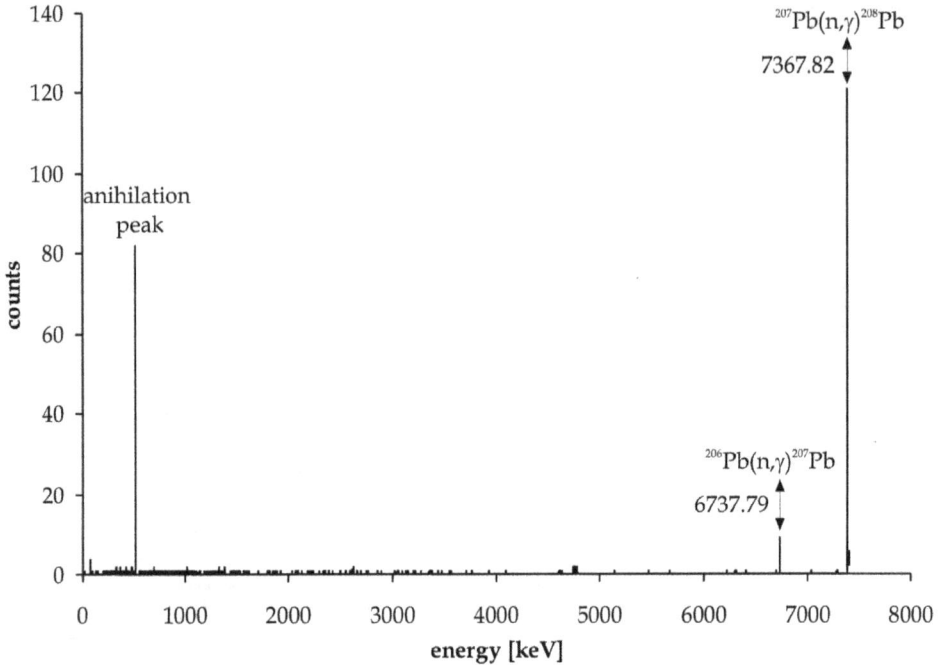

Fig. 5.8. The spectrum of gammas (6737.79 keV and 7367.82 keV) originating from the simple captures reactions: $^{206}Pb(n,\gamma)^{207}Pb$ and $^{207}Pb(n,\gamma)^{208}Pb$. The spectrum was calculated by the Monte Carlo method, using computer simulations based on the GEANT4 code.

6. Conclusions

The radioactivity induced inside the radiotherapy facility is a consequence of the photonuclear and electronuclear reactions as well as the neutron reactions. The following radioisotopes originate as a result of the (γ,n), $(\gamma,2n)$ and $(e,e'n)$ reactions: ^{13}N, ^{15}O, ^{57}Co, ^{57}Ni, ^{59}Ni, ^{178}W, ^{179}W, ^{181}W, ^{185}W, ^{196}Au, ^{202}Pb, ^{203}Pb and ^{205}Pb. The (n,γ) and (n,n') reactions give the radioisotopes of ^{28}Al, ^{56}Mn, ^{63}Ni, ^{123}Sn, ^{187}W, ^{198}Au, ^{205}Pb, ^{209}Pb and the metastable state of lead - ^{204m}Pb. Most of the activated materials are the building materials of the head of the commercial accelerators used in the teleradiotherapy treatment.

The radioactivity inside the radiotherapy facility can be accumulated because in the majority the half-lives of the induced radioisotopes are between tens of minutes to several days. Moreover, the level of this induced radiation is connected with the kind of the therapeutic beams and the nominal potential of the beams [see 19], because the cross sections of the (γ,n) and $(e,e'n)$ reactions depend on energy of gammas and electrons.

The good solution of the problem of the air activation is an uninterrupted ventilation of the radiotherapy facility, which makes it possible to reduce amount of the radioisotopes of ^{13}N and ^{15}O. The longer lasting emission of the high-energy X-ray beams should be avoided if it possible.

The knowledge of the neutron reactions taking place inside the radiotherapy facility is of great worth for designers of accelerators because elements with the large cross section of the neutron capture reactions can be eliminated from the accelerator construction. The information on the radioisotopes induced inside the radiotherapy treatment can be used in the radiological protection of the staff operating the medical accelerators.

7. References

[1] International Commission on Radiation Units and Measurements, Gamma-Ray Spectrometry in the Environment, „ICRU Report 53", Maryland, December 1994.

[2] International Electrotechnical Commission, In-situ photon spectrometry using a germanium detector for measuring discrete radionuclides in the environment, „IEC 209 FDIS", May 1992.

[3] http://geant4.web.cern.ch

[4] Konefał, A., 2006, Monte Carlo simulations with the use of the GEANT4 code. „Postępy Fizyki", 57(6): 242–251 (in Polish).

[5] Carrier J.-F., Archambault L., Beaulieu L., 2004, Validation of GEANT4, an object-oriented Monte Carlo Toolkit, for simulation in medical physics, „Med. Phys. ", 31, 484-492.

[6] Konefał, A., Szaflik P., Zipper, W., 2010, Influence of the energy spectrum and the spatial spread of the proton beams used in the eye tumor treatment on the depth-dose characteristics. „Nukleonika", 55(3): 313–316.

[7] Evdokimoff, V., Willins, J., Richter, H., 2002, Induced radioactive potential for a medical accelerator., „Health Physics", Operational Radiation Safety, 83, Sup.5:S68–S70.

[8] Price, W.J. Nuclear Radiation Detection. (USA: McGraw-Hill Book Company) (1964).

[9] Beckurc, K. and Wirtc, K. (eds), 1968, Niejtronnaja fizika. „Moscow: Atomizdat".

[10] Konefał, A., Polaczek-Grelik, K., Orlef, A., Maniakowski, Z., Zipper, W., 2006, Background neutron radiation in the vicinity of Varian Clinac-2300 medical accelerator working in the 20 MV mode., „Polish Journal of Environmental Studies", 15(4A): 177-180.

[11] Berman, B.L., Dietrich, S.S., (eds), 1998, Atlas of photonuclear cross sections obtained with monoenergetic photons., „Atom Data Nucl. Data Tables", 38: 199-338.

[12] National Nuclear Data Center, Brookhaven National Laboratory, web retrieval system, http://www. nndc.bnl.gov/sigma/search.jsp

[13] Firestone, R.B., (eds), 1996, „Table of Isotopes", 8th ed. version 1.0. Lawrence Berkeley National Laboratory, University of California, USA.

[14] Macklin, R.L., Pomerance, H.S., 1955, Resonance capture integrals., „Proc. 1st Intern. Conf. Peaceful Uses Atomic Energy", Geneva, P/833.

[15] Gostkowska B., 2005, Ochrona radiologiczna. Wielkości, jednostki i obliczenia., „Centralne Laboratorium Ochrony Radiologicznej", Warsaw (in Polish).

[16] Price, W.J., 1964, Nuclear Radiation Detection., „USA: McGraw-Hill Book Company".

[17] Facure, A., Falcao, R.C., Silva, A.X., Crispim, V.R., Vitorelli, J.C., 2005, A study of neutron spectra from medical linear accelerators., „Applied Radiation and Isotopes", 62: 69–72.

[18] Esposito,A., Bedogni, R., Lembo, L., Morelli M., 2008, Determination of the neutron
      spectra around an 18 MV medical LINAC with a passive Bonner sphere
      spectrometer based on gold foils and TLD pairs., „Radiation Measurements", 43:
      1038 – 1043.

[19] Konefał, A., Orlef, A., Dybek, M., Maniakowski, Z., Polaczek-Grelik, K., Zipper, W.
      2008, Correlation between radioactivity induced inside the treatment room and the
      undesirable thermal/resonance neutron radiation produced by linac., „Physica
      Medica", 24: 212 – 218.

9

Boron Studies in Interdisciplinary Fields Employing Nuclear Track Detectors (NTDs)

László Sajo-Bohus[1], Eduardo D. Greaves[1] and József K. Pálfalvi[2]
[1]Universidad Simón Bolívar, Valle de Sartenejas, Caracas,
[2]HAS KFKI Atomic Energy Research Institute, Budapest,
[1]Venezuela
[2]Hungary

1. Introduction

The aims of this study is the dissemination of the major experimental results in applied fields involving the boron-10 neutron capture with emphasis on advances of interdisciplinary fields employing passive nuclear track techniques. The stable isotope boron-10 has found many applications in various fields of physics and other branches of science. In this study details on applications and methodology are given.

2. Elemental Boron nuclear properties

Boron is a natural element having two stable isotopes. ^{11}B (80.1%) and ^{10}B (19.9%). Only the latter has a high thermal neutron capture cross section (3832 b). Due to its nuclear characteristics e.g. being a non radioactive element and readily available, the isotope boron-10 is often employed in application where the (n, α) reaction is of advantage and where other analytical techniques could not be employed satisfactorily. The probability for absorption of a neutron by this stable isotope via the $^{10}B(n,\alpha)^7Li$ capture reaction (^{10}BNC-reaction), is given by the absorption cross section. Its value is a function of the impinging neutron energy, (www.nndc.bnl.gov). The energetic fragments emitted in the ^{10}BNC-reaction produce a high value of "Linear Energy Transfer" (LET) or dE/dx, that is, a measure of the number of ionisations per unit distance as they traverse the absorbing material. Their combined path lengths are of short distance making them quite suitable where localized damage is of advantage. Industrial processes have been devised to modify the natural boron isotopic composition in order to obtain high values for ^{10}B concentration. Among the known boron compounds, several hundred are employed in today's applications and a growing list exists that enter the literature as related research progresses, Wikipedia (2011). Often natural boron is contained in biological material, in cells and it would be convenient to replace it with isotope enriched element. The possibility to introduce the ^{10}B isotope in any cell culture opened a new technique to treat e.g. cancer by Boron Neutron Capture Therapy (or ^{10}BNC Therapy). Chemical compounds containing^{10}B isotope are employed in purpose made specific applications such as neutron radiation shielding, nuclear reactors reactivity control, emergency shut-down systems, as occurred in the Fukushima No-1 nuclear power plant. Boron fibbers such as BN-nanotube material find the most convenient application as structural material for spacecraft and radiation shielding. The [10]BNC-reaction to take place requires a sample containing, even at ppb level, [10]B, a source set for irradiation with thermal or lower neutron energy (0.025eV or less) and a reaction fragment detecting device. The reaction phenomenon is related to a neutron interacting with boron nucleus, followed by breakup in two fragment of the [10]B+n compound nucleus (that survives a short time in the order of picoseconds). The two fragment nuclei depart acquiring kinetic energy due to a strong Coulomb field moving in opposite direction under the momentum conservation law, synthesized by the following process:

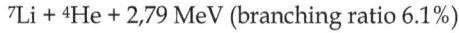

$$^7Li + {}^4He + 2,79 \text{ MeV (branching ratio 6.1\%)}$$

$$\nearrow$$

$$^{10}B + n \dashrightarrow [\,{}^{11}B] \hspace{5cm} (1)$$

$$\searrow$$

$$^7Li + {}^4He + \gamma \ (0,48 \text{ MeV}) + 2,31 \text{ MeV (branching ratio 93.9 \%)}$$

The reaction occurs with different branching ratio: the first has a relatively low frequency occurrence (6.1%) but has the advantage that the reaction is photon less and therefore the induced damage leads to a higher "Linear Energy Transfer" (LET) or dE/dx. The other, with higher occurrence is accompanied by a 0,48 MeV photon. If the alpha particle ($^4He^+$) leave the sample surface, with sufficient kinetic energy, then it can be detected e.g. by nuclear track techniques. The alpha particle fingerprint given by a suitable detecting material, provides information on the boron presence and it is recognized as a powerful analytical method for boron studies.

3. Experimental set up

In general terms the experimental system consists of a neutron source with a collimator to obtain the required beam characteristics, radiation shielding, sample and detecting device. Often it is necessary to take into account diffusion effects due to neutron scattering taking place due to the geometry and location of moderating materials, either in a material external to the sample under scrutiny, or in the sample itself. Charged particles emitted by the [10]BNC-reaction have path lengths, in most absorbing material either a cell or a nuclear track detector (NTD), around 12 μm, being sufficient to induce large molecular damage.

3.1 Thermal and cold neutron beam production

Neutron irradiation is obtained by different systems in which thermal neutron flux is a prevailing component. The selection of the experimental arrangement depends mainly on the required beam intensity (e.g. for [10]BNC-therapy > 5×10^8 ncm^{-2}s^{-1}) and its spectral characteristics (high % for thermal-epithermal fraction). Some aspects for selected application as well as sample size, geometry, chemical compound and density should be taken into account. If the application requires a high intensity beam, a major facility either a neutron reactor or an accelerator driven neutron source should be considered. However, for most applications it is sufficient to employ small, even portable, neutron generators (NG) or radioisotope neutron sources (RINS) offered commercially (QSA Global Inc., 40 North Avenue, Burlington, MA 01803). Some rather complex NG has been devised e.g. highly compact fusion neutron source with a small sub-critical fission assembly to multiply the neutron source intensity. Leung et al. (2002) have patented a high current density NG employing coaxial RF-driven plasma ion source, particularly advantageous in medical applications since it provides a neutron flux of 1.2 x 10^{12} n/cm^2 s (by 2.4 MeV D-D reaction). Persaud et al., (2011) developed recently a new concept for ionizing deuterium atoms to obtain neutrons from the D-D reaction. Ingenious NG device developed by Naranjo, et al., (2005) envisioned a tabletop device suitable for many industrial applications. Other physical solutions have been adopted to obtain the thermal neutron beam and some expensive technical approaches are currently employed. For instance, to extend the RINS operational efficiency, liquid nitrogen is used to lower the thermal neutron energy. Another method to produce a pure thermal beam is to employ a neutron guide as available at the Budapest Neutron Centre (Budapest, Hungary, http://www.bnc.hu/). Primary neutron field, either from fission chain, nuclear reaction involving spallation, beam fragmentation or RINS, have an energy distribution from which suitable neutron beams in thermal region may be derived. This process requires some technical arrangement involving hydrogen or deuterium rich materials (heavy water, PVC and others) that reduce the energy spectrum to the required energy window (close or lower to the thermal neutron group). On the other hand due to restrictions imposed by scattering and other physics laws, it is difficult to obtain a pure thermal beam, even in the case of an accelerator produced monoenergetic beams. The most employed technique to select neutrons of a given energy group, is time of flight. Recently Stevenato et al. (2010), for non-destructive identification, developed a time-tagged ^{252}Cf source. Based on these techniques, the experimenter can take advantage to obtain the maximum ^{10}BNC-reaction yield.

4. Basic aspects of nuclear track techniques

4.1 Linear energy transfer and latent nuclear track formation

Solid state nuclear track detectors (SSNTD) or more briefly NTDs are capable to detect charged particles within certain circumstances. They can be natural crystals or special plastics. Their theory was developed more than 40 years ago, the basic fundamentals can be found in Somogyi (1973) and in more details in Durrani et al. (1987). Even more details for detecting alpha particles, which is important from BNCT point of view, can be found in Nikezic (2003). Therefore, here we touch some aspects of interest, only. Popularly saying, an ionizing particle produces a narrow damaged zone in the plastic, 10-100 nm in diameter, which can be enlarged and visualized by a chemical treatment, so that the particle movement in the detector material, let us say the footprint of the particle or its track can be followed under optical microscope. Depending on the chemical treatment (called etching) and observation method there are basically two requirements: the range and energy deposition of the particle should be adequate. For instance, if we apply a 6 N NaOH solution at 70 °C for 2 hours and an optical microscope, the range of the alpha particle should be at least 2.8 μm, which requires minimum 400 keV kinetic energy. In this case, the energy deposition, the linear energy transfer or LET, is quite enough, ~ 200 keV/μm, to provide enough damage to be visualized. The essence of the visualization is that the chemical removes the damaged material from the zone more quickly than the undamaged, bulk material, from the surfaces of the plastic. Two quantities are responsible for this process: the track etch velocity, V_T and the bulk etch velocity, V_B. Evidently, V_T should be > than V_B. The length and diameter of the track formed characterize the incident particle: its type and energy. These are what we need to know for studying the physical aspects of the boron study.

If we approach the alpha particle detection mechanism by the attitude of a physicist some more details should be given, as follow below.

Ionizing radiation may induce atomic or molecular modifications that are observed as structural defects in the absorber such as polyallyl-diglicol-carbonate, PADC or similar materials. In particular, energetic alpha particles through Coulomb interaction induce ionization and consequently provide high charge density to displace Atoms. Two kinds of processes are involved:

i.   The impinging alpha particles, having energy above a value of ~100 keV, transfer their kinetic energy to electrons encountered on their path; this process is referred as Electronic Energy Loss, (Se),

ii.   The energy is transferred directly to the atomic nucleus; in this case we have Nuclear Energy Loss (Sn). This kind of process is observed for energies in the energy window below 100 keV.

Both of them independently contribute to the total rate of energy loss. Representing the detector density with ρ (kg/m³), dE/dx (keV/μm) for the incoming particle energy deposition per unit length, the following equations applies:

$$S_e = \frac{1}{\rho}\left(\frac{d}{dx}E\right)_x \tag{2}$$

$$S_n = \frac{1}{\rho}\left(\frac{d}{dx}E\right)_n$$

A variation of the energy-loss taking place during charged particles absorption exists and follows a curve in which S_e and S_n have a maximum at a given incident energy. These curves can be obtained using values estimated by SRIM-2010 code (Ziegler et al, 1985); for alpha particles absorption in PADC, a typical result is given in Fig. 1,

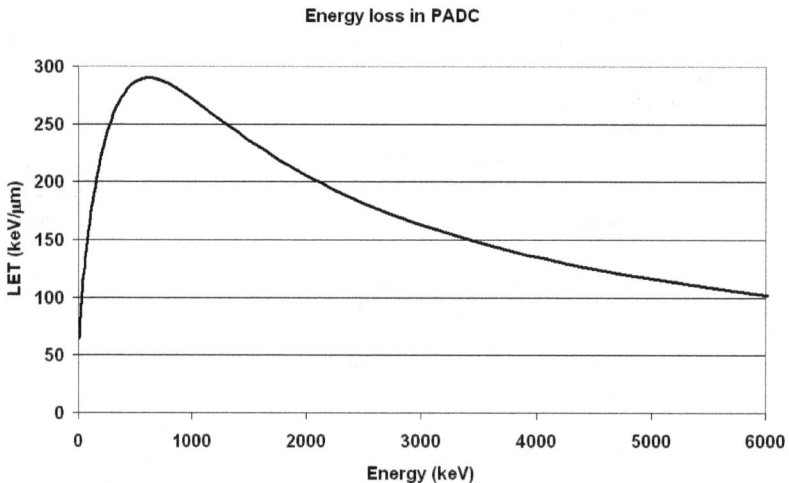

Fig. 1. Specific energy-loss (keV/μm) vs. alpha particle energy taking place during alpha particles absorption. The curve is the sum of S_e and S_n. Similar behaviour for any other absorbing material is expected.

The track etch rate along the particle path shows a similar feature according to the Bragg low as presented in Fig. 2.

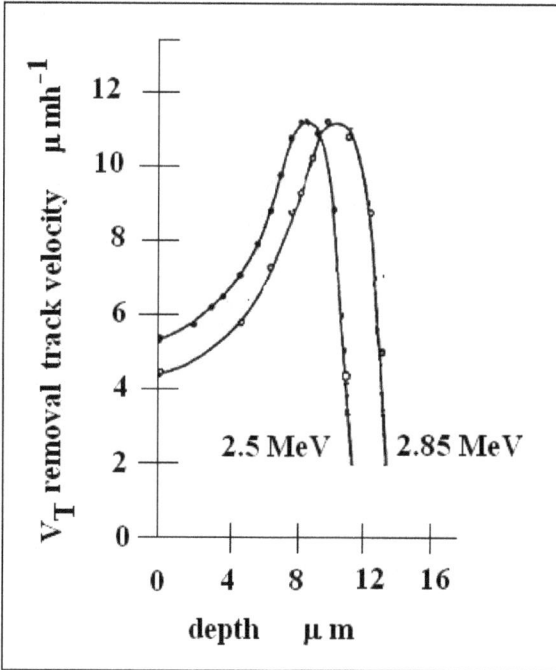

Fig. 2. Bragg ionizing curve for alpha particles of different energies in PADC as observed in the track etch rate, V_T, vs depth in the detector (adapted from Dörschel at al., 2002).

We may observe from Figure 2 that the charged particle entering the detector surface initially transfers energy with a low rate then with an increasing rate, after reaching a maximum value near the end of its path, it drops abruptly to zero. Along the alpha particle path the induced charge density is not uniform due to the Coulomb explosion.   This produces a non linear damage in the detector and it has to be taken into account if a precise etching model has to be developed.

4.2 Passive NDT response to charged particle and chemical etching

Due to the dielectric property of PADC detector, molecular polarization and ion diffusion takes place under a high gradient charge concentration. Particles move along an imaginary radius ionizing more molecules and even destroying layers cross-links. As the electrons has a major mobility drift immediately, the heavy positive ions moves at lower velocity and under the action of Coulomb force repulsion intervene and may displace and further ionize atoms favouring interstitial positions. During the short time of the process, also a relaxation effect (ion-hole recombination) takes place. For instance, 6 MeV alpha particle produce 150,000 ion pairs so that the PADC damaged zone that may reach atoms farther that 35 - 40 μm. The ion production is estimated to be around 3–4 ions pairs/nm, that may explain the ionization-density of molecules on its pathway.

The NTD surface absorbs during the process also oxygen from air interfering with relaxation process and therefore has an observable effect on the latent track formation. Several authors suggest that oxygen has an important role. It has already been observed that also the CO_2 gas influences has to be taken into account in making the latent track model. During the energy transfer, temperature gradient (abrupt cooling) increasing the entropy favouring disorder and possibly atomic cluster (amorphous state) formation.

The damage induced on the detector surface can be visualized also by Atomic Force Microscope and the surface opening of the latent track can be measured. (see Figure 3)

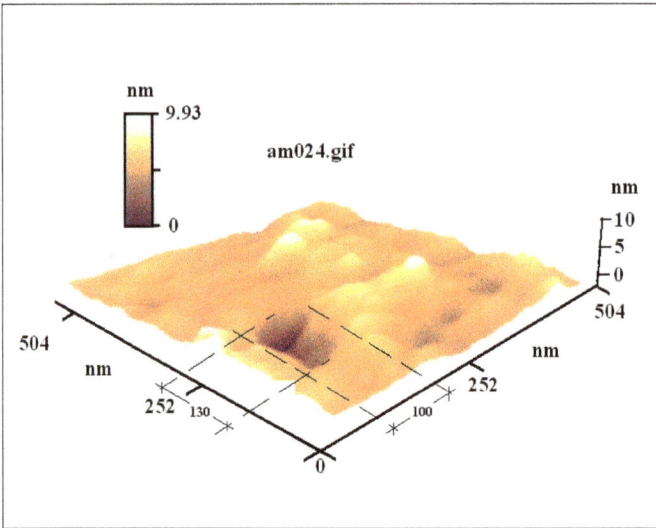

Fig. 3. AFM obtained image of a alpha induced latent track on the detector surface. Dimensions are in nm; alpha source is [241]Am, (picture taken by the authors)

Consequently having a proper detector material and analytical equipment with a convenient resolving power (e.g. AFM), the alpha particle fingerprint can be analyzed to obtain physical information. Since this method to analyze the latent track directly is time consuming, consequently unpractical, the latent track opening should be enlarged by chemical process which is the most often employed method.

Among a large group of NTD materials useful to detect charged particles we mention only one of the most sensitive material: called CR-39 ™ discovered by Cartwright et al., (1978), its chemical composition is polyallyl-diglycol-carbonate, PADC. Later, based on this PADC monomer many other products have been developed, more or less with the same properties. NTDs have an important property: that the damaged and the undamaged volume in the detecting materials, under chemical etching behave differently. This is an important property since it provides the possibility to recognize the region where alpha particle impact occurred. To determine the deposited energy it is essential to determine at first the bulk etch rate, V_B. To determine the V_B several experimental techniques have been suggested and a comprehensive report is given by Nickezic et al. (2004). For instance, assuming that the track etch rate is not changing quickly along the path of the perpendicularly incident particle, the following equation for the track diameter, D, is valid (Somogyi, 1973):

$$D = 2h\sqrt{\frac{V-1}{V+1}} \tag{3}$$

In which $V = V_T/V_B$ and h is the thickness of the removed layer during etching. Exposing the detector to very high LET particles as fission fragments we may apply the approximation V $\gg$ 1, and then the equation simplifies:

$$D \cong 2h = 2V_B t \tag{4}$$

Here t is the etching time and the V_B parameter can be determined.

Figure 4 shows examples of etched tracks incident on the detector perpendicularly and obliquely and the important, measurable track parameters as: major and minor axes (a, b) and the depth of the peak of the track (L). Also the projected track length is measurable (not marked). They can be used to determine the track etch rate, the incident angle and the real track length. From which, applying appropriate calibration the LET of the incident particle can be deduced (Pálfalvi, 2009).

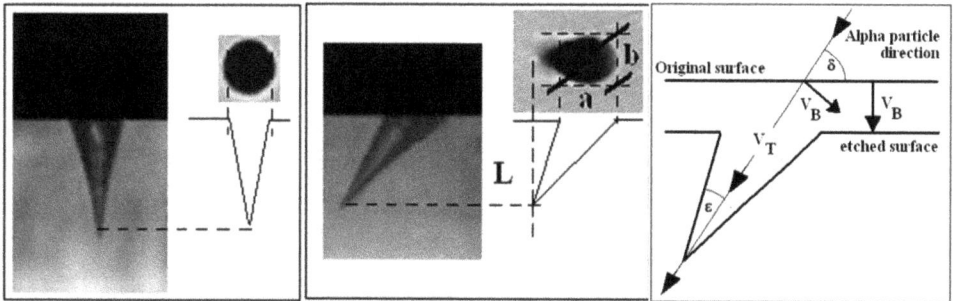

Fig. 4. Side and top views of etched tracks visualized by a light transmission microscope: a and b are the major and minor axes, respectively, L is track depth. V_T and V_B are the track and bulk etch rates, δ and ε are the incident and cone angles.

The size of the track opening (or pit size) depends on the detector material and etching conditions. For a real case when PADC detector (TASTRAK, TASL Co Ltd, Bristol, UK) is used for detecting alpha particles from the [10]BNC reaction, usually as control, a standardized, and collimated [210]Po alpha source is used, having an energy of 4.6 MeV (Szabó et al, 2002). If the etching is done in 6N NaOH at 70 °C, which results in a bulk etch velocity of V_B =1.34 ± 0.06 µm/h (Hajek et al. 2008), then the track parameters are as summarized in Table 1.

Etching time, h	6
Removed layer, µm	8.0
Track length, µm	18.0
Diameter, µm	10.53
Area, µm^2	87.04
V= (V_T/V_B)	2.55
Range in H$_2$O, µm	32.2
LET$_\infty$ in H$_2$O, keV/µm	96.9
Range in PADC, µm	25.8
LET$_\infty$ in PADC, keV/µm	121.5

Table 1. Reference data for control exposures by a standardized [210]Po alpha source.

While the averaged track etch rate, V_T, can be determined from Equ (5), for perpendicularly incident particles, when the incident angle, $\Theta = 0°$, in the case of oblique incidence the situation is not so simple. If the incidence angle measured is too high, then the track can be removed during the etching process, therefore, to establish the real number of incident alpha particles the critical detection angle, Θ_c, is introduced which is used then for correction purpose.

According to the "classical method" it is assumed that the V_T is constant along the particle path inside the detector material. If we accept this then the track and bulk etch rate ratio, $V=V_T/V_B$, can be calculated by the following formula (Durrani et al., 1987):

$$V = \frac{\sqrt{\left(1-B^2\right)^2 + 4A^2}}{1-B^2} , \tag{5}$$

where $A = \frac{a}{2h}$ and $B = \frac{b}{2h}$ (a is the measured major-, b is the minor axis, h is the etched off layer thickness, $h=V_B t$, t is the etching time). For the critical angle then the following formula is valid:

$$\cos^2 \Theta_c = \frac{V^2 - 1}{V^2} \tag{6}$$

In the case of isotropic radiation field, the differential alpha particle fluence at a given point x is defined by the next formula :

$$\Phi(\Omega, x) = \frac{d^2 N}{dA d\Omega} , \tag{7}$$

then considering the relation between the critical and the detection angles as shown in Figure 5, the fluence can be deduced (in detail see Benton 2004)

$$\Phi(\Omega, x) = \left(A2\pi \cos^2 \Theta_c\right)^{-1} N , \tag{8}$$

here A is the area of detector where the detected particle number is N.

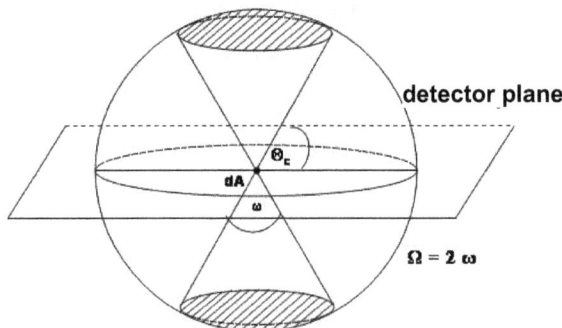

Fig. 5. Assuming isotropic radiation field solid angle (Ω) is determined by the critical angle (Θ_c). Those particles crossing the shaded spherical area and incident on detector area dA will be detected (Palfalvi, 2009).

We draw attention on the fact that PADC detector sensitivity may suffer also from environmental effects. In fact its recording properties change with storage time. It is observed that the registration property is influenced by the presence of oxygen or electron donor gases, even too low or high temperature is influencing it as shown in Figures 6 and 7, where the PADC detectors were exposed to alpha particles. In Figure 6, it is clearly seen from the increasing track area distribution that cooling enhances the registration ability. This is because of the cooling "freezes" the free radicals preventing their recombination. In opposite in Figure 7, the vacuum, the lack of oxygen, decreases the track area, because the oxygen cannot block the recombination of radicals and the damaged zone becomes smaller.

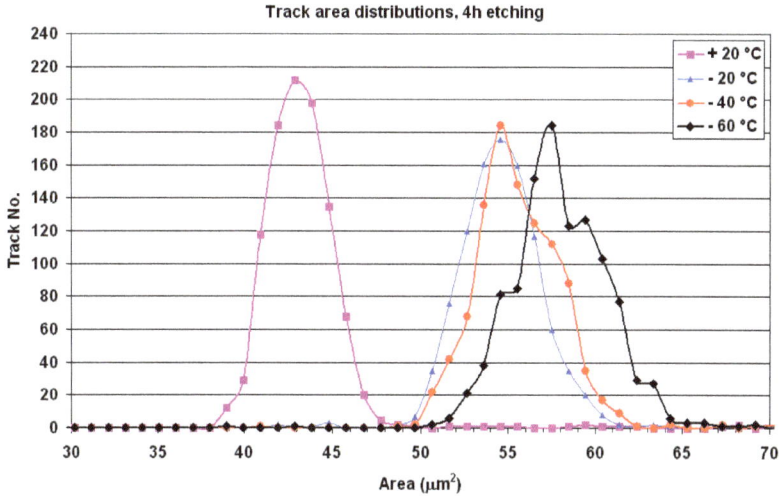

Fig. 6. The effect of the detector cooling on track area distribution (Palfalvi, 2011).

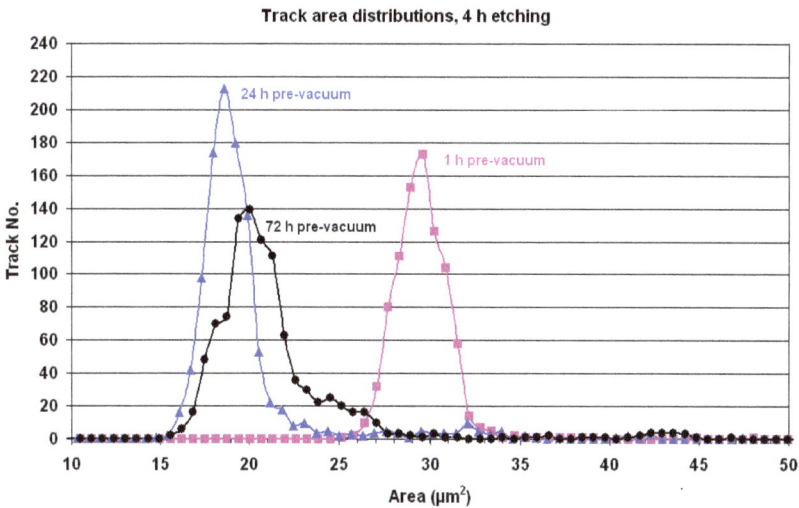

Fig. 7. ^{210}Po alpha track area distributions when vacuum of different durations were applied before exposure (Palfalvi, 2011).

It is important to consider that alpha particles emerge from a thick source with different energies due to self absorption and the track length and diameter depend on it.

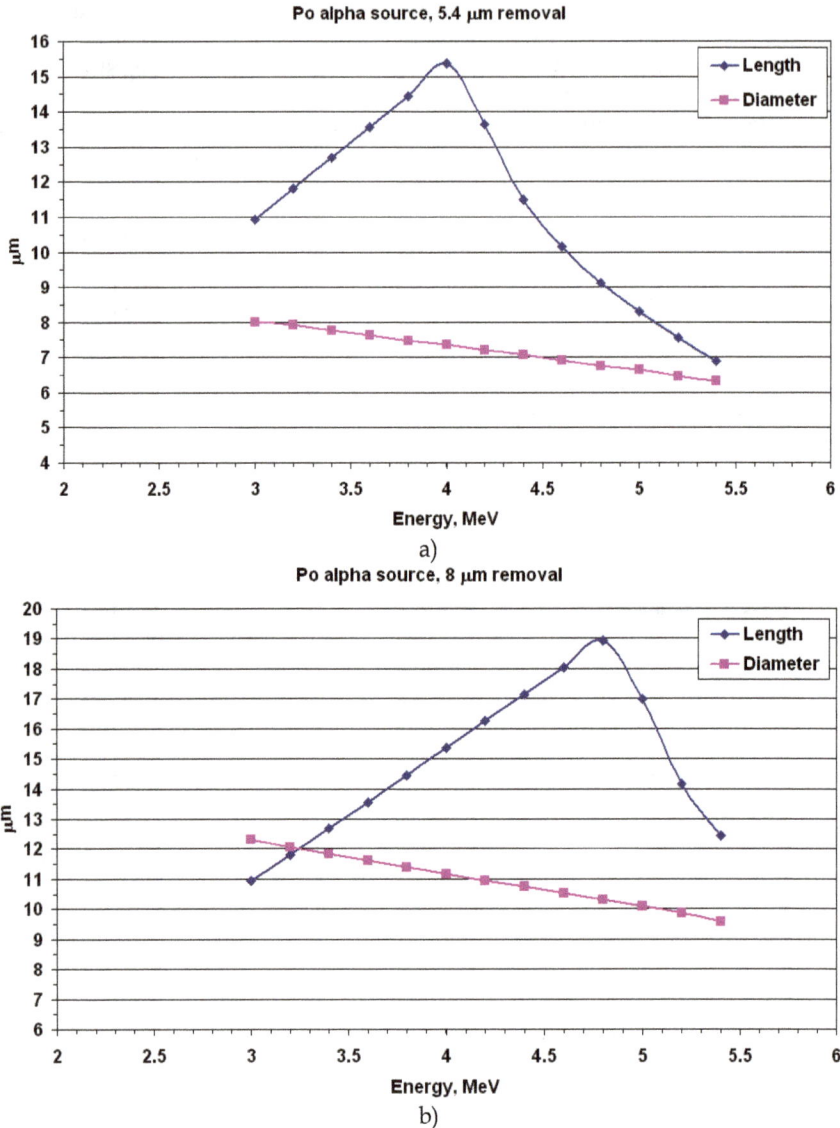

Po alpha source, 5.4 µm removal

a)

Po alpha source, 8 µm removal

b)

Fig. 8. a and b. PADC response curve to alpha particles with different energies. Curves were obtained after removing by etching 5.4 and 8 mN layers (Palfalvi, 2011).

The data on Figure 8 a and b, have an important role when the source is embedded in a material and the determination of the average depth from where the alpha particle comes is necessary. Sato (2008).

4.3 Automated track counting and software

Commercial and purpose made programs for automated track counting and analysis have been developed and upgraded regularly. A digital image analyser system essentially is a multi component unit, consisting of an optical microscope equipped with a CCD camera, a frame grabber and an analogue-to-digital converter to digitalize the image for further analysis. The microscope stage, the autofocus function and image analyzer are commanded by a purpose made software. Track parameters are obtained from these digital images employing a sequence of functions including mathematical approach for decomposing of multiple overlapping tracks. In most cases the track analysis process is to replace the human eye and its recognition capability, however, operator practice and training is needed even if sophisticated shape recognition method is available. Several custom made automated track counting and related software are reported in the literature, Espinosa et al. (1996), Palfalvi et al. (1997), Patiris et al. (2006), Coppola et al. (2009). A general overview on this subject is given by Hulber (2009).

A typical image of etched tracks obtained by a commercially available system is given in Figure 9. Together with the digitalized image, on the left side of the screen the macro for image analysis is shown. The tracks identification sequence for track parameterization (or macro) was assembled by Gonzalez (2010) employing LabView operating tool.

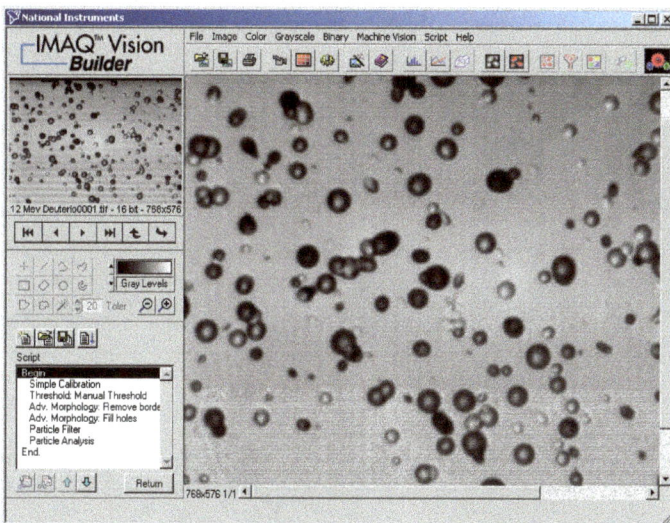

Fig. 9. Digitalized etched tracks as viewed by a commercially available NDT analyzer, Gonzalez (2010).

Once the digitalized image is obtained and unwanted background is eliminated the track perimeter, area, major and minor axes among others quantities are determined. For a given etching time the track position is defined by the (x, y) coordinates (Figure 10.), if etching is continued the same but enlarged track can be located and the new track opening may be observed and the increase of the track length along the third coordinate z can be measured which provides information on the track etch rate. This is relevant when reactions occurs inside the PADC e.g. (n,p) scattering on H atoms or (n,3α) reaction with C nuclei.

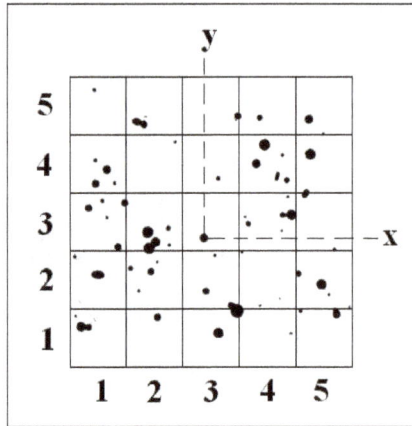

Fig. 10. Typical alpha particle etched tracks in PADC detector. Track diameter is related to the impinging energy, hence the difference in size. Track position is given by (x,y).

The software classifies the tracks on the basis of galleries composed of sample digital images with pre-selected shape and size. In Figure 11 a typical classification of track-openings are shown when the detector was exposed to a plane [241]Am alpha source being in a few cm

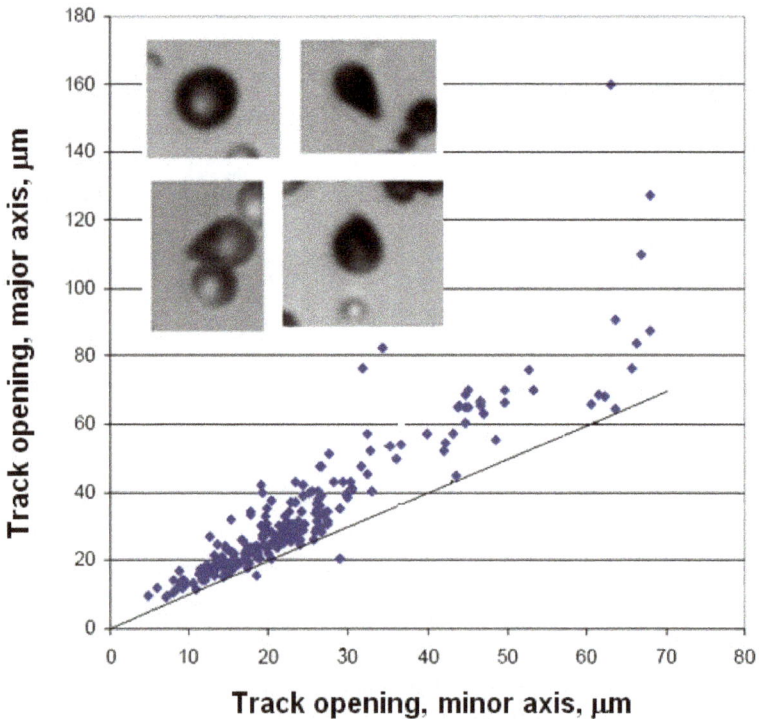

Fig. 11. Elliptical etched tracks distribution with related micrographs.

distance from the detector. The track size covers a large range between 10-70 μm. We may observe also, that a large number of particles have similar energy (15-30 μm) and only few are out of this range with incident angle near the critical one. The continuous line on Figure 14 is drawn to show where the circular tracks would be located. Below the mentioned line we still observe the presence of few tracks (negligible number), however their minor axis seems to be larger than the major axis indicating a false track registration. Similar picture can also be seen when studying alpha emission from a boron loaded sample exposed to neutrons if the [10]B bulk density is not uniformly distributed in the volume of the sample, but most atoms can be found in few μm below the surface.

A common feature among software based analyzers exists: all of them contain different solutions to similar problems. These are related to track identification and to pattern characterization including geometric parameters, approximations and software correction. To collect for example tracks into specific groups (discriminated by shape, size, area), chained mathematical and morphologic operations are necessary.

4.4 Control of etching with an alpha irradiation facility

Both the bulk and track etch rate depends on the etching conditions: concentration and temperature. To keep these quantities constant requires special attention, since few tenths of degree change in the temperature can affect the measure of the removed layer, which is responsible for the track etch rate, as seen from the Equ. (5). Usually, the control of the etching procedure is performed by exposing few detectors to alpha particles and measuring the track diameter after the etching. A simple facility can do the exposure with varying alpha energy as shown in Figure 12. The distance between the source and the detector can be adjusted and by this way the alpha energy can be changed. The source here is [210]Po, deposited onto a silver plate and placed into a copper holder and covered by a collimator which ensures the perpendicular incidence on the detector (Szabo et al., 2002).

A standard series of track diameters on the dependence of alpha energy and for different etching conditions can be experimentally obtained, which helps to check whether an etching is performed as required or planned.

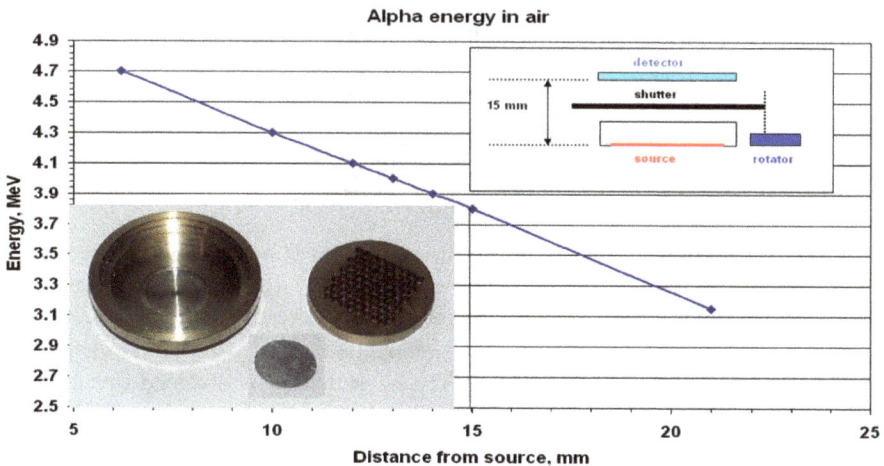

Fig. 12. Alpha exposure facility and the relationship between source-to-detector distance.

For instance, assuming a following relationship between the removed layer, h, and the measured circular track diameter, d, h = c * d, where c is a constant,and applying the data from Table 1, valid for a given detector material and etching, we may obtain for c a value of 0.7567, hence, h = 0.7567*d. So, for a given etching if the h differs from the pre-calculated one, here 8 µm, then the etching was not normal (Palfalvi, 2011).

Several kind of [210]Po-based alpha-particle irradiator either for calibration experiments or research have been reported in the literature, Sovland et al., (2000),. Zareena et al., (2008) reported an irradiation procedure and geometry that allows easier dose calculations for blood cells, biological culture or media in a sterile environment. Garty et al., (2005)proposed a novel microbeam facility for biological irradiation studies. Dosimetric characterisation of a simple irradiation device is given by Szabo et al., (2002). A calibration protocol for boron bio-distribution by autoradiography has been developed by Skvarc et al., (2002). Digital images of borated chicken liver and freeze-dried mouse tissue samples were evaluated for boron concentration on the basis of etched track densities in NTD type. Alfassi et al., (1999).

5. Selected examples of [10]BNC-reaction applications with PADC

PADC track etch detector (e.g. TASTRAK http://www.tasl.co.uk/tastrak.htm) are frequently applied for alpha particle detection. All the theoretical and experimental investigations resulting from the NDT method, is of great advantage not only in science and technology development but also for industrial application, Pálfalvi et al., (2010). Here we report some examples where the nuclear alpha-radiography by [10]BNC-reaction importance is highlighted.

5.1 Neutron field detection, monitoring and dosimetry

Monitoring and detecting of neutron fields, e.g. for nuclear reactor studies, consists of a (n,α) converter material, such as the binary glass metal strip $Fe_{75}B_{25}$, superposed on an alpha particle NTD detector LR 115 film wrapped in cadmium sheets. These NTD-converters are assembled in a stack on acrylic support, Pálfalvi, (1984) Detectors, purchased from DOSIRAD, France (http://dosirad.pagespro-orange.fr/soc_a.htm), consist of an active layer of red cellulose nitrate of 12 µm on a clear polyester base substrate of 100 µm. Although this detector is notcapable of providing information on track parameters mentioned earlier, it can be still useful because only the etched track density determination is needed.

The thermal neutron flux Φnt is determined by the well-known equation:

$$\Phi_{nt} = \frac{W(\rho - \rho_{Cd})}{K\rho_{Fe_{75}B_{25}} N_A foe\sum^x Bt_{irr}} \tag{9}$$

where K is the calibration coefficient, W the atomic weight of boron, ρ and ρ_{Cd} are the track densities without and with cadmium cover, respectively, ρ $_{Fe75B25}$ is the binary glass density, N_A Avogadro's number, f the [10]B isotopic abundance in the $Fe_{75}B_{25}$, σ the cross section of the [10]B(n,α)[7]Li reaction, $e^{-\Sigma x}$ the neutron average attenuation in the $Fe_{75}B_{25}$, B the yield quotient of the reaction and t_{irr} the irradiation time. The track density is obtained by etched track analysis Pálfalvi, (1982); it therefore provides the neutron field intensity near the energy region of thermal or cold neutrons Pálfalvi, et ali., (2001). The method is particularly useful to determine anisotropy of thermal neutron flux of graphite, heavy water or polyethylene moderated RINS such [252]Cf, [241]Am-Be or others (Alvarado et al., 2010).

The dose due to presence of high energy neutrons can be determined by PADC detectors Sainz et al., (2004), Santa Cruz et al., (2004). The technique requires several etching steps and individual track evaluation. In this case an automated analyser system is required. The technique provides neutron ambient dose equivalent (H*) between 200 keV and 20 MeV. It was applied with success at the International Space Station where values were reported in the range of 39 to 73 μSv/d, with an uncertainty estimated to be around ± 30 %, Pálfalvi (2004). PADC- CR39 coupled with a boron converter is employed currently as passive neutron "rem counter" as reported by Agosteo et al., (2010); essentially the device is a polythene designed to determine ambient dose equivalent conversion coefficients, H*(10)/F, for a broad energy window 0.025 eV - 1 GeV. Dose related study and its importance is given by Stap (2008).

5.2 Isotopic ^{10}B density determination

For low particle fluencies that induce well defined circular etched track without overlapping, a strong linear correlation exists between track density and particle yield. As the track density increases overlap cannot be avoided and density determination is less precise. However assuming an intermediate interval where track density and the overlapping frequency optimize, the etched track method demonstrated to be reliable and a non destructive procedure e.g. for:

i.   Determination of the boron either surface or bulk concentration in any given sample.

- ^{10}BNC-therapy. In this case, ^{10}BNC-reaction to be an effective technique in cancer treatment ^{10}B –bio distributions has to be known. The method of alpha-radiography is applied e.g. in sliced whole-body of tumour analysis as well as in cancerous cells in rat's brain, kidneys and liver. Occasionally a mixture of boric acid and borax compound medium is employed as carriers and ^{10}B is determined in microtome samples positioned between two pieces of NTD (Wittig et al., 2008).

    Etched track provides a remarkably good image of the boron density in tissue; see Figure 13 and17.

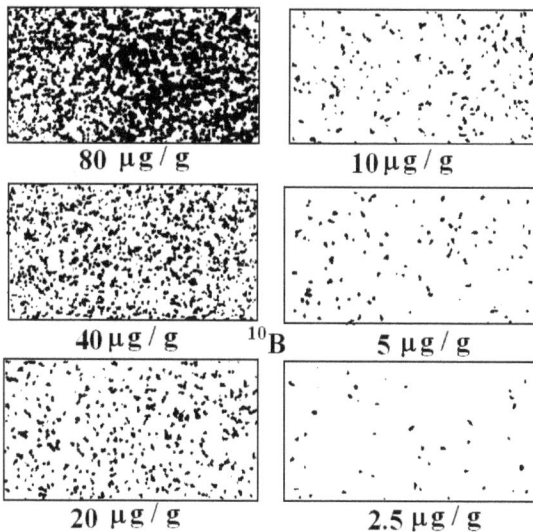

Fig. 13. Alpha radiography of boron in malignant melanoma by NTD. (adapted from Takagaki et al. 1990)

- In vitro study of radiobiological samples under low fluencies of alpha particles. Precise localization of each particle track in a monolayer cell is provided by alpha radiography, as demonstrated by Gaillard et al., (2005). Results evidence that living cell cultures exposed to alpha-radiation (dose rate 2-3 particle/cell/hour) suffer morphological changes Selmeczi et al., (2001). It has been demonstrated also that tracks give important information on the cell migration dynamics.

ii.   Study of boron spatial distribution.

Material science studies specifically binary glass metal ($Fe_{75}B_{25}$) have benefited from detailed knowledge of boron distribution. In Figure 14 etched tracks in LR-115 show the non uniform boron distribution where dislocations and its sparse agglomerations can be observed.

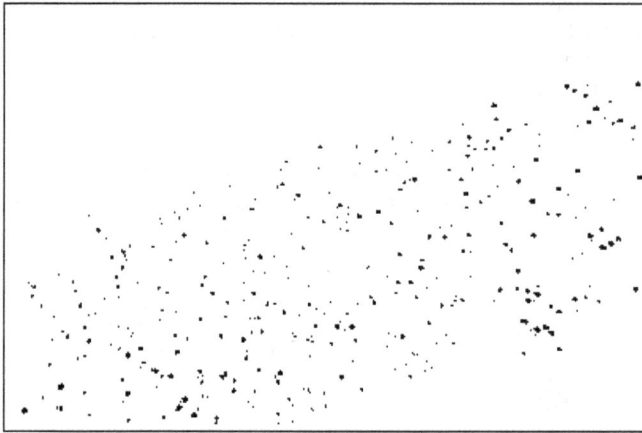

Fig. 14. Glass metal $Fe_{75}B_{25}$, irradiated with neutrons from ^{252}Cf. Spots are the etched alpha tracks (NTD-LR-115) and the metal edges are well recognized.

iii.   Lithium-Helium doping of a given Boron-structured material.

Megusar (1983), studied lithium and boron doping techniques to produce displacement or damage in bulk specimens of non-nickel bearing materials employed in fast fission reactors. It was demonstrated that lithium doping has an advantage over boron. Alloys with a stable boron compound uniformly distributed have been already developed and the (n,α) reaction could be a new technique to convert boron to lithium applying the reaction detailed in Equ. 1.

5.3 Gradient of boron concentration in thin film deposition

Local boron concentration and its micro distribution on materials such as polycrystalline beta-silicon carbide, laser ablated B_4C thin films or Mo-Si-B alloys among others have been studied or they will be scrutinized in the future using the (n,α) reaction and employing the NTD PADC, LR-115 or similar passive detectors. Its importance is related to the boron deposition technique and useful information can be derived by alpha auto radiography which may exhibit a large gradient of boron density. The area distribution of alpha tracks (Table 2) can be visualized in 3D as shown on Figure 15. To avoid high density overlapped tracks, it is convenient to calculate the irradiation time (t_{irr}) to obtain around $\rho=10^4$ tracks/cm^2. This can be estimated by the following equation:

$$t_{irr} = \rho(g\sigma_a \Phi N\, C_B R)^{-1} \tag{10}$$

Where: g is a geometrical factor, with an empirical value of 0.25; σ_a thermal neutron capture cross section for ^{10}B (7.6×10^{22} cm^2); Φ neutron flux (1.67×10^8 n/cm^2 s); N atom density of the sample, for boron steel (8.5×10^{22} cm^3); C_B fraction of boron atoms (1.7×10^4 for boron standard); $R = R_\alpha \tau_\alpha + R_{Li}\tau_{Li}$ in that R_α, R_{Li} are the ranges of alpha particles and lithium ions respectively; τ_α, τ_{Li} track registration efficiencies for alpha and lithium ions (assumed to be unity).

y / x	1	2	3	4	5	6	7	8	9	10	11	12	13	14
1	3	4	7	6	15	6	3	1	2	2	0	0	3	5
2	0	4	1	3	14	5	4	8	3	3	5	1	0	1
3	4	1	1	1	10	6	4	24	10	9	4	5	5	3
4	3	1	0	0	13	4	1	39	4	5	5	9	3	0
5	6	0	1	0	14	5	8	21	9	3	7	1	5	1
6	1	4	0	0	17	8	6	38	2	4	3	3	3	1
7	5	3	2	0	21	28	13	30	5	6	5	4	0	2
8	6	3	0	1	31	20	17	39	4	2	5	3	5	2
9	5	1	0	1	21	6	41	44	6	0	2	2	2	0
10	5	2	1	0	20	11	29	40	3	2	5	3	4	1
11	1	2	0	0	35	6	1	31	6	5	2	1	3	11
12	1	1	0	0	23	3	9	10	28	4	1	3	4	6
13	1	1	0	1	14	12	4	7	46	22	4	2	7	3
14	1	4	0	0	46	4	4	22	44	3	2	4	1	2
15	1	1	0	0	46	3	18	29	6	4	4	2	0	0
16	1	0	0	0	6	9	2	3	31	39	4	3	3	4
17	4	2	1	0	0	4	2	4	18	26	2	1	1	2
18	4	0	0	0	0	3	5	7	39	12	11	3	2	2
19	1	0	2	1	2	0	3	4	9	26	3	3	2	3
20	2	2	1	0	0	4	4	3	2	5	5	5	2	1
21	3	0	0	0	1	3	5	3	6	3	5	3	2	1
22	2	3	0	0	3	1	2	0	4	5	2	2	1	1
23	0	2	0	0	2	1	6	0	1	4	2	2	1	1

Table 2. Measured track density (given by bolded numbers) localized by numerical coordinates (x,y)

The coordinate number refers to a microscope field view of 100μm x 150 μm area positioned on a plastic detector of the size of 10 mm x 21 mm area. These data are plotted as 3D map by a software program. The resulting hill like map is shown on Figure 15.

In Figure 15 etched track spatial distribution of track density, $\rho(x,y)$ and its gradient is given. 3D map was rendered by Surfer ™. On the left, disposition of glass metal strip as positioned on the NTD is shown. The other information, discerned from the picture or map on the right side, is that the sample edges are sharp and where two samples superimpose the track density does not show an abrupt increase as it would be expected. The main reason is because the alpha particle range is comparable to the sample thickness and self absorption effect reduces sharply track recording.

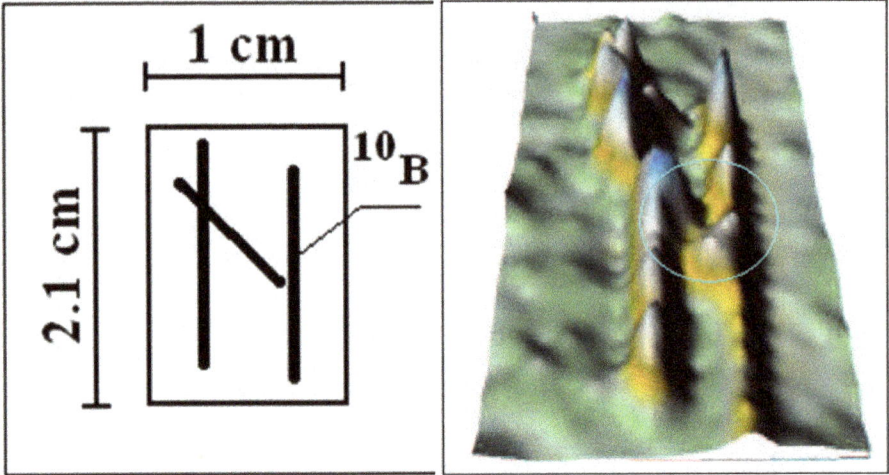

Fig. 15. Overall view of alpha tracks spatial distribution resulting from glass-metal strips $(Fe_{75}B_{25})$ shown as heavier black lines on the left side, detector dimensions are in cm. Circle indicate the place where auto absorption effect is dominant.

It would be interesting to apply the autoradiography methodology to study e.g. micro cracks or fissures due to material fatigue, manufacturing imperfections including micro structural temporal evolution. Where these structural in-homogeneities occur, we would observe track density variation of the type shown in the encircled region of Figure 15.

Important results were obtained studying toughness and ductility in the high-temperature intermetallic compound by etched track mapping of the location of the boron. Results (see Figure 16) supports that it is distributed randomly in (Al–Ru–B) alloys, (Middlemas et al., 2010). Fleisher (2005) observed from alpha radiography on PADC that it is concentrated at grain boundaries, effect that lowers the alloy's shear modulus (see Figure 16)

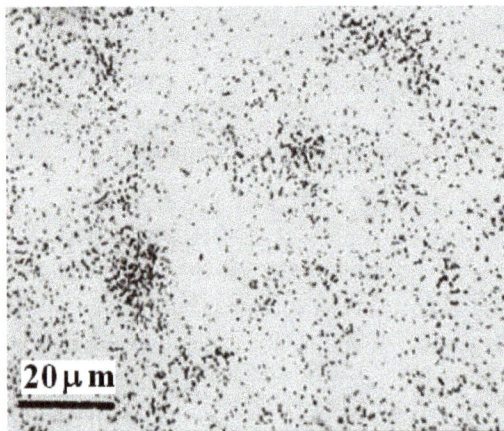

Fig. 16. The etched detectors viewed in a Leitz Ortholux microscope for track distributions at magnifications of 100x (adapted from Fleisher, 2005)

5.4 [10]B-Carrier deposition effectiveness in biomedical sample

[10]B isotope finds application in nuclear medicine and generally in the biophysical field. It has been demonstrated to be an ideal isotope for cancer therapy. The cancerous cells absorb more boron compound than normal cells. Thus applying the [10]BNC-reaction the alpha products have the possibility to destroy the tumour without or with low damage to normal tissues. The therapy taking advantage of the method is briefly referred as [10]BNC-therapy or simply BNCT. In the literature often it is also referred to as a binary radiation therapy modality. This technique is based on carrier-modality that requires a molecule rich in boron content, having a capability to enter selectively targeted cells. For instance carboranyl groups, called *carrier*, have attached boron atoms e.g. $B_{10}H_{10}$ and $B_{12}H_{12}$ and these compounds can be synthesized to contain only the requested [10]B isotope. Other appropriate chemical compounds tagged with [10]B (e.g. BPA (boron-phenylalanine) BSH (sulphydril-borane); BOPP (protoporfirina boronated) are currently employed. To take advantage of the disruptive power of the [10]BNC-reaction product appropriate neutron irradiations is required. A relatively low dose of thermal neutrons is applied inducing energetic alpha particles and therefore a high density atomic ionization (LET). During the atomic and molecular recombination, chemical structure is modified which in turn, inhibits or disrupts the cell activity. Further advantage is the boron element low toxicity (300 $\mu g/g$) and the fact that in bacteria the concentration may already be around 150 ($\mu g/g$) B_{nat} suggesting the applicability of [10]BNC-reaction as a new analytical technique in oncology. We give some results of the mentioned carrier-modality in biomedical field and later we refer to another method called internal boron-modality. Well understood chemistry of boron allows synthesizing borated drugs or purpose made molecules to provide incorporation techniques into e.g. high-grade gliomas and generally any tumour, (Altieri et al., 2008). In order to establish the effectiveness of these boron carriers it is necessary to determine their distribution e.g. in GS-9L gliosarcoma cells. Several studies point that carrier agents, like p-boronophenylalanine (BPA) distributes relatively homogeneously (Bennett et al., 1994). Recently an important study was completed with [10]BNC-therapy at RA-3 reactor facility by Portu et al. (2011), in which histological features have been accurately visualized by alpha etched track density in NTD-Lexan, see Figure 17.

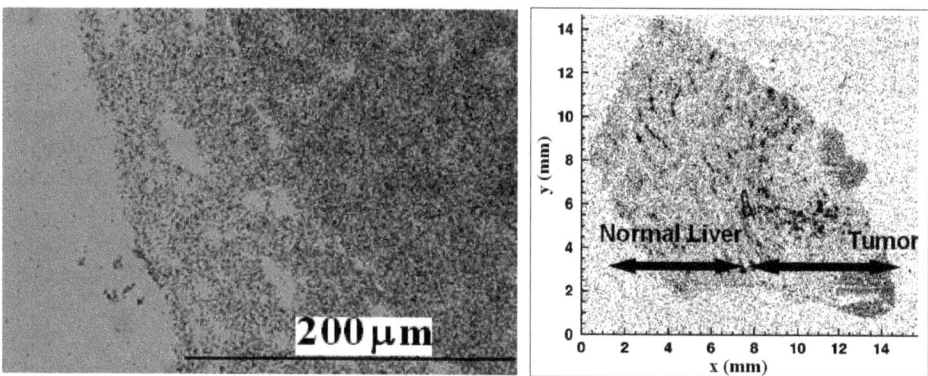

Fig. 17. Alpha autoradiography image of a tumour section (left) adapted from Portu et al. (2011) and on the right [10]B distributions in VX-2 tumour that shows selective uptake by intra-arterial injection, (adapted from Mikado (2009)

The other so called internal boron-modality is related to boron content in many naturally occurring bio-samples. Just to mention one, the natural antibiotic such as boromycin, contains elemental boron. The [10]BNC-reaction technique could be employed to determine the concentration and its bio-efficiency. As mentioned at the beginning several other substances occur in our environment and the given NTD methodology could be applied in revealing the boron content in many environmental samples.

5.5 Boron mapping in plant physiology

An interesting application for [10]BNC-reaction method is related to plant nutrients. It has been recognized that boron is essential for growth performance; in fact then concentration should be in the range of 0.8 - 1.8 ppm. Otherwise poor plant grows or boro soil boron toxicity shows up. Therefore it would be convenient to employ the alpha radiography technique for plant physiology studies; alpha track mapping could detect boron concentrations with values lower than few ng/g. It could be also employed for uptake determination in different parts of the plant. A few studies have already been reported and it can still be considered, from the NDT point of view, as a little explored field.

It seems that the isotopic ratio [11]B/[10]B in this study field, departs from the naturally occurring value. It has been observed that ratio values may change significantly from the expected value of [11]B/[10]B =4.1 depending on the sample taken from different parts of the plants. This kind of isotopic effect related to boron biochemistry foliar processes, indicates that the boron uptake should be studied further (Vanderpool et al., 1992).

5.6 Topics for further application of [10]BNC-reaction relevant for environmental monitoring, industry and commercial certifications

Boron is dissolved in surface and underground water other than the sea as a chemical compound e.g. $B(OH)_3$ (aq) or $B(OH)_4^-$ (aq); its concentration is in the range of 4-15 ppm (dry mass) (Water, 2011).

Boron concentration in surface waters comes from minerals with boron concentrations between 5 and 80 ppm; such as kernite, borax, ulexite, colemanite among others. It can be measured even in households as sodium perborate in detergents, eventually damaging the environment. However the boron has an important role in environmental studies; it is an indicator to detect the presence of other hazardous substances. For example, the boron-trifluoride-ethyl-ether complex reacts with water, forming diethyl ether BF_3, and releasing some highly flammable gases. A number of boron compounds, such as boron tri iodide, are hydrolysed in water. Therefore detecting boron by [10]BNC-reaction and PADC technique could be important for environmental monitoring.

On the other hand industrial plants with significant B-concentration such as that found in fumaroles pollute the environment with compounds such as boric acid, borates and boron minerals, (Environment, 1998). Often boron compounds are also found in food; for concentration measurement the colorimetric curcumin method is used. That could be replaced by the [10]BNC-reaction or autoradiography being simpler and a direct method. It could be applied also in the field of molecular synthesis, (Solozhenko et al., 2009) where this technique could be of advantage including far apart fields such as nano-tube manufacturing (Yu et al., 2006) and human genetics (Vithana et al., 2006).

6. Conclusions

Distribution of ^{10}B in samples raging from medicine to industry from environment to food certification are determined using passive nuclear track detectors and a neutron beam for irradiation. This technique will spread among a larger community as the demand for boron determination increases. For the near future the most important field of application will be environmental monitoring and cancer therapy. Tumour tissue destruction is a health priority and studies will be further extended to improve ^{10}B carriers and boron imaging to visualize microscopic uptake and its distribution in cell or any other material.

7. Acknowledgements

The support received from the INFN (Sezione di Padova) and the University of Studies of Padova is acknowledged. This work was also partly supported by the European Commission in the frame of the FP7 HAMLET project (Project # 218817).

8. References

Agosteo S., Caresana M., Ferrarini M., Silari M.,(2010). A passive rem counter based on CR39 SSNTD coupled with a boron converter. Rad. Meas., 44, 985–987.

Alfassi Z. B., Probst T. U., (1999). On the calibration curve for determination of boron in tissue by quantitative neutron capture radiography. Nucl. Instr. Meth. A 428, 502-507.

Alvarado R., Palacios D., Sajo-Bohus L., Greaves E.D., Barros H., Nemeth P. Goncalves I.F. (2010). Neutron flux characterization using LR-115 NTD and binary glass metal as converter. Rev. Mexicana de Fisica S 56, 5–8.

Altieri, S., Braghieri, A., Bortolussi, S., Bruschi, P., Fossati, F., Pedroni, P., Pinelli, T., Zonta, A., Ferrari, C., Prati, U., Roveda, L., Barni, S., Chiari, P., Nano, R., (2008). Neutron autoradiography imaging of selective boron uptake in human meta- static tumours. Appl. Rad. Isot. 66, 1850–1855.

Bennett B. D., Mumford-Zisk J., Coderre J. A. and Morrison G. H. (1994). Subcellular Localization of p-Boronophenylalanine-Delivered Boron-10 in the Rat 9L Gliosarcoma: Cryogenic Preparation In Vitro. Rad. Res. 140, 72-78.

Benton, E. R. (2004). Radiation Dosimetry at Aviation Altitudes and in Low-Earth Orbit, Ph.D. Thesis, University College, Dublin, Ireland.

Cartwright, B. G., Shirk E. K. and Price, P. B., (1978). A nuclear-track-recording polymer of unique sensitivity and resolution. Nucl. Instr. Meth. 153, 457–460.

Coppola F., Durante M., Gialanella G., Grossi G., Manti L., Pugliese M. and Scampoli P., (2009). Development of an automated scanning system for the analysis of heavy ions' fragmentation reaction by nuclear track detectors. Rad. Meas., 44, 802-805.

Dörschel B., Hermsdorf D. and Kadner K., (2002). Studies of experimentally determined etch-rate ratios in CR-39 for ions of different kinds and energies. Rad. Meas. 35, 183-187.

Durrani, S.A., Bull, R.K., (1987). Solid State Nuclear Track Detection: Principles, Methods and Applications. Pergamon Press, Oxford. p. 284.

Environmental Health Criteria 204: Boron the IPCS. (1998).
    http://www.inchem.org/documents/ehc/ehc/ehc204.htm.

Espinosa G., Gammage R.B., Meyer K.E. and Dudney C.S., (1996). Nuclear Track Analysis by Digital Imaging. Rad. Prot. Dos., 66, 363-366.

Fleischer R.L., (2005). The distribution of boron in AlRu: Effect on ductility and toughness. Acta Materialia 53, 2623–2627.

Gaillard S., Armbruster V., Hill M.A., Gharbi T., Fromm M., (2005). Production and validation of CR-39-based dishes for alpha-particle radiobiological experiments. Rad. Res. 163, 343-350.

Garty G., Rossi G. J., Bigelowa A., Randers-Pehrson G., Brenner D. J. (2005). A microbeam irradiator without an accelerator. Nucl. Instr. Meth. B, 231, 392-396.

Gonzalez (2010), Private Communication.

Hajek, M., Berger, T., Vana, N., Fugger, M., Pálfalvi, J.K., Szabó, J., Eördögh, I., Akatov, Y. A., Arkhangelsky, V. V. And Shurshakov, V. A., (2008). Convolution of TLD and SSNTD measurements during the BRADOS-1 experiments onboard ISS (2001). Rad. Meas. 43, 1231-1236.

Hulber E. (2009), Overview of PADC nuclear track readers. Recent trends and solutions Rad. Meas., 44, 821-825.

Leung, K. (2002) U.S. Patent #6,907,097.

Lounis Z., Djeffal S., Morsli K. and Alla M. (2001). Track etch parameters in CR-39 detectors for proton and alpha particles of different energies. Nucl. Instr. Meth. B: 17, 543-550.

Megusar J., Harling O.K. and Grant N.J. (1983). Lithium doping of candidate fusion reactor alloys to simulate simultaneous helium and damage production. J. Nucl. Mat., 115, 192-196

Middlemas M.R. (2010), http://www.lbl.gov/ritchie/Programs/Moly/Moly.html.

Mikado S., Yanagie H., Yasuda N., Higashi S., Ikushima I, Mizumachi R., Murata Y., Morishita Y., Nishimura R., Shinohara A., Ogura K, Sugiyama H., Iikura H., Ando H., Ishimoto M., Takamoto S., Eriguchi M, Takahashi H., Kimura M., (2009). Application of neutron capture autoradiography to Boron Delivery seeking techniques for selective accumulation of boron compounds to tumor with intra-arterial administration of boron entrapped water-in-oil-in-water emulsion. Nucl. Instr. Meth. A 605, 171–174.

Naranjo B., Gimzewski J.K. and Putterman S., (2005). Observation of nuclear fusion driven by a pyroelectric crystal. Nature 434, 1115-1117.

Nikezic D. and Yu K.N., (2003). Calculations of track parameters and plots of track openings and wall profiles in CR-39 detector. Rad. Meas. 37, 595-601.

Nikezic D. Yu K.N., (2004). Formation and growth of tracks in nuclear track materials Mat. Sci. and Eng. R 46 51–123.

Pálfalvi, J. K., (1982). Neutron Sensitivity of LR-115 SSNTD Using Different (n,alpha) Radiators, Nucl. Instr. Meth., 203, 451-457.

Pálfalvi, J. K. (1984), Neutron Sensitivity Measurements of LR-115 Tracks Detector with Some (n,alpha) Converters, Nuclear Tracks, 9, 47-57.

Pálfalvi J. K., Eordog, I., Szabo,J. and Sajo-Bohus L., (1997). A new generation image analyser for evaluating ssntds. Rad. Meas. 28, 849-852.

Pálfalvi J. K., Sajó-Bohus L., M. Balaskó and I. Balásházy (2001). Neutron field mapping and dosimetry by CR-39 for radiography and other applications. Rad. Meas. 34, 471-475.

Pálfalvi, J. K., Akatov, Y. A., Szabó, J., Sajó-Bohus, L. and Eördögh, I., (2004). Evaluation of SSNTD Stacks Exposed on the ISS. Rad. Prot. Dos. 110, 393-397.

Pálfalvi J. K., (2009). Fluence and dose of mixed radiation by SSNTDs: achievements and constrains. Rad. Meas. 44, 724-728.

Pálfalvi J. K., Szabó J. and Eördögh I., (2010). Detection of high energy neutrons, protons and He particles by solid state nuclear track detectors. Rad. Meas., 45, 1568-1573.

Pálfalvi J. K., (2011). Unublished, private communication.

Patiris D.L., Blekas K., Ioannides K.G., (2006). TRIAC: A code for track measurements using image analysis tools. Nucl. Instr. Meth. 153, B 244, 392–396.

Persaud A., I. Allen, M. R. Dickinson, T. Schenkel, R. Kapadia, K. Takei and A.Javey, (2011). Development of a compact neutron source based on field ionization processes. American Vacuum Society. DOI: 10.1116/1.3531929.

Portu A., Carpano M., Dagrosa A., Nievas S., Pozzi E., Thorp S., Cabrini R., Liberman S., SaintMartin, G., (2011). Reference systems for the determination of [10]B through autoradiography images: Application to a melanoma experimental model. Appl. Radiat.Isotopes, DOI:10.1016/j.apradiso.2011.02.049.

Sainz C., Soto, J., Casado, J., Bueren, J., (2004). Effects on CD34+ haematopoietic stem cells induced by low doses of alpha irradiation. Int'l J. Low Radiation 2, 208 – 215.

Santa Cruz GA, Zamenhof L., (2004). The microdosimetry of the [10]B reaction in boron neutron capture therapy: a new generalized theory. Rad. Res. 162, 702-10.

Sato F., Kuchimaru T., Kato Y. and Iida T., (2008). Digital Image Analysis of Etch Pit Formation in CR-39 Track Detector. Jpn. J. Appl. Phys. 47 269-272.

Selmeczi D., Szabo B., Sajo-Bohus L., Rozlosnik N., (2001). Morphological changes in living cell cultures following α-particle irradiation studied by optical and atomic force microscopy. Rad. Meas., 34, 549-553.

Skvarc, J., Giacomelli, M., Yanagie, H., Kuehne, G., (2002). Selective radiography of 10B distribution in organs using cold and thermal neutron beams. Cell Mol Biol Lett.,7, 162-4.

Solozhenko, V.L., Kurakevych, O.O., Le Godec, Y., Andrault D., Mezouar, M., (2009). "Ultimate Metastable Solubility of Boron in Diamond: Synthesis of Superhard Diamondlike BC5". Phys. Rev. Lett. 102: 015506.

Somogyi G. and Szalay, S.A., (1973). Track-diameter kinetics in dielectric track detectors. Nucl. Inst. Meth. 109, 211-232.

Søyland C, Hassfjell S. (2000). A novel [210]Po-based alpha-particle irradiator for radiobiological experiments with retrospective alpha-particle hit per cell determination. Rad. Environ. Biophys. 39, 125-30.

Stap J., Krawczyk P., Van Oven C, Barendsen G, Essers J, Kanaar R, Aten J.,.(2008).Induction of linear tracks of DNA double-strand breaks by alpha-particle irradiation of cells. Nat. Meth.. 5, 261-266.

Stevanato L., Caldogno M., Dima R., Fabris D., Hao X., Lunardon M., Moretto S., G. Nebbia G., Pesente S., Sajo-Bohus L., and Viesti G., (2010). A new facility for Non-Destructive Assay with a time-tagged [252]Cf source. AIP Conf. Proc., 4, 1265, 431-434.

Szabo J., Feher I., Pálfalvi J., Balashazy I., Dam A.M., Polonyi I., Bogdandi E.N., (2002). In vitro cell irradiation systems based on 210Po alpha source: construction and characterisation. Rad. Meas., 35, 575-8.

Takagaki M. and Mishima Y., (1990). Boron-10 quantitative analysis of neutron capture therapy on malignant melanoma by spaectrophotometric alpha-track reading. Nucl. Tracks Rad. Meas. 17, 531-535.

Vanderpool R.A., Johnson P.E. (1992). Boron isotope ratios in commercial produce and boron-forliar and hydroponic enriched plants. J. Agric. Food Chem. 40 (3), pp 462-466

Vithana E., Morgan P., Sundaresan P., Ebenezer N., Tan D., Mohamed M., Anand S., Khine K., Venkataraman D., Yong V., Salto-Tellez M., Venkatraman A., Guo K., Hemadevi B., Srinivasan M., Prajna V., Khine M., Casey J., Inglehearn C., Aung T. (2006). "Mutations in sodium-borate cotransporter SLC4A11 cause recessive congenital hereditary endothelial dystrophy. Nat. Genet. 38, 755 – 757.

Water (2011), http://www.lenntech.com/periodic/water/boron/boron-and-water.htm#ixzz1LDAyFUVO and references therein.

Wikipedia (2011), http://en.wikipedia.org/wiki/Boron, and refrences therein.

Wittig A., Michel J., Moss R.L., Stecher-Rasmussen F., Arlinghaus H.F., Bendel P., Mauri P.L., Altieri S., Hilger R., Salvadori P.A., Menichetti L., Zamenhof R., Sauerwein W.A., (2008). Boron analysis and boron imaging in biological materials for boron neutron capture therapy (BNCT). Crit. Rev. Oncol. Hematol. 68, 66–90.

Yu J., Ying C., Elliman R. G., Petravic M., (2006). Isotopically enriched [10]B nanotubes. Adv. Mat., 18, 2157-2160.

Zareena Hamza, V., Vivek Kumar P. R., Jeevanram R. K., Santanam R., Danalaksmi B. and Mohankumar M. N., (2008), A simple method to irradiate blood cells in vitro with radon gas, Rad. Prot. Dos. 1–8.

Ziegler James F., Ziegler M.D., and Biersack J.P., (2010). SRIM - the stopping and range of ions in matter. Nucl. Instr. Meth. B 268, 1818–1823.

Part 2

Radioisotopes and Radiology in Medical Science

10

Radiolabelled Nanoparticles for Diagnosis and Treatment of Cancer

Dimple Chopra

Department of Pharmaceutical Sciences, Punjabi University,
India

1. Introduction

Cancer is one of the leading cause of death worldwide.[1] In 2010, a total of 1,529,560 new cancer cases and 569,490 cancer deaths were estimated in the United States alone.[1]

Despite advances in our understanding of tumor biology, cancer biomarkers, surgical procedures, radio- and chemotherapy, the overall survival rate from cancer has not improved significantly in the past two decades. Early detection, pathological characterization, and individualized treatments are recognized as important aspects for improving the survival of cancer patients. Many novel approaches, such as imaging for the early detection of molecular events in tumors, comprehensive and personalized treatments, and targeted delivery of therapeutic agents to tumor sites, have been developed by various research groups; and some of these are already in clinical trials or applications for cancer patients. Radiation therapy, in conjunction with chemotherapy and surgery, is an effective cancer treatment option, especially for radiation-sensitive tumors. Radiation therapy utilizes high dose ionizing radiation to kill cancer cells and prevent progression and recurrence of the tumor. Traditionally, radiation therapies fall into one of three categories: external radiation, internal radiation and systemic radiation therapy. External radiation therapy delivers high-energy x-rays or electron or proton beams to a tumor from outside the body, often under imaging guidance. Internal radiation therapy (also called brachytherapy) places radiation sources within or near the tumor using minimally invasive procedures. Systemic radiation therapy delivers soluble radioactive substances, either by ingestion, catheter infusion, or intravenous administration of tumor-targeting carriers, such as antibodies or biocompatible materials, which carry selected radioisotopes. Although systemic radiation offers desirable advantages of improved efficacy as well as potentially reducing radiation dosage and side effects, in vivo delivery of radioisotopes with tumor targeted specificity needs to address many challenges that include: (i) the selection of radioisotopes with a proper half life; (ii) a delivery vehicle that can carry an optimal amount of radioisotopes and has favorable pharmacokinetics; (iii) suitable tumor biomarkers that can be used to direct the delivery vehicle into cancer cells; and (iv) specific tumor targeting ligands that are inexpensive to produce and can be readily conjugated to the delivery vehicles. In addition, a multifunctional carrier that not only delivers radioisotopes but also provides imaging capability for tracking and quantifying radioisotopes that have accumulated in the tumor is highly desirable.[2]

Recent advances in nanotechnology have led to the development of novel nanomaterials and integrated nanodevices for cancer detection and screening, in vivo molecular and cellular imaging,[3] and the delivery of therapeutics such as cancer cell killing radio-isotopes.[4,5] An increasing number of studies have shown that the selective delivery of therapeutic agents into a tumor mass using nanoparticle platforms may improve the bioavailability of cytotoxic agents and minimize toxicity to normal tissues.[6-8] Radiolabelled nanoparticles represent a new class of agents which has enormous potential for clinical applications. This book chapter provides deep insight into designing radiolabeled nanocarriers or nanoparticles tagged with appropriate radionuclides for cancer diagnosis and therapy. The combination of newer nuclear imaging techniques providing high sensitivity and spatial resolution such as dual modality imaging with positron emission tomography/computed tomography (PET/CT) and use of nanoscale devices to carry diagnostic and therapeutic radionuclides with high target specificity can enable more accurate detection, staging and therapy planning of cancer.

2. Molecular imaging with radiolabeled nanoparticles

The visualization, characterization and measurement of biological processes at the molecular and cellular levels in humans and other living systems is termed as molecular imaging.[9] Molecular imaging includes molecular magnetic resonance imaging (mMRI), magnetic resonance spectroscopy (MRS), optical bioluminescence, optical fluorescence, targeted ultrasound, single photon emission computed tomography (SPECT) and positron emission tomography (PET).[10] The availability of scanners for small animals provide similar vivo imaging capability in mice, primates and humans. This facilitates correlation of molecular measurements between species.[11,12] Molecular imaging gives whole body readout in an intact system which is more relevant and reliable than in vitro/ ex vivo assays.[13] Non-invasive detection of various molecular markers of diseases lead to earlier diagnosis, earlier treatment and better prognosis. Radionuclide-based imaging includes SPECT and PET, where internal radiation is administered through a low mass amount of pharmaceutical labeled with a radioisotope. The major advantages of radionuclide-based molecular imaging techniques (SPECT and PET) over other modalities (optical and MRI) are that they are very sensitive, quantitative without any tissue penetration limit.[10,15]But the resolution of SPECT or PET is same as that of MRI. Mostly, nanoparticles are labeled with a radionuclide for non-invasive evaluation of its biodistribution, pharmacokinetic properties and/or tumor targeting efficacy with SPECT or PET.[16]

Radioisotopes used for SPECT imaging include ^{99m}Tc ($t_{1/2}$: 6.0 h), ^{111}In ($t_{1/2}$: 2.8 days) and radioiodine (^{131}I, $t_{1/2}$: 8.0 days). The source of SPECT images are gamma ray emissions. The radioisotope decays and emits gamma rays, which can be detected by a gamma camera to obtain 3-D images.[17,18] The pharmacokinetics, tumor uptake and therapeutic efficacy of an ^{111}In-labeled, chimeric L6 (ChL6) monoclonal antibody linked iron oxide (IO) nanoparticle was studied in athymic mice bearing human breast cancer tumors.[19] The ^{111}In-labeled ChL6 was conjugated to the carboxylated polyethylene glycol (PEG) on dextran-coated IO nanoparticles (~ 20 nm in diameter), with one to two ChL6 antibodies per nanoparticle. It was proposed that the nanoparticles remained in the circulation for long period of time which provides them the opportunity to access the cancer cells. Inductively heating the nanoparticle by externally applied alternating magnetic field (AMF) caused tumor necrosis at 24 h after AMF therapy. In a follow-up study, different doses of AMF was delivered at 72 h after nanoparticle injection.[20]SPECT imaging was carried out to quantify the nanoparticle uptake in the tumor, which was about 14 percentage injected dose per gram (%ID/g) at 48 h post-injection. A delay in tumor growth occurred after the AMF treatment, which was statistically significant when compared with the untreated group. Subsequently, similar nanoparticles with diameters of 30 and 100 nm were also studied.[21] Although the heating capacity of these large nanoparticles is several times greater, the tumor targeting efficacy was significantly less than that of the 20 nm-sized counterparts. In another report, recombinant antibody fragments were tested for tumor targeting of these nanoparticles. Pharmacokinetic and whole-body autoradiography studies demonstrated that only 5% of the injected dose was targeted to the tumor after 24 h.[22]

As cancer cells undergo metastasis ie. they invade and migrate to a new tissue. They penetrate and attach to the target tissue's basal matrix. This allows the cancer cell to pull itself forward into the tissue. The attachment is mediated by cell-surface receptors known as integrins, which bind to components of the extracellular matrix. Integrins are crucial for cell invasion and migration, not only for physically tethering cells to the matrix, but also for sending and receiving molecular signals that regulate these processes.[23] Till date 24 integrins have been discovered, integrin $\alpha_v\beta_3$ is the most intensively studied.[24,25] It is expressed in many types of tumor and plays a critical role in tumor angiogenesis.[26] Integrin $\alpha_v\beta_3$-targeted ^{111}In-labeled perfluorocarbon (PFC) nanoparticles were tested for the detection of tumor angiogenesis in New Zealand white rabbits. The PFC nanoparticles bearing approximately 10 ^{111}In per particle was found to have better tumor-to-muscle ratio than those containing approximately 1 ^{111}In per particle. At 18 h post-injection, the mean tumor radioactivity in rabbits receiving integrin $\alpha v\beta 3$-targeted PFC nanoparticle was about 4-fold higher than the non-targeted control. Biodistribution studies revealed that nanoparticles were principally cleared from spleen.[27] Carbon nanotubes are promising carriers for use in biomedical and pharmaceutical sciences. Wang et al.(2004) studied its biological properties in vivo.[28] They labeled water-soluble hydroxylated carbon single-wall nanotubes with radioactive [125]In atoms, and then the tracer was used to study the distribution of hydroxylated carbon single-wall nanotubes in mice. They moved easily among the compartments and tissues of the body, behaving as small active molecules though their apparent mean molecular weight is tremendously large. This study gave a quantitative analysis of carbon nanotubes accumulated in animal tissues. Singh et al.(2006) functionalized water-soluble SWNTs with the chelating molecule diethylentriaminepentaacetic (DTPA) and labeled them with ^{111}In for imaging purposes.[29] Both the studies suggested that SWNTs were not retained in any of the RES organs (e.g. liver or spleen) and were cleared rapidly from the circulation through the renal route. Villa et al. (2008) synthesized and studied the biodistribution of oligonucleotide functionalized tumor targetable carbon nanotubes.[30] Recently Mehmet Toner have developed a microfluidic device composed of carbon nanotubes which can detect cancer cells in 1ml of patient's blood.[31]

SPECT and PET are extremely valuable technologies in nuclear medicine. SPECT has superior spatial resolution, it can potentially allow for simultaneous imaging of multiple radionuclides, since the gamma rays emitted from different radioisotopes can be differentiated based on energy.[32,33] PET on the other hand has much higher detection efficiency.[34] The biodistribution of [64]Cu ($t_{1/2}$: 12.7 h)-labeled SWNTs in mice has been investigated by PET imaging and Raman spectroscopy. It was found that these SWNTs are highly stable in vivo. PEGylated SWNTs exhibit relatively long circulation half-life (about 2

h) and low uptake by the RES. Most importantly, efficient targeting of integrin $\alpha_v\beta_3$-positive tumor in mice was achieved with SWNTs coated with PEG chains linked to cyclic RGD peptides. Good agreement of biodistribution data obtained by PET and ex vivo Raman measurements confirmed the in vivo stability and tumor-targeting efficacy of SWNT-RGD. [35] Molecular imaging of living subjects continues to rapidly evolve with bioluminescence and fluorescence strategies, in particular being frequently used for small-animal models. Keren *et al.*(2008) demonstrated noninvasive molecular imaging of small living subjects using Raman spectroscopy. Surface-enhanced Raman scattering nanoparticles and single-wall carbon nanotubes were used to demonstrate whole-body Raman imaging, nanoparticle pharmacokinetics, multiplexing, and *in vivo* tumor targeting, using an imaging system adapted for small-animal Raman imaging. This imaging modality holds significant potential as a strategy for biomedical imaging of living subjects.[36] An optimized noninvasive Raman microscope was used to evaluate tumor targeting and localization of single walled carbon nanotubes (SWNTs) in mice. Raman images were acquired in two groups of tumor-bearing mice. The control group received plain-SWNTs, whereas the experimental group received tumor targeting RGD-SWNTs intravenously. Raman imaging commenced over the next 72 h and revealed increased accumulation of RGD-SWNTs in tumor ($p < 0.05$) as opposed to plain-SWNTs. These results support the development of a new preclinical Raman imager.[37]Photoacoustic imaging of living subjects offers higher spatial resolution and allows deeper tissues to be imaged compared with most optical imaging techniques. Many diseases do not exhibit a natural photoacoustic contrast, especially in their early stages, so it is necessary to administer a photoacoustic contrast agent. De la Zerda et al (2008) showed that single-walled carbon nanotubes conjugated with cyclic Arg-Gly-Asp (RGD) peptides can be used as a contrast agent for photoacoustic imaging of tumours. Intravenous administration of these targeted nanotubes to mice bearing tumours showed eight times greater photoacoustic signal in the tumour than mice injected with non-targeted nanotubes. These results were verified ex vivo using Raman microscopy. Photoacoustic imaging of targeted single-walled carbon nanotubes may contribute to non-invasive cancer imaging and monitoring of nanotherapeutics in living subjects.[38]

Carbon nanotubes are promising new materials for molecular delivery in biological systems. The long-term fate of nanotubes intravenously injected into animals in vivo is currently unknown, an issue critical to potential clinical applications of these materials. Liu et al (2008) using the intrinsic Raman spectroscopic signatures of single-walled carbon nanotubes (SWNTs), measured the blood circulation of intravenously injected SWNTs and detected SWNTs in various organs and tissues of mice ex vivo over a period of three months. Functionalization of SWNTs by branched polyethylene-glycol (PEG) chains was developed, to prolong SWNT residence time in blood up to 1 day, relatively low uptake in the reticuloendothelial system (RES), and near-complete clearance from the main organs in approximately 2 months. Raman spectroscopy detected SWNT in the intestine, feces, kidney, and bladder of mice, suggesting excretion and clearance of SWNTs from mice via the biliary and renal pathways. No toxic side effect of SWNTs to mice was observed in necropsy, histology, and blood chemistry measurements. These findings pave the way to future biomedical applications of carbon nanotubes.[39]Liu et al. (2008) further conjugated paclitaxel to branched polyethylene glycol chains on SWNTs via a cleavable ester bond to obtain a water-soluble SWNT-PTX conjugate. SWNT-PTX affords higher efficacy in suppressing tumor growth than clinical Taxol in a murine 4T1 breast cancer model, owing to prolonged blood circulation and 10-fold higher tumor PTX uptake by SWNT delivery likely through enhanced permeability and retention. Drug molecules carried into the reticuloendothelial system are released from SWNTs and excreted via biliary pathway without causing obvious toxic effects to normal organs. Thus, nanotube drug delivery is promising for high treatment efficacy and minimum side effects for future cancer therapy with low drug doses.[40] Selective tumor targeting with a soluble, nanoscale SWNT construct mediated by appended specific antibodies was also achieved. The soluble, reactive SWNT platform was used as the starting point to build multifunctional constructs with appended antibody, metal-ion chelate, and fluorescent chromophore moieties to effect specific targeting, to carry and deliver a radiometal-ion, and to report location, respectively.[41]These constructs were found to be specifically reactive with the human cancer cells they were designed to target, both in vitro and in vivo. In a follow-up study, PET imaging was carried out to determine the tissue biodistribution and pharmacokinetics of ^{86}Y (t1/2: 14.7 h)-labeled SWNTs in a mouse model. It was found that ^{86}Y cleared from the blood within 3 hours and distributed predominantly to the kidneys, liver, spleen, and bone. Although the activity that accumulated in the kidney cleared with time, the whole-body clearance was quite slow.[42]

Most of the molecular imaging modalities detect nanoparticle only, whereas radionuclide-based imaging detects the radiolabel rather than the nanoparticle. The nanoparticle distribution is measured indirectly by assessing the localization of the radionuclide, which can provide quantitative measurement of the tumor targeting efficacy and pharmacokinetics only if the radiolabel on the nanoparticle is stable enough under physiological conditions. However, dissociation of the radionuclide (usually metal) from the chelator, and/or the radionuclide-containing polymer coating from the nanoparticle, may occur which can cause significant difference between the nanoparticle distribution and the radionuclide distribution. Thus, the biodistribution data of radiolabeled nanoparticles based on PET/SPECT imaging should always be interpreted with caution.[7]

No single molecular imaging modality is perfect and sufficient to obtain all the necessary information for a particular study.[2] For example, it is difficult to accurately quantify fluorescence signal in living subjects, particularly in deep tissues; MRI has high resolution yet it suffers from low sensitivity; Radionuclide-based imaging techniques have very high sensitivity but they have relatively poor resolution. So, combination of molecular imaging modalities can offer synergistic advantages over any modality alone. Multimodality imaging using a small molecule-based probe is very challenging due to the limited number of attachment points and the potential interference with its receptor binding affinity. For this nanoparticles can be investigated as they have large surface areas where multiple functional moieties can be incorporated for multimodality molecular imaging.[7]

Quantum dots (QDs) are inorganic fluorescent semiconductor nanoparticles with many desirable optical properties for imaging applications, such as high quantum yields, high molar extinction coefficients, strong resistance to photobleaching and chemical degradation, continuous absorption spectra spanning the ultraviolet (UV) to near-infrared (NIR, 700–900nm) range, narrow emission spectra, and large effective Stokes shifts.[41-43] However, in vivo targeting and imaging of QDs is very challenging due to the relatively large overall size (typically > 20 nm in hydrodynamic diameter) and short circulation half-lives of most QD conjugates.[41-44]Radioactive cadmium telluride/zinc sulfide ($Cd^{125m}Te/ZnS$) nanoparticles were targeted to mouse lung with antibody to mouse lung endothelium and quantified using radiological histology in order to test the *in vivo* targeting efficacy of a nanoparticle-antibody (NP–mAb) system. The nanoparticles were linked to either a monoclonal antibody to mouse lung thrombomodulin (mAb 201B) or a control antibody (mAb 33), and injected into groups of 6-week-old Balb/C female mice. Animals were sacrificed at 1, 4, 24, 72 and 144 h post-injection, and biodistribution in major organs was determined. Full body microSPECT/CT imaging was performed on a pair of mice (experimental and control) providing visual confirmation of the biodistribution. The Cd^{125m}Te/ZnS NPs conjugated to mAb 201B principally target the lungs while the nanoparticles coupled to mAb 33 accumulate in the liver and spleen. These data provide, for the first time, a quantitative measurement of the *in vivo* targeting efficacy of an inorganic nanoparticle–mAb system.[45] In a follow-up study it was found that CdTe NP, either targeted or untargeted, interact with the reticuloendothelial system very soon after intravenous injection. This interaction promotes uptake in the liver and spleen and limits even very rapid targeting efforts. For the first several hours after injection, the CdTe NP are subject to interaction with the reticuloendothelial system of the animal but then become refractory to removal. Temporary depletion of phagocytic cells can increase targeting efficiency and retention of the CdTe NP at the target site. The elimination of CdTe NP from the body is complex, and at least, some of the injected NP remain in body tissues for weeks after injection. Long whole-body retention times can lead to increased organ toxicity and radiotoxicity.[46] Quantum dots (QDs) can be used to perform multicolor images with high fluorescent intensity and are of a nanosize suitable for lymphatic imaging via direct interstitial injection. Kobayashi *et al.* (2007) showed simultaneous multicolor in vivo wavelength-resolved spectral fluorescence lymphangiography using five quantum dots with similar physical sizes but different emission spectra. This allows noninvasive and simultaneous visualization of five separate lymphatic flows draining and may have implications for predicting the route of cancer metastasis into the lymph nodes.[47]Combination of the multiplexing capabilities of both SPECT (with different isotopes) and QDs may be worth exploring in the future for multiple-event imaging. A few other reports have focused on radiolabeling QDs with PET isotopes such as ^{18}F (t1/2: 110 min) and ^{64}Cu.[48-50] However, neither incorporation of a targeting moiety nor optical imaging was carried out in these studies. Due to the difficulties in quantifying the fluorescence signal in vivo and many other technical challenges which remain to be solved, in vivo imaging of QDs is so far mostly qualitative or semi-quantitative.[51-53] PET has been routinely used in the clinic for staging and evaluating many types of cancer.[54] Development of a dual-modality agent containing both a NIR QD and a PET isotope will allow for sensitive, accurate assessment of the pharmacokinetics and tumor targeting efficacy of NIR QDs by PET, which may greatly facilitate future translation of QDs into clinical applications.[55] Vascular endothelial growth factor (VEGF)/VEGF receptor (VEGFR) signaling pathway plays a pivotal role in regulating tumor angiogenesis.[56] Many therapeutic agents targeting VEGF or VEGFR are currently in preclinical and clinical development.[57,58] Since the radiolabeled QDs primarily targeted the tumor vasculature rather than the tumor cells, we investigated VEGFR targeting of QDs in a follow-up study.[59] Tumor uptake of ^{64}Cu-labeled DOTA-QD was significantly lower than that of ^{64}Cu-labeled DOTA-QD-VEGF. Most importantly, good correlation was also observed between the results measured by ex vivo PET and NIRF imaging of excised major organs. In clinical settings, optical imaging is relevant for tissues close to the surface of the skin, tissues accessible by endoscopy, and during intraoperative visualization.[41] Combination of PET and optical imaging overcomes the tissue penetration limitation of NIRF imaging and enables quantitative in vivo targeted imaging in deep tissue, which will be crucial for future image-guided surgery through sensitive, specific, and real-time intra-operative visualization of the molecular features of normal and diseased processes. One scenario where a QD-based dual-modality PET/NIRF agent will be particularly useful is that an initial whole body PET scan can be carried out to identify the location of tumor(s), and optical imaging can be subsequently used to guide tumor resection.[15]

MRI is a non-invasive diagnostic technique based on the interaction of protons (or other nuclei) with each other and with surrounding molecules in a tissue of interest.[60] Different tissues have different relaxation times which can result in endogenous MR contrast. The major advantages of MRI over radionuclide-based imaging are the absence of radiation, higher spatial resolution (usually sub-millimeter level), and exquisite soft tissue contrast. The major disadvantage of MRI is its inherent low sensitivity, which can be partially compensated for by working at higher magnetic fields (4.7–14 T in small animal models), acquiring data for a much longer time period, and using exogenous contrast agents. IO nanoparticles, consisting of a crystalline IO core surrounded by a polymer coating such as dextran or PEG, are the most widely used nanoparticle-based MR contrast agents.[61] The presence of thousands of iron atoms in each particle can give very high T2 relaxivity.[62]

Accurate localization of PET probe uptake is very difficult in cases where anatomical structures are not identifiable, particularly in the abdomen.[63,64] MRI has exquisite soft tissue contrast and combination of PET/MR can have many synergistic effects. PET/MR imaging, acquired in one measurement, has the potential to become the imaging modality of choice for various clinical applications such as neurological studies, certain types of cancer, stroke, and the emerging field of stem cell therapy.[65] The future of PET/MR scanners will greatly benefit from the use of dual-modality PET/MR probes. Recently, an ^{124}I ($t_{1/2}$: 4.2 days)-labeled IO nanoparticle was also reported as a dual-modality PET/MR probe for lymph node imaging in rats.[66] This nanoparticle may be useful in the clinic for accurate localization and characterization of lymph nodes, which is critical for cancer staging since the lymphatic system is an important route for cancer metastasis.[68]

3. Radiation therapy with radiolabeled nanoparticles

Radiation therapy (radiotherapy) has been quite effective in the treatment of different types of cancer and minimizing the risk of local recurrence after surgical removal of the primary tumor.[76,77] Radiation kills cells largely through the generation of free radicals, which deposits a large amount of energy that can cause single- and double-strand breaks in the DNA. Generally, tumor cells are less capable of repairing DNA damage than normal cells since the tumor cells are more frequently in a sensitive cell-cycle phase, such as mitosis.[78,79] The radiation dose is divided into a number of treatment fractions to allow DNA repair to take place within the normal cells and let proliferating tumor cells redistribute through the cell cycle and move into more radiosensitive phases. The main goal of radiotherapy is to kill tumor cells selectively, without damaging the normal cells.[15]Radiation therapy utilizes radiation energy to induce cell death. By directly delivering external radiation beams to a tumor in the patient, external radiation therapy offers a relatively simple and practical approach to cause radiation damage in the tumor. Although the intensity, location and timing for external radiation can be well controlled and modulated, its main disadvantages include: 1) the destruction of normal tissue adjacent to tumors and in the path of the beam; 2) the need of high radiation doses for penetrating tissues with a large field or volume; 3) prolonged treatment with the requirement of daily hospital visits for 5–6 weeks; and 4) the use of only selected radiation sources due to the technical requirements and limitations of radiation devices and radiation sources (e.g. high energy x-rays). Therefore, external radiation treatment may not be applicable to certain cancers and not effective in the improvement of clinical symptoms.[68] In contrast to external radiation treatment, systemic radiotherapy delivers radiation energy from the radioisotopes that are conjugated to a suitable delivery carrier, such as antibodies, liposome emulsions or nanoparticles with tumor targeting ligands, and transported to the tumor site as illustrated in Figure 1.

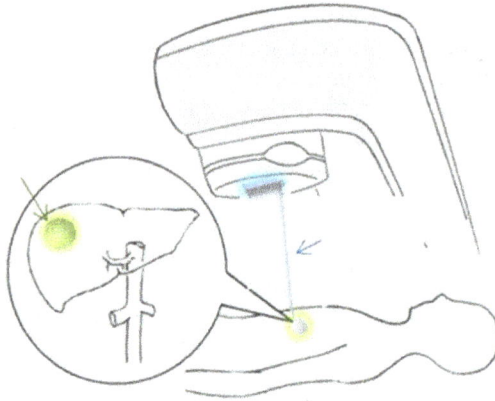

External radiation

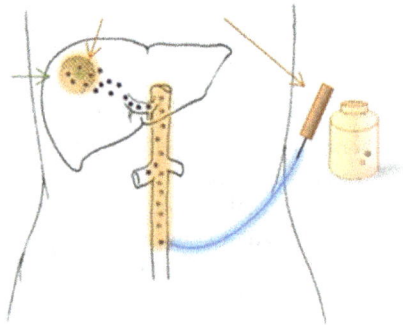

Systemic radiation

Fig. 1. External radiation therapy and systemic radiation therapy (Reference 2).

Since tumor targeted and localized delivery of radiation enhances the treatment effect and reduces the toxicity to normal tissue, systemic radiotherapy is considered to be a promising approach for personalized oncology. Although systemic radiotherapy presents major challenges in the design and production of delivery vehicles, it offers great opportunities for the application of novel nanomaterials and nanotechnologies.

Although many radioisotopes can be used as radiation sources, only a few have been developed and applied in preclinical and in vivo studies. When selecting a candidate for experimental and clinical studies, the advantages and disadvantages of radioisotopes should be evaluated based on their physical and chemical properties, patient and environmental safety, specific requirements for in vivo applications and technical feasibility. Table 1 summarizes the physical and radiation properties of therapeutic radioisotopes that have been used in previous studies. These radioisotopes can be categorized into three types, ie, α, β and auger particles.

Radioisotopes	Particle(s) emitted	Half-life	Particle energy (keV)	Maximum particle range
α-particle				
^{211}At	α	7.2 hours	6,000	0.08
^{225}Ac	α , β	10 days	6-8,000	0.1 mm
^{212}Bi	α , β	60.6 minutes	6,000	0.09 mm
^{213}Bi	α , β	46 minutes	6,000	<0.1 mm
^{223}Ra	α , β	11.4 days	6-7,000	<0.1 mm
^{212}Pb	α , β	10.6 hours	7,800	<0.1 mm
^{149}Tb	α	4.2 hours	400	<0.1 mm
β-particle				
^{131}I	β, Υ	193 hours	610	2.0 mm
^{90}Y	β	64 hours	2,280	12.0 mm
^{67}Cu	β, Υ	62 hours	577	1.8 mm
^{186}Re	β, Υ	91 hours	1,080	5.0 mm
^{177}Lu	β, Υ	161 hours	496	1.5 mm
^{64}Cu	β	12.7 hours	1,670	2.0 mm
Auger-particle				
^{67}Ga	Auger, Υ	78.3 hours	90	10 nm
^{123}I	Auger, Υ	13.3 hours	159	10 nm
^{125}I	Auger, Υ	60.5 days	27	10 nm

Abbreviation: keV, kilo electron volts.

Table 1. Characteristics of some therapeutic radioisotopes (Reference2)

3.1 α-emitters

Many radioisotopes emit α-particles but most of them decay too quickly to be considered for therapeutic use. Only a few α-emitters, including actinium-225 (^{225}Ac), astatine-211 (^{211}At), bismuth-213 (^{213}Bi) and bismuth-212 (^{212}Bi), have therapeutic potential and have been investigated in animal models or humans. α-particle emitters can eject high energy (4 – 8 MeV) helium nuclei (i.e. α-particle) which can cause severe cytotoxicity, however their ejection range is quite short (typically 40 – 80 μm). α-particles have linear energy transfer

(LET) of 100 keV/μm. LET refers to the average radiation energy deposited in tissue per unit length of track (keV/μm). Cell death occurs only when α-particles traverse the cell nucleus. By virtue of these properties, α-particles are highly efficient and specific in treatment of microscopic and small-volume tumors, or residual tumors in a variety of cancer types, including leukemia, lymphoma, glioma, melanoma, and peritoneal carcinomatosis.[76-81] However, poor radionuclide supply, complicated methodologies for calculating the radiation dosimetry and the need for relevant data relating to normal organ toxicity limit the applications of α-emitter radioisotopes and impede the development of targeted α-emitters.[82]

3.2 β-emitters

β-emitters are the most widely used radioisotopes in cancer therapy. These radioisotopes can release electrons which have lower energy and cause lower cytotoxicity than the α-particle emitters do, but they can travel a much longer distance and kill cells by indirect damage to the DNA.[15] Commonly used β-emitter radioisotopes are iodine-131 (^{131}I), yttrium-90 (^{90}Y); copper-67 (^{67}Cu), rhenium-186 (^{186}Re), lutetium-177 (^{177}Lu), and copper-64 (^{64}Cu). ^{131}I and ^{90}Y are the most popular candidates since these two isotopes are readily available and inexpensive.[131] I has a long half-life (8 days) and also provides ϒ-emissions that can be used in imaging for tracking and quantifying the radioisotope in vivo. It can be easily attached to tumor targeted antibodies.[83,84] ^{131}I gets rapidly degraded and has a short retention time in the tumor.[79] Additionally, the high energy ϒ-emission presents some safety concerns to patients and the environment. ^{90}Y has fewer environmental radiation restrictions than ^{131}I because of its pure β-emitter nature, higher energy and low-range (12 mm), and a longer residence time in the tumor, making it more suitable for the irradiation of large tumors that require a higher radiation dose and a stable link between radioisotopes and the tumor targeting antibody.[79] β -particles have lower LET and longer radiation ranges than α-particles. Because of their long radiation range (several millimeters), β-particles can destroy tumor cells through the "crossfire effect," even though the radioimmunoconjugate is not directly bound to the cells. Therefore, they are particularly useful in overcoming treatment resistance. β-particles are considered to be most suitable for the treatment of bulky or large volume tumors.[85]

3.3 Auger-emitters

An auger is a low energy (≤1.6 keV), and short- range (≤150 nm) electron derived from inner-shell electron transitioning. During the decay of these radioisotopes, the vacancy formed in the K shell as a consequence of electron capture or internal conversion is rapidly filled by electrons dropping in from higher shells, resulting in a cascade of atomic electron transitions and emitting a characteristic X-ray photon or an auger. Auger emitters, such as gallium-67 (^{67}Ga), iodine-123 (^{123}I) and iodine-125 (^{125}I), have been used for cancer radiotherapy. Auger emitters deposit high linear energy transfer (LET) over extremely short distances and are therefore most effective when the decay occurs in the nucleus and less so when the decay occurs in the cytoplasm. The dimensions of many mammalian cell nucleus components, such as chromatin fiber (30-nm), fall in the range of the auger emitter (<150 nm); therefore, auger emitters are more damaging to these cellular structures. As a result, the use of auger emitters has been relatively restricted because of the extreme toxicity of such radioisotopes.[86]Electrons from Auger electron emitters travel the shortest range (< 1 µm) and are cytotoxic only when they are very close to the nucleus, thus Auger electron emitters are generally not applicable to nanoparticle-based radiotherapy.

The selection of radioisotopes for cancer therapy should take into account the specific cancer types, characteristics of the tumor, toxicity and safety of radioisotopes, availability and production of radioisotopes, and the chemistry that is involved in assembling the radioisotopes to the delivery carriers. It has been shown that a combination of radioisotopes with different energies can be more beneficial than using a single radioisotope. It was experimentally found that the combination of a high energy and long tissue range radioisotope with a medium energy and shorter tissue range radioisotope is able to destroy both large-volume tumors and micrometastases.[79] Tumor-targeted nanocarriers have been designed to deliver radionuclide payloads in a selective manner to improve the efficacy and safety of cancer imaging and therapy.[1,73-75]

4. Antibodies conjugated radioisotopes for tumor targeted radiation therapy

Most of the anticancer drugs are unable to differentiate between cancerous and normal cells, leading to systemic toxicity and adverse effects. A simple tumor targeting strategy is the use of monoclonal antibodies (mAb) interacting with cancer cell surface markers.[87]

Small, high affinity antibody fragments, such as single chain antibodies and affibodies, which are expressed as recombinant proteins in prokaryotic cells, are cost effective targeting ligand.[88] They have been extensively investigated for the delivery of radioisotopes, as internal radiation sources. [89-91] This method of using mAbs conjugated with radioisotopes for internal or systemic radiation treatment is known as radioimmunotherapy.

There are several limitations in using antibody conjugated radioisotopes for the delivery of radiation therapy. Firstly, mAbs may bind to cell surface markers on normal tissues, causing potential systemic toxicity. Secondly, mAbs have only a few sites available for conjugating radioisotopes. Therefore, delivery of a large dose of therapeutic radioisotopes may require a larger amount of antibodies. Thirdly, the use of mAbs presents potentially unwanted immune responses. Additionally, antibodies may be susceptible to protease degradation. Attempts are being made to overcome these limitations using nanoparticulate delivery systems.

5. Biocompatible nanoparticles

Nanoparticles are colloidal materials that can be fabricated with a variety of compositions and morphologies using special techniques and chemistries. Nanomaterials currently used in biomedical applications include fluorescent CdSe nanoparticles known as quantum dots (QDs), photosensitive gold nanoparticles, magnetic nanoparticles, as well as polymeric nanoparticles and nanoscale liposomes. Nanoparticles, especially metallic and metal oxide nanoparticles, in the "mesoscopic" size range of 5–100 nm in diameter often exhibit unique chemical and physical properties that are not possessed by their bulk or molecular counterparts.[92] For example, QDs made from CdSe exhibit photoluminescence with a controllable wavelength ranging from the visible to near infrared depending on their size. Colloidal gold nanoparticles exhibit unique surface plasmon resonance (SPR) properties derived from the interaction of electromagnetic waves with the electrons in the conduction band.[93] Magnetic nanoparticles, such as iron III oxide (Fe_3O_4), are superparamagnetic and exhibit high magnetization and yet no residual magnetization in the absence of an externally applied magnetic field.[94] Both the chemical properties and reactivity of the nanoparticles are controlled by the surface chemistries offered by functionalized polymer coatings or blocks, which are also important to the stability and biocompatibility of the nanoparticles, interactions between particles, biomolecules and cells, in addition to tissue distributions of the nanoparticles. Nanoparticles provide a large surface area and various types of functional groups that allow for chemical reactions taking place on the nanoparticle surface and to assemble or load bioactive ligands or small molecular agents.

6. Biocompatibilities and functionalization of nanoparticles

Metal oxide nanoparticles are coated with polymers to stabilize them from aggregating and precipitating in physiological conditions while maintaining the desired physical properties. These polymers improve biocompatibility of metallic nanoparticles by minimizing toxicity and modulating interactions between nanoparticles and biomolecules, cells and tissues. They alter secretion and biodistribution of nanoparticles. Coating polymers are functionalized with reactive functional groups, such as $-COOH$, NH_2 and $-SH$, for conjugation with tumor targeting ligands. For carrying and delivering therapeutic radioisotopes, coating polymers with reactive functional groups allows for covalent cross-linking or non-covalently incorporating chelates of radioisotopes. There are a variety of polymers and their derivatives, such as dextran, polyethylene glycol (PEG), and dendrimer, developed for ensuring the biocompatibility and functionalization of nanoparticles.[95,96] For instance, Zhang and co-workers modified the surface of iron oxide nanoparticles with trifluoroethyl ester-terminal-PEG-silane, which was then converted to an amine-terminated PEG.[97] The terminal amine groups were used for the conjugation of Cy5.5, a near infrared (NIR) optical probe, and chlorotoxin, a targeting peptide for glioma tumors. In vitro MRI and confocal fluorescence microscopy showed a strong preferential uptake of the multimodal nanoparticles by glioma cells compared to the control nanoparticles and noncancerous cells. To reduce nonspecific uptake of nanoparticles by normal tissues and extend the blood circulation time of nanoparticles to allow particle accumulation at the target site, polymer coatings must be specifically designed to meet such applications. The physical characteristics of polymer coated nanoparticles affect their in vivo performance. Surface morphology, overall particle size and surface charge are all considered important factors that determine toxicity and biodistribution. The overall particle size must be small enough to evade uptake by reticuloendothelial system (RES) but large enough to avoid renal clearance, leaving a window of between 5.5 and 200 nm.[98] However, it has also been demonstrated that for particles smaller than 40 nm in diameter, both the biodistribution and the half-life of nanoparticles are determined by the coating material rather than the mean size.[99] The surface charge of nanoparticles depends on the nature of the coating material, which in turn plays an important role in determining blood half-life. Nanoparticles with strong positive or negative charge tend to bind to cells. [100,101] Thus, nanoparticles with a neutral surface charge are recommended to extend circulation times. Nanoparticles with neutral surfaces resist protein binding and provide steric hindrance for preventing aggregation after in vivo administration.[98] New coating materials composed of zwitterionic polymers have been developed to provide a biocompatible surface with both positive and negative charges, which exhibit high resistance to nonspecific protein adsorption and uptake by macrophages in liver and spleen.[102-114] Chen and colleagues developed an antibiofouling copolymer PEO-b-P$_T$MPS for coating nanocrystals.[115] This new copolymer made hydrophobic iron oxide nanoparticles mono-disperse in physiological conditions with great stability. This amphiphilic blocked coating polymer can be functionalized with reactive amine groups on the particle surface, making it readily available for the conjugation of tumor targeting ligands and therapeutic agents such as radioisotope chelates. These composite nanoparticles showed a reduced nonspecific uptake.

7. Advantages of nanoparticulate drug delivery system

Radioisotopes undergo rapid elimination due to their widespread distributions into normal organs and tissues. One common solution to this problem is to administer large quantity of radioisotope, which is not cost effective and often results in undesirable toxicity. Nanoparticles with proper biocompatible polymer coatings provide better platforms for carrying radioisotopes and subsequently delivering the agents to the tumor. There are several advantages of using nanoparticles to deliver therapeutic radioisotopes:

i.   Nanoparticles have prolonged blood retention time, ranging from 30 minutes to 24 hours, depending on the morphology and size of the particle, coating materials and compositions of nanoparticle conjugates.
ii. Nanoparticulate carriers used for targeting cancer cells exhibit high tumor retention time and thus enhance the concentration of therapeutic agents.
iii. Nanoparticles have high loading capacity, they can even carry more than one type of radioisotope.
iv. Internalization of receptor targeted nanoparticles leads to the uptake of large amounts of radioisotopes into the target cells, resulting in effective killing of tumor cells with a relatively low level of receptor-expression.
v.   The unique chemical and physical properties of nanoparticles, such as magnetization and photosensitizing provide additional capabilities and functions for improving delivery of the radioisotopes and monitoring the response to radiotherapy.

With the controlled formulation and optimized drug carrying strategies, nanoparticle platforms may offer appropriate pharmacokinetics for optimal delivery of radioisotopes for cancer treatment. Radioisotopes can be conjugated on to hydrophilic functional groups present on the surface of micelles for better transport.[106]Nanoparticulate drug delivery system often possess multi-functional capacity which enables it to load multiple moieties like targeting ligands and therapeutic agents. This is of immense importance to tumor targeted delivery of radioisotopes in vivo. It has been reported that nanoparticles consisting of streptavidin that linked three biotinylated components: the antiHer2 antibody trastuzumab, the tat peptide and the ^{111}In-labeled antiRIa messenger RNA antisense morpholino (MORF) oligomer, produce significant radiation-induced antisense mediated cytotoxicity of tumor cells in vitro.[107]

8. Strategies for targeting nanoparticles to cancer

Principally, two mechanisms are used for targeting nanoparticles to tumors, passive and active targeting. In passive targeting, nanoparticles reach the tumor through highly permeable tumor vasculature. They get accumulated in the tumor and subsequently remain their due to its lack of lymphatic drainage. In active targeting, nanoparticles are engineered to target specific biomarker molecules that are unique and over populated in a tumor or cancer cell surface. Differences in the expression of cellular receptors between normal and tumor cells provides an opportunity for targeting nanoparticles to cancer cells. Active targeting nanoparticles carrying radioisotopes to tumors are the current research focus and the subject of intensive investigations. The surface coating polymer of nanoparticles is conjugated to ligands like antibodies, peptides and small molecules targeting the receptors highly expressed on tumor cells. Extensive reviews and discussions on the mechanisms of targeting nanoparticles to tumors have been published.[3,108-109] In vivo tumor targeting have been achieved using folic acid modified dendrimers,[111] synthetic small-molecule modified iron oxide nanoparticles,[112] PEGylated arginine-glycine-aspartic acid peptide modified carbon nanotubes[113] and PEGylated single chain variable fragment antibody modified gold nanoparticles.[93] Tumor targeted nanoparticles are believed to be a promising platform for nanobiotechnology.

Antibodies have been extensively studied as tumor targeting ligand for magnetic or photosensitive nanoparticles in the area of cancer imaging. Conjugates of nanoparticles and antibodies were found to retain the properties of both the antibody and the nanoparticle. Herceptin, a well-known antibody against HER2/neu receptors which are over-expressed in breast cancer cells when conjugated with magnetic iron oxide nanoparticles showed in vivo cancer targeting and imaging of HER2/neu with a high sensitivity. The smallest parts of the antibody, the so called ScFv, are among the frequently used ligands. Nanoparticles conjugated with mAb fragments have increased circulation times in the blood compared to nanoparticles conjugated with whole mAbs. Because mAb fragments lack the Fc domains which binds to Fc receptors on phagocytic cells.[114]

Besides mAbs and antibody fragments, small molecule ligands can be readily obtained from chemical synthesis in a large quantity. Small peptide ligands, such as Arg-Gly-Asp (RGD) has high affinity for tumor integrins $\alpha_v\beta_3$ or $\alpha_v\beta_5$ in its conformationally constrained cyclic form than its linear form, have been extensively investigated for their applications in delivering tumor targeted nanoparticles carrying imaging and therapeutic agents. This increases the probability of RGD-targeted nanoparticles to act on tumor endothelial cells and produce anti-angiogenesis effect.[115,116] The folate receptor (FR) is an attractive molecular target for tumor targeting because it is over expressed by most of tumor cells like ovarian, colorectal, breast, nasopharyngeal carcinomas and has limited expression in normal tissues.[117,118]FR-mediated tumor delivery of drugs, gene products, radionuclides and nanoparticles for imaging have been reported.[119-121] Folic acid, attached to polyethyleneglycol-derivatized, distearyl-phosphatidylethanolamine, was used to target in vitro liposomes to folate receptor (FR)-overexpressing tumor cells. Confocal fluorescence microscopic observations demonstrated binding and subsequent internalization of rhodamine-labeled liposomes by a high FR-expressing, murine lung carcinoma line (M109-HiFR cells), with inhibition by free folic acid. Additional experiments tracking doxorubicin (DOX) fluorescence with DOX-loaded, folate-targeted liposomes (FTLs) indicate that liposomal DOX is rapidly internalized, released in the cytoplasmic compartment, and, shortly thereafter, detected in the nucleus, the entire process lasting 1-2 h. FR-mediated cell uptake of targeted liposomal DOX into a multidrug-resistant subline of M109-HiFR cells (M109R-HiFR) was unaffected by P-glycoprotein-mediated drug efflux.[122]

9. Radioisotope loaded nanoparticles for tumor targeting

Radioisotopes are very powerful agents in diagnosis and treatment of solid tumors.[130-131] However they lack tumor selectivity, damage surrounding normal tissues and organs, which results severe toxicity that often outweighs their anti-tumor effects. Many researchers have focused on the direct administration of radioisotopes into the tumor site. This method appears to be very effective, but failed due to rapid clearance from the injected tumor site. There is a strong indication that radioisotope carriers can improve the efficiency of intratumoral administration.[126,127] Suzuki et al. described the biodistribution and kinetics of the Holmium 166–chitosan complex in rats and mice. They suggested that chitosan prolonged the retention time of Holmium 166 in the tumor site.[126] Nakajo et al. also designed a ^{131}I-labeled lipiodol for the treatment of liver cancer patients. They determined that the radioactive concentration in blood after ^{131}I-lipiodol administration could be maintained at levels as low as $10 \times 10^{-4}\%$ injected dose (ID)/mL for 8 days.[127] Stimuli-sensitive polymeric nano-carriers are another potential candidate for intratumoral radioisotope administration. They are readily administered due to their favorable biocompatibility, small size, and low viscosity. Alterations in their properties (hydrophilic to hydrophobic) help them accumulate in tumor sites, which results in prolongation of radioisotope retention time.[128] Very few studies have been conducted using polymeric nanoparticles labeled radioisotopes for anti-tumor treatment.[129] Conventional polymeric nanoparticles from amphiphilic block or random copolymers possess insufficient functional groups for radioisotope labeling, which results in lower labeling efficiency. Whereas, self assembled nanoparticles from polysaccharide derivatives facilitate the tagging process, as a consequence of their abundant functional groups, which enable the direct labeling of radioisotopes. Park et al. utilized ionic strength (IS)-sensitivity in the development of new radioisotope carriers for intratumoral administration. A polysaccharide derivative, pullulan acetate nanoparticle (PAN)was prepared via dialysis. The PAN had a spherical shape with size range of 50–130 nm and a low critical aggregation concentration (CAC) (<8_g/mL). With increases in the IS of the dialysis media, the CAC of PAN was reduced gradually and the rigidity of the hydrophobic core in PAN was increased. This suggests that the property of PAN was altered more hydrophobically at high IS values. PAN evidenced a high degree of [99m]Technetium (^{99m}Tc) labeling efficiency (approximately 98%). The percentage retention rate (%RR) of the ^{99m}Tc-labeled PAN was significantly longer than that of the free [99m]Tc ($p < 0.05$), due largely to PAN's IS-sensitivity. Thus, PAN may constitute a new approach to the achievement of maximal radioisotope efficiency with regard to intratumoral administration.[122] Noninvasive, focused hyperthermia can be achieved by using an externally applied alternating magnetic field (AMF) if effective concentrations of nanoparticles can be delivered to the target cancer cells. Monoclonal antibodies or peptides, linked to magnetic iron oxide nanoparticles (NP), represent a promising strategy to target cancer cells. A new radioconjugate NP ((111)In-DOTA-di-scFv-NP), using recombinantly generated antibody fragments, di-scFv-c, for the imaging and therapy of anti-MUC-1-expressing cancers was developed by Natarajan et al.[130]

10. Future implications

With the advent of nanotechnology, researchers world over are interested in designing a magic bullet which would detect the malignant tissue and destroy it. Radionuclides can be targeted to malignant tissue by coupling them to antibodies or their parts. These radioimmunoconjugates are being developed to meet the challenges facing cancer detection and therapy today and in the future. Several radiolabeled multifunctional and multimodality nanoparticles have been effectively demonstrated in detecting and treating cancer in animal models. However, further preclinical and clinical efficacy and toxicity studies are required to translate these advanced technologies to the health care of cancer patients.

11. References

[1] Ting G, Chang CH, Wang HE. Cancer nanotargeted radiopharmaceuticals for tumor imaging and therapy. *Anticancer Res.* 2009 ;29(10):4107-18.

[2] American Cancer Society, Inc., Surveillance and Health Policy Research, 2010.

[3] Zhang et al, Nanotechnology, Science and Applications 2010:3: 159–170.

[4] Jarzyna PA, Skajaa T, Gianella A, et al. Iron oxide core oil-in-water emulsions as a multifunctional nanoparticle platform for tumor targeting and imaging. *Biomaterials.* 2009;30:6947–6954.

[5] Peng X, Qian X, Mao H, et al. Targeted magnetic iron oxide nanoparticles for tumor imaging and therapy. *Int J Nanomed.* 2008;3:311–321.

[6] Yang L, Peng X, Wang A, et al. Receptor-targeted nanoparticles for in vivo imaging of breast cancer. *Clin Cancer Res.* 2009;15:4722–4732.

[7] Shenoy D, Little S, Langer R, Amiji M. Poly(ethylene oxide)-modified poly(beta-amino ester) nanoparticles as a pH-sensitive system for tumor-targeted delivery of hydrophobic drugs. 1. In vitro evaluations. *Mol Pharm.* 2005;2:357–366.

[8] Bae Y, Diezi TA, Zhao A, Kwon GS. Mixed polymeric micelles for combination cancer chemotherapy through the concurrent delivery of multiple chemotherapeutic agents. *J Control Release.* 2007;122:324–330.

[9] Lee H, Lee E, Kim K, Jang NK, Jeong YY, Jon S. Antibiofouling polymer-coated superparamagnetic iron oxide nanoparticles as potential magnetic resonance contrast agents for in vivo cancer imaging. *J Am Chem Soc.* 2006;128:7383- 7389.

[10] Mankoff DA. A definition of molecular imaging. *J Nucl Med.* 2007;48:18N. 21N.

[11] Massoud TF, Gambhir SS. "Molecular imaging in living subjects: Seeing fundamental biological processes in a new light"*Genes Dev* 2003;17:545-580.

[12] Koo V, Hamilton PW, Williamson K. Non-invasive in vivo imaging in small animal research. *Cell. Oncol.* 2006;28:127–139.

[13] Pomper MG, Lee JS. Small animal imaging in drug development. *Curr Pharm Des.* 2005;11:3247–3272.

[14] Cai W, Rao J, Gambhir SS, Chen X. How molecular imaging is speeding up anti-angiogenic drug development. *Molecular Cancer Therapeutics*, 2006, 5, 2624-2633.

[15] Gambhir SS. Molecular imaging of cancer with positron emission tomography. *Nat Rev Cancer* 2002;2:683-693.

[16] Hong H, Zhang Y, Sun J, Cai W. Molecular imaging and therapy of cancer with radiolabeled nanoparticles. *Nano Today* 2009; 4: 399-413.

[17] Peremans K, Cornelissen B, Van Den Bossche B, Audenaert K, Van de Wiele C. A review of small animal imaging planar and pinhole spect Gamma camera imaging. *Vet Radiol Ultrasound* 2005;46:162-170.

[18] Kjaer A. Molecular imaging of cancer using PET and SPECT. *Adv Exp Med Biol* 2006;587:277-284.

[19] DeNardo SJ, DeNardo GL, Miers LA, Natarajan A, Foreman AR, Gruettner C, Adamson GN, Ivkov R. *Clin Cancer Res* Development of tumor targeting bioprobes

(^{111}In-chimeric L6 monoclonal antibody nanoparticles) for alternating magnetic field cancer therapy. 2005;11:7087s-7092s.

[20] DeNardo SJ, DeNardo GL, Natarajan A, Miers LA, Foreman AR, Gruettner C, Adamson GN, Ivkov R. Thermal dosimetry predictive of efficacy of ^{111}In-ChL6 nanoparticle AMF--induced thermoablative therapy for human breast cancer in mice. *J Nucl Med.* 2007 48(3):437-444.

[21] Natarajan A, Gruettner C, Ivkov R, DeNardo GL, Mirick G, Yuan A, Foreman A, DeNardo SJ. NanoFerrite particle based radioimmunonanoparticles: binding affinity and in vivo pharmacokinetics. *Bioconjug Chem* 2008;19:1211-1218.

[22] Natarajan A, Xiong CY, Gruettner C, DeNardo GL, DeNardo SJ. Development of multivalent radioimmunonanoparticles for cancer imaging and therapy. *Cancer Biother Radiopharm* 2008;23:82-91.

[23] Hood JD, Cheresh DA. Role of integrins in cell invasion and migration. *Nat Rev Cancer.* 2002;2:91-100.

[24] Cai W, Chen X. Anti-angiogenic cancer therapy based on integrin $\alpha_v\beta_3$ antagonism. *Anti-cancer Agents Med Chem*, 2006; 6: 407-428.

[25] Cai W, Niu G, Chen X. Imaging of integrins as biomarkers for tumor angiogenesis. *Curr Pharm Des* 2008;14:2943-2973.

[26] Hu G, Lijowski M, Zhang H, Partlow KC, Caruthers SD, Kiefer G, Gulyas G, Athey P, Scott MJ, Wickline SA, Lanza GM. Imaging of Vx-2 rabbit tumors with alpha(nu)beta3-integrin-targeted 111In nanoparticles. *Int J Cancer.* 2007; 1;120:1951-1957.

[27] Wang H, Wang J, Deng X, Sun H, Shi Z, Gu Z, Liu Y, Zhao Y. Biodistribution of single-wall carbon nanotubes in mice. *J Nanosci Nanotechnol* 2004;4:1019-1024.

[28] Singh R, Pantarotto D, Lacerda L, Pastorin G, Klumpp C, Prato M, Bianco A, Kostarelos K. *Proc Natl Acad Sci USA* 2006;103:3357-3362.

[29] Villa CH, McDevitt MR, Escorcia FE, Rey DA, Bergkvist M, Batt CA, et al. Synthesis and Biodistribution of Oligonucleotide-Functionalized, Tumor-Targetable Carbon Nanotubes. *Nano Lett.* 2008; 8: 4221-4228.

[30] Chen GD, Fachin F, Fernandez-Suarez M, Wardle BL, Toner M. Nanoporous Elements in Microfluidics for Multiscale Manipulation of Bioparticles. *Small*, 2011; DOI: 10.1002/smll.201002076.

[31] Shokouhi S, Metzler SD, Wilson DW, Peterson TE. Multi-pinhole collimator design for small-object imaging with SiliSPECT: a high-resolution SPECT. *Phys Med Biol* 2009;54:207-225.

[32] Berman DS, Kiat H, Van Train K, Friedman JD, Wang FP, Germano G. Dual-isotope myocardial perfusion SPECT with rest thallium-201 and stress Tc-99m sestamibi. *Cardiol Clin* 1994;12:261-270.

[33] Visser EP, Disselhorst JA, Brom M, Laverman P, Gotthardt M, Oyen WJ, Boerman OC. Spatial resolution and sensitivity of the Inveon small-animal PET scanner. *J Nucl Med* 2009;50:139-147.

[34] Liu Z, Cai W, He L, Nakayama N, Chen K, Sun X, Chen X, Dai H. In vivo biodistribution and highly efficient tumour targeting of carbon nanotubes in mice. *Nat Nanotechnol* 2007;2:47-52.

[35] Keren S, Zavaleta C, Cheng Z, de la Zerda A, Gheysens O, Gambhir SS. Noninvasive molecular imaging of small living subjects using Raman spectroscopy. *Proc Natl Acad Sci USA* 2008;105:5844-5849.

[36] Zavaleta C, de la Zerda A, Liu Z, Keren S, Cheng Z, Schipper M, Chen X, Dai H, Gambhir SS. Noninvasive Raman spectroscopy in living mice for evaluation of tumor targeting with carbon nanotubes. *Nano Lett* 2008;8:2800-2805.

[37] De la Zerda A, Zavaleta C, Keren S, Vaithilingam S, Bodapati S, Liu Z, et al. Carbon nanotubes as photoacoustic molecular imaging agents in living mice. *Nat Nanotechnol* 2008;3:557-562.

[38] Liu Z, Davis C, Cai W, He L, Chen X, Dai H. Circulation and long-term fate of functionalized, biocompatible single-walled carbon nanotubes in mice probed by Raman spectroscopy. *Proc Natl Acad Sci* USA 2008;105:1410-1415.

[39] Liu Z, Chen K, Davis C, Sherlock S, Cao Q, Chen X, Dai H. Drug delivery with carbon nanotubes for in vivo cancer treatment. *Cancer Res* 2008;68:6652-6660.

[40] McDevitt MR, Chattopadhyay D, Kappel BJ, Jaggi JS, Schiffman SR, Antczak C, Njardarson JT, Brentjens R, Scheinberg DA. Tumor targeting with carbon nanotubes. *J Nucl Med* 2007;48:1180-1189.

[41] McDevitt MR, Chattopadhyay D, Jaggi JS, Finn RD, Zanzonico PB, Villa C, Rey D, Mendenhall J, Batt CA, Njardarson JT, Scheinberg DA. *PLOS ONE* 2007;2:e907.

[42] Cai W, Hsu AR, Li ZB, Chen X. Are quantum dots ready for *in vivo* imaging in human subjects?. *Nanoscale Res Lett.* 2007;2:265-281.

[43] Michalet X, Pinaud FF, Bentolila LA, Tsay JM, Doose S, Li JJ, Sundaresan G, Wu AM, Gambhir SS, Weiss S. Quantum dots for live cells, in vivo imaging, and diagnostics. *Science* 2005;307:538.

[44] Medintz IL, Uyeda HT, Goldman ER, Mattoussi H. Quantum dot bioconjugates for imaging, labelling and sensing *Nat Mater* 2005;4:435-446.

[45] Li ZB, Cai W, Chen X. Semiconductor quantum dots for in vivo imaging. *J Nanosci Nanotechnol* 2007;7:2567-2581.

[46] Woodward JD, Kennel SJ, Mirzadeh S, Dai S, Wall JS, Richey T, Avenell J, Rondinone AJ. *In vivo* SPECT/CT imaging and biodistribution using radioactive Cd125mTe/ZnS nanoparticles. *Nanotechnology* 2007;18:175103.

[47] Kennel SJ, Woodward JD, Rondinone AJ, Wall J, Huang Y, Mirzadeh S. The fate of MAb-targeted Cd[125m]Te/ZnS nanoparticles in vivo. *Nucl Med Biol* 2008;35:501-514.

[48] Kobayashi H, Hama Y, Koyama Y, Barrett T, Regino CA, Urano Y, Choyke PL. Simultaneous multicolor imaging of five different lymphatic basins using quantum dots. *Nano Lett* 2007;7:1711-1716.

[49] Schipper ML, Cheng Z, Lee SW, Bentolila LA, Iyer G, Rao J, Chen X, Wu AM, Weiss S, Gambhir SS. microPET-based biodistribution of quantum dots in living mice. *J Nucl Med.* 2007;48:1511-1518.

[50] Schipper ML, Iyer G, Koh AL, Cheng Z, Ebenstein Y, Aharoni A, Keren S, Bentolila LA, Li J, Rao J, Chen X, Banin U, Wu AM, Sinclair R, Weiss S, Gambhir SS. Particle size, surface coating, and PEGylation influence the biodistribution of quantum dots in living mice. *Small* 2009;5:126-134.

[51] Duconge F, Pons T, Pestourie C, Herin L, Theze B, Gombert K, Mahler B, Hinnen F, Kühnast B, Dollé F, Dubertret B, Tavitian B. Fluorine-18-labeled phospholipid

quantum dot micelles for in vivo multimodal imaging from whole body to cellular scales. *Bioconjug Chem* 2008;19:1921-1926.

[52] Cai W, Shin DW, Chen K, Gheysens O, Cao Q, Wang SX, Gambhir SS, Chen X. Peptide-labeled near-infrared quantum dots for imaging tumor vasculature in living subjects. *Nano Lett* 2006;6:669-676.

[53] Gao X, Cui Y, Levenson RM, Chung LWK, Nie S. In vivo cancer targeting and imaging with semiconductor quantum dots. *Nat Biotechnol* 2004;22:969-976. 2007;67:1138.

[54] Tada H, Higuchi H, Wanatabe TM, Ohuchi N. In vivo real-time tracking of single quantum dots conjugated with monoclonal anti-HER2 antibody in tumors of mice. *Cancer Res* 2007;67:1138-44.

[55] Gambhir SS, Czernin J, Schwimmer J, Silverman DH, Coleman RE, Phelps ME. A tabulated summary of the FDG PET literature. *J Nucl Med* 2001;42:1S-93S.

[56] Cai W, Chen K, Li ZB, Gambhir SS, Chen X. Dual-function probe for PET and near-infrared fluorescence imaging of tumor vasculature. *J Nucl Med* 2007;48:1862-1870.

[57] Cai W, Chen X. Multimodality imaging of vascular endothelial growth factor and vascular endothelial growth factor receptor expression. *Front Biosci* 2007;12:4267-4279.

[58] Ellis LM, Hicklin DJ. VEGF-targeted therapy: mechanisms of anti-tumour activity. *Nat Rev Cancer* 2008;8:579-591.

[59] Ferrara N, Hillan KJ, Gerber HP, Novotny W. Discovery and development of bevacizumab, an anti-VEGF antibody for treating cancer. *Nat Rev Drug Discov* 2004;3:391-400.

[60] Chen K, Li ZB, Wang H, Cai W, Chen X. Dual-modality optical and positron emission tomography imaging of vascular endothelial growth factor receptor on tumor vasculature using quantum dots. *Eur J Nucl Med Mol Imaging* 2008;35:2235-2244.

[61] Pathak AP, Gimi B, Glunde K, Ackerstaff E, Artemov D, Bhujwalla ZM. Molecular and functional imaging of cancer: advances in MRI and MRS. *Methods Enzymol* 2004;386:3-60.

[62] Thorek DL, Chen AK, Czupryna J, Tsourkas A. Superparamagnetic iron oxide nanoparticle probes for molecular imaging *Ann Biomed Eng* 2006;34:23-38.

[63] Bertini I, Kowalewski J, Luchinat C, Parigi G. Cross correlation between the dipole-dipole interaction and the Curie spin relaxation: the effect of anisotropic magnetic susceptibility. *J Magn Reson* 2001;152:103-108.

[64] Ruf J, Lopez Hanninen E, Oettle H, Plotkin M, Pelzer U, Stroszczynski C, Felix R, Amthauer H. Detection of recurrent pancreatic cancer: comparison of FDG-PET with CT/MRI. *Pancreatology* 2005;5:266-272.

[65] Pannu HK, Cohade C, Bristow RE, Fishman EK, Wahl RL. PET-CT detection of abdominal recurrence of ovarian cancer: radiologic-surgical correlation. *Abdom Imaging* 2004;29:398-403.

[66] Wehrl HF, Judenhofer MS, Wiehr S, Pichler BJ. Pre-clinical PET/MR: technological advances and new perspectives in biomedical research. *Eur J Nucl Med Mol Imaging* 2009;36:S56-S68.

[67] Choi JS, Park JC, Nah H, Woo S, Oh J, Kim KM, Cheon GJ, Chang Y, Yoo J, Cheon J. A hybrid nanoparticle probe for dual-modality positron emission tomography and magnetic resonance imaging. *Angew Chem Int Ed Engl* 2008;47:6259-6262.

[68] Bogenrieder T, Herlyn M. Axis of evil: molecular mechanisms of cancer metastasis. *Oncogene* 2003;22:6524-6536.

[69] Korb L. Radiotherapy for the palliation of prostate cancer. *Semin Surg Oncol.* 2000;18:75-79.

[70] Sharkey RM, Goldenberg DM. Novel radioimmunopharmaceuticals for cancer imaging and therapy. *Curr Opin Investig Drugs* 2008;9:1302.

[71] Buchholz TA. Radiation therapy for early-stage breast cancer after breast-conserving surgery. *N Engl J Med* 2009;360:6370.

[72] Sakata K, Someya M, Matsumoto Y, Hareyama M. Ability to repair DNA double-strand breaks related to cancer susceptibility and radiosensitivity. *Radiat Med* 2007;25:433-438.

[73] Sankaranarayanan K, Chakraborty R. Cancer predisposition, radiosensitivity and the risk of radiation-induced cancers. I.Background. *Radiat Res* 1995;143:121-143.

[74] Mitra A, Nan A, Line BR, Ghandehari H. Nanocarriers for Nuclear Imaging and Radiotherapy of Cancer. *Curr Pharm Res* 2006;12:4729-4749.

[75] Elbayoumi TA, Torchilin VP. Current trends in liposome research. *Methods Mol Biol.* 2010;605:1-27.

[76] Torchilin VP. Targeted Pharmaceutical Nanocarriers for Cancer Therapy and Imaging. *AAPS* 2007; 9; E128-E147.

[77] Hartmann F, Horak EM, Garmestani K, et al. Radioimmunotherapy of nude mice bearing a human interleukin 2 receptor alpha-expressing lymphoma utilizing the alpha-emitting radionuclide-conjugated monoclonal antibody [213]Bi-anti-Tac. *Cancer Res.* 1994;54:4362–4370.

[78] Kennel SJ, Stabin M, Roeske JC, et al. Radiotoxicity of bismuth-213 bound to membranes of monolayer and spheroid cultures of tumor cells. *Radiat Res.* 1999;151:244-256.

[79] Ballangrud AM, Yang WH, Charlton DE, et al. Response of LNCaP spheroids after treatment with an alpha-particle emitter ([213]Bi)-labeled anti-prostate-specific membrane antigen antibody (J591). *Cancer Res.* 2001;61:2008-2014.

[80] Goldenberg DM. Targeted therapy of cancer with radiolabeled antibodies. *J Nucl Med.* 2002;43:693-713.

[81] Salvatori M, Indovina1 L, Mansi L. Targeted α-particle therapy: a clinical overview. *Curr. Radiopharm.* 2008;1:251-253.

[82] Zalutsky MR. Targeted a-particle therapy of microscopic disease: providing a further rationale for clinical investigation. *J Nucl Med.* 2006;47:1238-1240.

[83] Mühlhausen U, Schirrmacher R, Piel M, et al. Synthesis of [131]Ilabeled glucose-conjugated inhibitors of O6-methylguanine- DNA methyltransferase (MGMT) and comparison with nonconjugated inhibitors as potential tools for in vivo MGMTimaging. *J Med Chem.* 2006;49:263-272.

[84] Schipper ML, Riese CG, Seitz S, et al. Efficacy of 99mTc pertechnetate and ^{131}I radioisotope therapy in sodium/iodide Symporter (NIS)-expressing neuroendocrine tumors in vivo. *Eur J Nucl Med Mol Imaging.* 2007;34:63 :638-650.

[85] Mulford DA, Scheinberg DA, Jurcic JG. The promise of targeted α-particle therapy. *J Nucl Med*. 2005;46:199S–204S.

[86] Mitra A, Nan A, Line BR, et al. Nanocarriers for nuclear imaging and radiotherapy of cancer. *Curr Pharm Des*. 2006;12:4729–4749.

[87] Saravanakumar G, Kim K, Park JH, et al. Current status of nanoparticle based imaging agents for early diagnosis of cancer and atherosclerosis. *J Biomed Nanotechno*. 2009;5:20–35.

[88] Marcucci F, Lefoulon F. Active targeting with particulate drug carriers in tumor therapy: fundamentals and recent progress. *Drug Discovery Today*. 2004;9:219–228.

[89] Li L, Wartchow CA, Danthi SN, et al. A novel antiangiogenesis therapy using an integrin or anti-FLK-1 antibody cotated ^{90}Y-labeled nanoparticles. *Int J Radiation Oncology Biol Phys*. 2004;58:1215–1227.

[90] DeNardo SJ, DeNardo GL, Miers LA, et al. Development of tumor targeting bioprobes (^{111}In-chimeric L6 monoclonal antibody nanoparticles) for alternating magnetic field cancer therapy. *Clin Cancer Res*. 2005;11:7087S–7092S.

[91] Chen J, Wu H, Han D, et al. Using anti-VEGF McAb and magnetic nanoparticles as double-targeting vector for the radioimmunotherapy of liver cancer. *Cancer Letters*. 2006;231:169–175.

[92] Nie S, Xing Y, Kim GJ, Simons JW. Nanotechnology applications in cancer. *Annu Rev Biomed Eng*. 2007;9:257–288.

[93] Qian X, Peng XH, Ansari DO, et al. In vivo tumor targeting and spectroscopic detection with surface-enhanced Raman nanoparticle tags. *Nat Biotechnol*. 2008;26:83–90.

[94] Gupta AK, Gupta M. Synthesis and surface engineering of iron oxide nanoparticles for biomedical applications. *Biomaterials*. 2005;26:3995–4021.

[95] Veiseh O, Sun C, Gunn J, et al. Optical and MRI multifunctional nanoprobe for targeting gliomas. *Nano Letters*. 2005;5:1003–1008.

[96] Xie J, Xu C, Kohler Y, et al. Controlled PEGylation of monodisperse Fe_3O_4 nanoparticles for reduced non-specific uptake by macrophage cells. *Adv Mater*. 2007;19:3163–3166.

[97] Veiseh O, Sun C, Fang C, et al. Specific targeting of brain tumors with an optical/magnetic resonance imaging nanoprobe across the blood brain barrier. *Cancer Res*. 2009;69:6200–6207.

[98] Sun C, Lee JS, Zhang M. Magnetic nanoparticles in MR imaging and drug delivery. *Adv Drug Deliv Rev*. 2008; 60:1252–1265.

[99] Briley-Saebo K, Bjørnerud A, Grant D, et al. Hepatic cellular distribution and degradation of iron oxide nanoparticles following single intravenous injection in rats: implications for magnetic resonance imaging. *Cell Tissue Res*.2004;316:315–323.

[100] Fujita T, Nishikawa M, Ohtsubo Y, et al. Control of in-vivo fate of albumin derivatives utilizing combined chemical modification. *J Drug Target*. 1994;2:157–165.

[101] Kairdolf BA, Mancini MC, Smith AM, et al. Minimizing nonspecific cellular binding of quantum dots with hydroxyl derivatizied surface coatings. *Anal Chem.* 2008;80:3029–3034.

[102] Ladd J, Zhang Z, Chen S, et al. Zwitterionic polymers exhibiting high resistance to nonspecific protein adsorption from human serum and plasma. *Biomacromolecules.* 2008;9:1357–1361.

[103] Breus VV, Heyes CD, Tron K, et al. Zwitterionic Biocompatible Quantum Dots for Wide pH Stability and Weak Nonspecific Binding to Cells. *ACS Nano.* 2009;3:2573–2580.

[104] Yang W, Zhang L, Wang S, et al. Functionalizable and ultra stable nanoparticles coated with zwitterionic poly(carboxybetaine) in undiluted blood serum. *Biomaterials.* 2009;30:5617–5621.

[105] Chen H, Wu X, Duan H, et al. Biocompatible polysiloxane-containing diblock copolymer PEO-b-P$_\text{r}$MPS for coating magnetic nanoparticles. *ACS Applied Materials and Interfaces.* 2009;1:2134–2140.

[106] Nasongkla N, Bey E, Ren J, et al. Multifunctional polymeric micelles as cancer-targeted, MRI-ultrasensitive drug delivery systems. *Nano Letters.* 2006;6:2427–2430.

[107] Liu X, Wang Y, Nakamura K, et al. Auger radiation – induced, antisense mediated cytotoxicity of tumor cells using a 3- component streptavidindelivery nanoparticle with [111] In. *J Nucl Med.* 2009;50:582–590.

[108] Wang MD, Shin DM, Simon JW, et al. Nanotechnology for targeted cancer therapy. *Expert Rev Anticancer Ther.* 2007;7:833–837.

[109] Brannon-Peppas L, Blanchette JO. Nanoparticle and targeted systems for cancer therapy. *Adv Drug Deliv Rev.* 2004;56:1649–1659.

[110] van Vlerken LE, Vyas TK, Amiji MM. Poly (ethylene glycol)-modified nanocarriers for tumor-targeted and intracellular Delivery *Pharm. Res.* 2007;24:1405–1414.

[111] Kukowska-Latallo JF, Candido KA, Cao Z, et al. Nanoparticle targeting of anticancer drug improves therapeutic response In animal model of human epithelial cancer. *Cancer Res.* 2005;65:5317–5324.

[112] Weissleder R, Kelly K, Sun EY, et al. Cell-specific targeting of nanoparticles by multivalent attachment of small molecules. *Nat Biotechnol.*2005;23:1418–1423.

[113] Liu Z, Cai W, He L, et al. In vivo biodistribution and highly efficient tumour targeting of carbon nanotubes in mice. *Nat Nanotechnol.* 2007;2:47–52.

[114] Yang L, Mao H, Wang YY, et al. Single chain epidermal growth factor receptor antibody conjugated nanoparticles for in vivo tumor targeting and imaging. *Small.* 2009;5:235–243.

[115] Zhang CF, Jugold M, Woenne EC, et al. Specific targeting of tumor angiogenesis by RGD-conjugated ultrasmall superparamagnetic iron oxide particles using a clinical 1.5-T magnetic resonance scanner.*Cancer Res.* 2007;67:1555–1562.

[116] Cai WB, Chen XY. Preparation of peptide-conjugated quantum dots for tumor vasculature-targeted imaging. *Nature Protocols.*2008;3:89–96.

[117] Antony AC. Folate receptors. *Annu Rev Nutr.* 1996;16:501–512. 117. Ke CY, Mathias CJ, Green MA. Folate-receptor-targeted radionuclide imaging agents. *Adv Drug Deliv Rev.* 2004;56:1143-1160.

[118] Das M, Mishra D, Dhak P, et al. Biofunctionalized, phosphonategrafted, ultrasmall iron oxide nanoparticles for combined Targeted cancer therapy and multimodal imaging. *Small.* 2009;5:2883–2893.

[119] Thomas M, Kularatne SA, Qi L, et al. Ligand-targeted delivery of small interfering RNAs to malignant cells and tissues. *Ann N Y Acad Sci.* 2009;1175:32–39.

[120] Setua S, Menon D, Asok A, et al. Folate receptor targeted, rare-earth oxide nanocrystals for bi-modal fluorescence and magnetic imaging of cancer cells. *Biomaterials.* 2010;31:714–729.

[121] Goren D, Horowitz AT, Tzemach D, Tarshish M, Zalipsky S, Gabizon A. Nuclear delivery of doxorubicin via folate- targeted liposomes with bypass of multidrug-resistance efflux pump.*Clin Cancer Res.* 2000;6:1949-1957.

[122] Seong SK, Ryu JM, Shin DH, Bae EJ, Shigematsu A, Hatori Y, Nishigaki J, Kwak C, Lee SE, Park KB, Biodistribution and excretion of radioactivity after the administration of 166Ho–chitosan complex (DW-166HC) into the prostate of rat *Eur. J. Nucl. Med. Mol. Imaging* 2005; 32: 910–917.

[123] Hafeli UD, Dasillas S, Dietz DW, Pauer GJ, Rybicki LA, Conzone SD, Day DE, Hepatic tumor radioembolization in a rat model using radioactive rhenium ([186]Re/[188]Re) glass microspheres, *Int. J. Radiat. Oncol. Biol. Phys.* 1999; 44:189–199.

[124] Lau WY, Ho S, Leung TW, Chan M, Ho R, Johnson PJ, Li AK, Selective internal radiation therapy for nonresectable hepatocellular carcinoma with intraarterial infusion of 90Y microspheres, *Int. J. Radiat. Oncol. Biol. Phys.* 1998; 40:583–592.

[125] Suzuki YS, Momose Y, Higashi N, Shigematsu A, Park KB, Kim YM, Kim JR, Ryu JM, Biodistribution and kinetics of holmium-166–chitosan complex (DW-166HC) in rats and mice, *J. Nucl. Med.* 1998; 39:2161–2166.

[126] Nakajo M, Kobayashi H, Shimabukuro K, Shirono K, Sakata H, Taguchi M, Biodistribution and in vivo kinetics of iodine-131 lipiodol infused via the hepatic artery of patients with hepatic cancer, *J. Nucl. Med.* 1988; 29:1066–1077.

[127] Na K, Park KH, Kim SW, Bae YH, Self-aggregated hydrogel nanoparticles from curdlan derivatives: characterization, anti-cancer drug release and interaction with a hepatoma cell line (HepG2), *J Control. Release* 2000; 69:225–236.

[128] Park YJ, Lee JY, Chang YS, Jeong JM, Chung JK, Lee MC, Park KB, Lee SJ, Radioisotope carrying polyethylene oxide–polycarprolactone copolymer micelles for targetable bone imaging, *Biomaterials* 2002; 23:873–879.

[129] Park KH, Song HC, Na K, Bom HS, Lee KH, Kim S, Lee DH, Ionic strength-sensitive pullulan acetate nanoparticles (PAN) for intratumoral administration of radioisotope: Ionic strength-dependent aggregation behavior and 99mTechnetium retention property, *Colloids and Surfaces B: Biointerfaces* 2007; 59:16–23.

[130] Natarajan A, Xiong CY, Gruettner C, DeNardo GL, DeNardo SJ, Development of multivalent radioimmunonanoparticles for cancer imaging and therapy. *Cancer Biother Radiopharm.* 2008; 23:82-91.

[131] Sofou S. Surface-active liposomes for targeted cancer therapy. *Nanomedicine* 2007; 2, 711–724.

[132] Ogihara I, Kojima S, Jay M. Differential uptake of gallium-67-labeled liposomes between tumors and inflammatory lesions in rats. *J Nucl Med* 1986, 27, 1300–1307.

Production and Selection of Metal PET Radioisotopes for Molecular Imaging

Suzanne V. Smith[1,2], Marian Jones[1] and Vanessa Holmes[1]
*[1]Centres of Excellence in Antimatter Matter Studies (CAMS) at
Australian Nuclear Science and Technology Organisation, and
[2]The Australian National University
Australia*

1. Introduction

Today advances in the design of modern drugs rely on an understanding of their mechanism of action. Positron Emission Tomography (PET) has proven to be an important imaging tool for diagnosis in oncology, cardiology and neurology and more recently in personalised medicine (Roach, M. et al., 2011; Smith, S.V., 2005). Its exquisite sensitivity ($<10^{-9}$ M) has made it a valuable tool in the screening and risk assessment of new agents in drug development programs (Cropley, V.L. et al., 2006; Gomes, C.M.F. et al., 2011; Gregoire, V. et al., 2007; Lancelot, S. & Zimmer, L., 2010; Langer, A.2010; Lonsdale, M.N. & Beyer, T., 2010; Roach, M. et al., 2011; Seam, P. et al., 2007; Truong, M.T. et al., 2011; Vyas, N.S. et al., 2011; Wahl, R.L. et al., 2009). It has been mainly focused on the use of the classical short lived PET isotopes, ^{11}C and ^{18}F, in the evaluation of small molecular weight molecules. PET contributions to the molecular imaging and drug development fields however would be significantly enhanced if longer lived PET radioisotopes were readily available. Generally metal PET radioisotopes are better suited for labelling larger molecular agents such as peptides, proteins and particles. This chapter will review a range of metal PET radioisotopes under development, assess their physical characteristics and suitability for PET imaging.

1.1 How does PET work?

The high sensitivity of PET is related to its mode of detection. The positron (β^+) which is emitted from the PET isotope must lose sufficient energy (to <100 eV) so it is able to attach to an electron and form the neutral, unstable positronium. The positronium annihilates rapidly (within less than 150 picoseconds (ps)) to give off two 511 keV gammas simultaneously, in opposite directions. If the two coincident gammas are detected by the circumferentially arranged detectors within a fixed time period, it is assumed that they arise from the same radioactive source (see Figure 1). This co-registration of the gamma signals allows one to determine the location and concentration of the PET probe in the body (Lonsdale, M.N. &Beyer, 2010 T.; Smith, S.V., 2005).

2. PET Imaging

The quality of information gained by PET imaging is governed by the spatial resolution and sensitivity of PET to the radioisotope used. While the sensitivity and flexibility of PET is superior to other existing imaging modalities it lacks the anatomical or structural detail achieved with CT and MRI (Smith, S.V., 2007). Today PET technology has evolved significantly in design, detector type, data processing and construction of hybrid systems. This section will give an overview of this evolution and future opportunities for PET. Table 1 summarises the latest relative spatial resolution of human and animal cameras (Bockisch, A. et al., 2009; Cai, W.B. &Chen, X.Y., 2008; Cañadas, M.E. et al., 2008; Kitajima, K. et al., 2011; Lewellen, T.K., 2008; Wang, C. et al., 2010). The reader is encouraged to consider a number of excellent reviews for more detailed discussion.

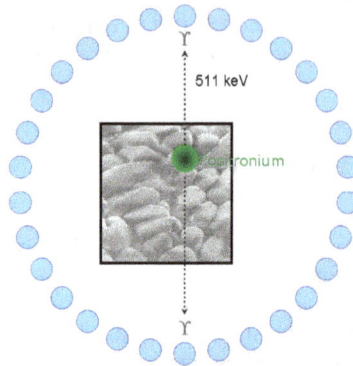

Fig. 1. Detection of coincident gamma rays from positronium annihilation

Human		Animal		
	Resolution (mm)		Resolution	Sensitivity (moles detected)
TOF-PET[a]	2 – 3	PET	1 – 2 mm	10^{-15}
SPECT[a]	6 – 10	SPECT	1 – 2 mm	10^{-14}
MRI[a]	< 1	MRI	50 μm	10^{-9} to 10^{-6}
CT[a]	< 1	CT	50 μm	10^{-6}

[a] Time of Flight Positron Emission Tomography; Single Photo Emission Computed Tomography; Magnetic Resonance Imaging; Computed Tomography

Table 1. Comparison of specification of Human and Animal Cameras (Baker, M., 2010; Pimlott, S.L. &Sutherland, A., 2011; Smith, S.V., 2005)

2.1 Human PET cameras

The first PET camera was developed in the 1950s and by the 90s they were well established imaging tools. Its detectors (up to 32 thousand) are arranged in a ring around the subject of interest. To obtain an image, the subject containing the PET agent is moved within the field of view of the detectors. The radioactivity present in the region of interest is measured and the data is processed to generate a 3 D image. The radial distribution of the detectors can be varied. For human whole body and brain imaging PET diameters are approximately 70–87 cm and 40-50 cm, respectively (Lee , J.S., 2010; Moses, W.W. et al., 1997). The further the detectors are placed apart the more likely the gamma signal will deviate from 180ºC (i.e.

become non-collinear) compromising spatial resolution. As a result, whole body and brain PET have optimum resolutions of approximately 4-6 mm and 2 mm (Lancelot, S. & Zimmer, L., 2010). For animal PET, the radial distance of the detector is smaller therefore the achievable spatial resolution is 1-2 mm (using [18F]). Because the detectors are closer, images can be significantly affected (i.e. become blurred) by high energy positrons and gammas (350 – 700 keV). This comparatively poor spatial resolution of PET created the need to combine it with high resolution anatomical data obtained with CT and MRI. Initially this was achieved by fusion imaging, where images are taken from independent CT and PET cameras and merged to produce a single image.

By 1998 CT and PET detectors were integrated into a single system (Winant, C.D. et al., 2010). The separate images were superimposed and produced images of superior quality than their respective stand-alone systems. Today, stand alone PET systems comprise less than 10% of the PET market and their sales are projected to cease by 2015 (Frost & Sullivan, 2010). There are almost 5000 PET-CT systems worldwide, while stand-alone PET tend to be purchased only when patient throughput is low or a diagnostic CT camera is in routine operation.

PET-CT is mostly used for oncology studies (95%), however both neurology (3%) and cardiology (2%) are expected to grow to 15 and 10%, respectively. Rubidium-82 is becoming the agent of choice for myocardial perfusion imaging (MPI). Generator produced [82Rb] can be made widely available. Compared to traditional SPECT agents, such as [99mTc](terefosmin) and [201Tl], it provides equivalent or better performance with 5-20 times lower radiation dose. The next generation of PET-CT scanners will have 64-slice CT scanners (64 slice) and are expected to improve their competitive edge (Winant, C.D. et al., 2010). In neurology, [18FDG] has already proven its ability in differential diagnosis of dementias, such as Alzheimer's disease from frontotemporal dementia and normal patients. The recent launch of the clinical trial of the [18F]-PET amyloid imaging agent ([18F]-AV-45) is expected to further expand the role of PET-CT in neurology.

Even though the PET and CT images must still be acquired sequentially, the PET-CT systems have increased the patient throughput by 30%, reducing time and cost to patient and healthcare providers. These systems have enhanced collaboration between the nuclear medicine and radiology communities, leading to improved quality of care to the patient and increased potential for reimbursements for the radiopharmaceuticals used.

More recently PET has been able to capitalise on Time of Flight (TOF) techniques to enhance image resolution. The first TOF-PET system was developed as far back as the 80s, but only in recent times have detectors and data processing capabilities been sufficiently cost–effective to incorporate this capability into commercial systems (Lee, J.S., 2010). TOF-PET can be used to reduce scan time or quantity of radiotracer required to obtain the image (Moses, W.W., 2007). This advancement has been particularly important for larger patients (>115 kg and BMI $\geq$ 38) (Lee , J.S., 2010) where small lesion detection can be difficult. Further advances in the timing resolution to less than 500 picoseconds is expected to improve signal to noise ratio and image contrast. This will provide more accurate tracer quantification in small structures. One drawback for PET-CT is the radiation dose and poor soft tissue contrast. It has created the impetus for the development of MRI and PET hybrid systems (Wolf, W., 2011).

PET-MRI is emerging as a disruptive technology. (Lancelot, S. & Zimmer, L., 2010). Still only used in research communities, the advancement of PET-MRI has relied on improved resolution and stability of solid state detectors for PET. These are smaller, cheaper and insensitive to magnetic field. Ultimately PET-MRI is expected to significantly reduce the radiation dose to patient, provide higher spatial resolution and superior soft tissue contrast. Such characteristics are invaluable for imaging systemic soft tissue disease and the brain. Areas such as Alzheimer's and Parkinson's disease and neuropsychiatric disorders such as schizophrenia and depression in particular, are expected to benefit from such developments.

2.2 Animal PET cameras

Dedicated animal PET (microPET) and PET-CT cameras have been available for some time. While they have proven to be useful for *in vivo* measurements of various physiological processes in small animals, there are a number of remaining challenges. Of these, spatial resolution, sensitivity and radiation dose are the most obvious. Currently high resolution microPET cameras have a maximum resolution of ~1 mm (with ^{18}F) and a maximum sensitivity of almost 10% (Lancelot, S. & Zimmer, L., 2010). For animal cameras, positron energy and specific activity (i.e. radiolabelled product vs unradiolabelled product) of the radiopharmaceutical can have a detrimental effect on imaging resolution and sensitivity. In addition, CT scanners expose the animal to unnecessary additional high radiation dose. In contrast the PET-MRI systems are predicted to reduce this radiation exposure by up to 50%. Unfortunately they are still under development and there are no commercial systems available at this time.

3. Cyclotrons – How PET radioisotopes are produced

PET radioisotopes are proton-rich and are predominantly produced using a cyclotron or linear accelerator (International Atomic Energy Agency, 2006; Schmor, P.W., 2010). The cyclotron was first conceived by Lawrence at Berkeley, USA in the early 1930s, and by the end of that decade it was producing radioisotopes for medical research. Over 15 years later there were cyclotrons in St Louis and Boston, USA and London, UK. The first company to produce cyclotrons, Scanditronix, was established in 1961, and today there are over 12 companies producing commercial cyclotrons. The design of a cyclotron can be quite varied and tailored to meet the specifications of the users.

Cyclotrons accelerate a beam of particles such as protons (p), deuterons (d), $^3He^{2+}$ or $^4He^{2+}$ (α) in a circular path, with protons being the most common particle used for radioisotope production. The particles need to reach certain energies (MeV) for nuclear reactions to take place. The particle beam must also have sufficient current ($\mu A.hr$ to $mA.hr$) to produce radioisotopes at reasonable quantities of radioactivity (e.g. 10s GBq to 10s TBq) (International Atomic Energy Agency, 2009).

Today over 350 cyclotrons operate worldwide and most are devoted to radioisotope production (International Atomic Energy Agency, 2006; Petrusenko, Y.T. et al., 2009). More than half of these cyclotrons produce p with energies from 10 to 20 MeV (majority of these (~75%) are designed to produce ^{18}F), and others produce energies of up to 70 MeV. The number of cyclotrons is expected to increase in conjunction with the expanding role of PET and SPECT in molecular imaging.

The production of radioisotopes involves bombarding a target material with p, d, $^3He^{2+}$ or α to cause a nuclear reaction to occur. Typically, these reactions involve a p entering the target nuclei and one or more neutrons (n) or alphas (α) exiting with various gamma emissions. These nuclear reactions may be expressed in the following manner, (p, 2n), (p, x) and (p, xn).

Reactions with d, $^3He^{2+}$ or α less utilised, because the achievable energies and currents of their respective beams are comparatively lower than for protons.

3.1 Cyclotron types

Commercially available cyclotrons can be characterised into three types based on the energy of the particles they produce: (Schmor, P.W., 2010)

3.1.1 Energies up to 20 MeV

Often termed "baby" cyclotrons, they are the dominant producers of short lived PET isotopes (i.e. ^{11}C, ^{18}F, ^{13}N, and ^{15}O). They are generally located in hospital centres, a short distance from the PET camera. They produce beams of low currents (< 50 μA) and can handle a number of target types (typically gas or liquid and, more recently, solid targets). These cyclotrons can produce sufficient PET radiopharmaceuticals to support a number of PET cameras. The solid targets generally used to produce metal PET isotopes can be positioned external or internal (under vacuum) to the cyclotron, requiring extensive cooling systems.

3.1.2 Energies 20-30 MeV

These cyclotrons are used to produce both SPECT (e.g. ^{67}Ga, ^{201}Tl, ^{111}In and ^{123}I) and PET radioisotopes. Gas, liquid and solid target systems are available. Beam currents are less than 50 μA for gas and liquid, and up to 250 μA for solid target systems. These cyclotrons are generally dedicated facilities, requiring highly skilled staff to maintain them. These production facilities can be located a substantial distance (i.e. different continents) from the nuclear medicine department.

3.1.3 Energy 30-70 MeV

High intensity and high energy cyclotrons are generally deployed for the production of large quantities of radioisotopes for sale. Those operating in the higher energy range are commonly used for research purposes. The recently commissioned 70 MeV cyclotron at ARRONAX in France (Haddad, F. et al., 2008) will be used to produce radioisotopes for imaging and radiotherapy. They include, ^{67}Cu, ^{211}At, ^{47}Sc ^{52}Fe, ^{55}Co, ^{76}Br and ^{82}Sr. These radioisotopes can not be produced economically in lower energy systems. The selective production of a radioisotope, limiting the production of long lived contaminating radioisotopes, can be achieved with this system. For example, ^{52}Fe can be produced via $^{55}Mn(p,4n)^{52}Fe$, and the long lived ^{55}Fe ($T_{1/2}$ = 2.74 years) is co-produced, however using a different reaction, such as ^{50}Cr (α,2n)^{52}Fe the production of ^{55}Fe is eliminated. The range of beam particles and their respective currents for this cyclotron are summarised in Table 2.

Beam	Accelerated particles	Energy range (MeV)	Intensity (μA)
p	H- HH+	30 – 70 17.5	< 350 (x2) < 50
d	D-	15 – 35	50[a]
α	He++	70	<35

[a] Local radioprotection authorities in France set the intensity limit for deuterons.

Table 2. Characteristics of the beams for 70 MeV Cyclotron ARRONAX, France

4. Factors governing selection of a radioisotope

Historically, the criteria governing the selection of a PET radioisotope has been relatively simple and largely dominated by the availability, production yield and ability to image. Significant advances in design and stability of cyclotrons, cameras and data processing, has increased the potential range of PET radioisotopes. Today, the choice of PET radioisotopes is influenced by many factors (see Figure 2). It is important to understand how they impact on the production and clinical setting as well as on patients when administered repeatedly.

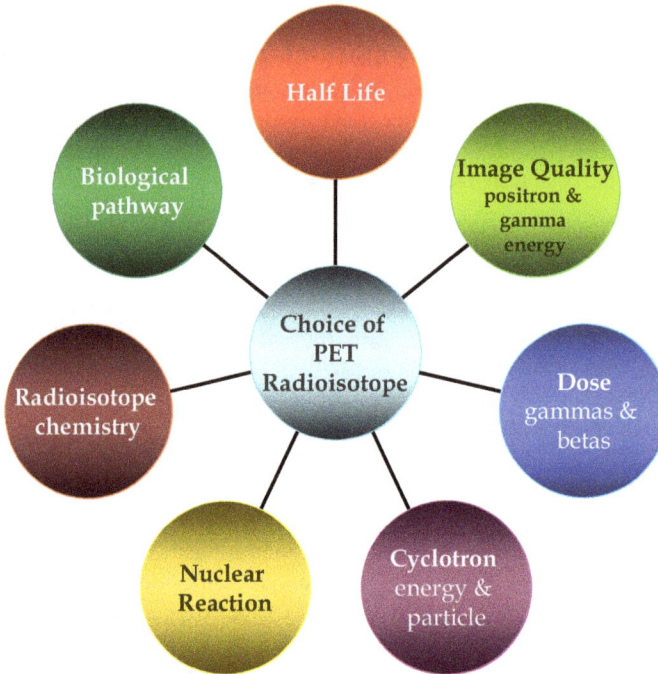

Fig. 2. Criteria governing selection of a PET radioisotope.

4.1 Half life

The half life of the PET radioisotope needs to match the biological half life of the carrier agent. For applications in diagnosis, this may simply be a radioisotope with a half life long enough to see the radiopharmaceutical accumulate in the diseased tissue and clear from blood. When imaging to determine prognosis of a patient (ability to respond to treatment) or assessing risk (therapeutic index) of an agent, the half life may need to be longer. For the latter, understanding uptake and clearance from all critical organs for a patient is important when determining its therapeutic index and or appropriate dose.

4.2 Image quality

The decay characteristics of the radioisotopes need to match the performance of the imaging camera. The resolution of PET remains dependent on three fundamental factors:

1. Non-collinearity of annihilating gammas (i.e. deviation from 180° angle on emission of 511 keV signal)
2. Positron energy spectrum
3. Gamma emissions

4.2.1 Positrons

Non-collinearity can be reduced by bringing the detectors closer together. The higher the positron energies, the further they need to travel to lose enough energy to form a positronium, and this can cause images to blur (Pimlott, S.L. & Sutherland, A., 2011). For human PET and particularly TOF-PET systems, the tolerance for the range of positron energies is higher than that of an animal PET camera. The spatial resolution of animal PET cameras is dependent on the maximum positron energy and its spectrum. Therefore in designing animal PET studies one needs to take into consideration the positron energy of the PET isotope.

4.2.2 Gamma emissions

Gamma emissions from the PET radioisotopes, of specified energies (and intensities), can interfere with the image quality, particularly those emitted within the PET discriminator window (ie. 350 and 700 keV). This leads to gamma coincidences that result in images of poor quality and flat unstructured background.

4.3 Dose

PET radioisotopes may decay via a number of pathways and give off x-rays, betas (i.e. electrons or positrons) and gammas. All of these emissions contribute to the total absorbed dose to the subject. From a production and imaging point of view it is the high intensity gamma emissions that cause concern for operators (i.e. production and animal handling staff) and applications in the clinical setting. Most PET facilities have been designed to handle ^{18}F and do not have sufficient shielding for high energy gammas. Where the patient is concerned, the presence of high energy gammas prevents re-injection (such as that used in cardiology) due to accumulated radiation dose.

4.4 Cyclotron energy

Cyclotrons need to produce a beam of suitable particles at a desired energy and current for the preferred nuclear reaction. Inappropriate beam characteristics will prevent operators from producing sufficient quantities for clinic applications.

4.5 Nuclear reaction

Most PET radioisotopes are produced by bombarding a target material, which often is isotopically enriched. An enriched target is expensive (up to $1000s per target irradiation). As only a negligible (~0.01%) portion of the target material is converted to the desired radioisotope, the target should be treated as an asset and recycled. Sometimes long lived gamma emitting radioisotopes can not be chemically removed. Their presence in the used target material can prevent its recycling in a timely manner. In such circumstances, a sizeable inventory of enriched target material needs to be purchased (e.g. >$250,000 US) to permit decay of contaminants prior to reuse.

4.6 Chemistry of the PET radioisotope

The ease, efficiency and cost of production of a PET radiopharmaceutical are affected by the chemistry of the radioisotope. The first aspect is the chemical separation of the PET radioisotope from the target material; this is more easily achieved if the target element is chemically different from the radioisotope. The separation process is generally conducted at the cyclotron site.

The second aspect of PET radiopharmaceutical production is the incorporation of the PET radioisotope into the carrier agent. Such processes must be performed relatively quickly and preferably quantitatively in order to produce high specific activity products. The production of the radiopharmaceutical may require hot-cells and/or an automated system. This can add considerable cost to the process and prevent production of the desired PET radiopharmaceutical at the site of use.

4.7 Biological pathway

Each radioisotope needs to be stably secured to a carrier. However, if the radiolabelled carrier breaks down *in vivo* it is likely to release the radioisotope. As radioisotopes mimic their naturally occurring metal ions, they can be expected to excrete from the body using normal biological pathways. Hence in choosing the radioisotope, one needs to consider these pathways and their potential contribution to the radiation dose to patient. For example, copper excretes via the liver and can be recirculated with ceruloplasmin. All the copper isotopes, ^{60}Cu, ^{61}Cu, ^{62}Cu and ^{64}Cu are expected to behave in a similar manner. In contrast, ^{86}Y, like other lanthanides, will accumulate in bone while the biological pathway of ^{89}Zr is not known and therefore its behaviour is difficult to predict.

5. Metal PET radioisotopes

As mentioned previously the greatest potential for the PET radioisotopes is to provide diagnostic information on pathological processes and any changes to them, prior to anatomical changes. They have a significant role to play in both the screening of new diagnostic and therapeutic target agents in animal studies and in clinical trials. By radiolabelling the target agent, such as antibodies, the PET radiopharmaceuticals are used to determine the therapeutic index (distribution) of the antibody, the appropriate dose for the patient and also assist in selecting patients likely to respond to treatment.

Radiometal isotopes can also be incorporated into a ligand or complexing agent and used to monitor various biological properties such as blood flow, hypoxia or glomerular filtration and or organ function (e.g. heart, kidney or liver) (Smith, S.V., 2004 and references within). In this case, the design of the ligand and the resultant metal complex dictates the function of the PET radiopharmaceutical.

If the ligand is *bi-functional*, one portion of the molecule will be responsible for complexing the metal radioisotope and the other will contain a reactive group (e.g. amino, carboxylic acid or isothiocyanate) that is responsible for coupling the metal complex to a target agent. For target agents with long biological half lives (i.e. hours and days) it is important to ensure the radiolabel is secured and stable *in vivo* until the target agent clears from the body. This ensures the PET signal *in vivo* is from the radiolabelled target agent and not the radioisotope in its ionic state or non-specifically bound to naturally occurring biological components. Furthermore, the half life of the radioisotope needs to be sufficiently long enough to monitor the localisation of its carrier to the disease site and attain a signal to noise ratio of at least 2 fold. Table 3 summarises metal PET radioisotopes reported for these applications.

Isotope	Half Life	Maximum β+ Energy (keV)	Mean β+ Energy (keV)	β+ Intensity (%)	γ energy (keV) (Intensity %)
48V	16.0 d	695	290	49.9	944 (7.9); 984 (100); 1312 (98.2)
52Fe	8.3 h	804	340	55.0	169 (99.2) ; 338 (1.6)
52mMn	21.1 m	2633	1174	94.8	1434 (98.3)
52Mn	5.6 d	576	242	29.6	744 (90.0) ; 848 (3.3); 936 (94.5) ; 1434 (100.0)
55Co	17.5 h	1021 1113 1499	436 476 649	25.6 4.3 46.0	477 (20.2); 931 (75.0); 1317 (7.1); 1409 (16.9)
60Cu	23.7 m	1835 1911 1981 2946 3772	805 840 872 1325 1720	4.6 11.6 49.0 15.0 5.0	467 (3.5); 826 (21.7); 1035 (3.7); 1333 (88.0); 1792 (45.4); 1862 (4.8); 3124 (4.8)
61Cu	3.3 h	559 932 1148 1215	238 399 494 524	2.6 5.5 2.3 51.0	283 (12.2) ; 656 (10.8) ; 1185 (3.7)
62Zn	9.3 h	605	259	8.4	508 (14.8) ; 548 (15.3); 597 (26.0)
62Cu	9.7 m	2926	1316	97.2	
63Zn	38.5 m	1382 1675 2345	600 733 1042	4.9 7.0 80.3	670 (8.2) ; 962 (6.5)
64Cu	12.7 h	653	278	17.6	
66Ga	9.5 h	924 4153	397 1904	3.7 51.0	834 (5.9) ; 1039 (37.0) ; 2190 (5.3); 2752 (22.7) ; 4295 (3.8)
68Ga	1.1 h	822 1899	353 836	1.2 87.9	1077 (3.2)
81Rb	4.6 h	580 1026	254 448	1.8 25.0	190 (64.0) ; 446 (23.3) ; 510 (5.3)
82Rb	1.3 m	2601 3378	1168 1535	13.1 81.8	777 (15.1)
83Sr	1.4 d	830 1212	363 531	3.1 9.0	382 (14.0) ; 418 (4.4); 763 (30.0)
86Y	14.7h	1221 1545 1988 3141	535 681 883 1437	11.9 5.6 3.6 2.0	515 (4.9) ; 581 (4.8); 628 (32.6) ; 646 (9.2); 703 (15.4); 777 (22.4) ; 836 (4.4) ; 1077 (82.5) ; 1153 (30.5); 1443 (16.9); 1854 (17.2); 1921 (20.8)
89Zr	78.4 h	902	396	22.7	909 (99.0)
94mTc	52.0 m	2439	1094	67.6	871 (94.2) ; 1522 (4.7) ; 1869 (5.7)
94Tc	4.9 h	811	358	10.5	703 (99.6) ; 850 (95.7) ; 871 (99.9) ; 916 (7.6)
110In	1.2 h	2260	1043	62.0	658 (98.0)

Table 3. Selection of Metal PET radioisotopes and their physical characteristics (http://www.nndc.bnl.gov/; International Atomic Energy Agency, 2009)

Their physical characteristics including mean positron energy, gamma emissions and half lives are given. Only gamma emissions of greater than 5% intensity, or that are within the PET detector window, or contribute significantly to operator dose have been included. All of the radioisotopes listed emit positrons of sufficient energy and intensity ($> \sim 10\%$) for imaging with available PET cameras. Further each radioisotope listed can be produced in commercially available cyclotrons.

5.1 Half life

The half lives of the radioisotopes presented in Table 3 vary from minutes to days. They may be classified into groups based on these half lives (see Figure 3). The timeframes chosen take into consideration the time to purify the radioisotopes, its incorporation into a carrier agent and the packaging and transport of the resultant radiopharmaceutical across continents.

The first class of radioisotope ($t_{1/2}$ = 1- 60 mins) is generally used as surrogates for cations *in vivo* or incorporated into complexing agents. A number of the radioisotopes can be produced from a generator system (see Figure 4) making them more accessible to the wider PET community. Radioisotopes with longer half lives (hours to days) are best suited for studying the pharmacokinetics or dynamics of large molecules such as peptides, proteins and particles.

There are 6 PET radioisotopes (^{82}Rb, ^{62}Cu, ^{52m}Mn, ^{60}Cu, ^{63}Zn and ^{94m}Tc), with half lives varying from 1.3 – 52 min. Three, ^{82}Rb, ^{62}Cu and ^{52m}Mn are particularly attractive because they can be produced from a generator utilising the decay of ^{82}Sr, ^{62}Zn or ^{52}Fe to produce the respective daughter PET radioisotopes (see Figure 4). The ^{82}Sr/^{82}Rb and ^{62}Zn/^{62}Cu generators have been commercially available for some time. The radioisotopes ^{60}Cu, ^{63}Zn and ^{94m}Tc must be produced and used at or within a short distance from the cyclotron production centres.

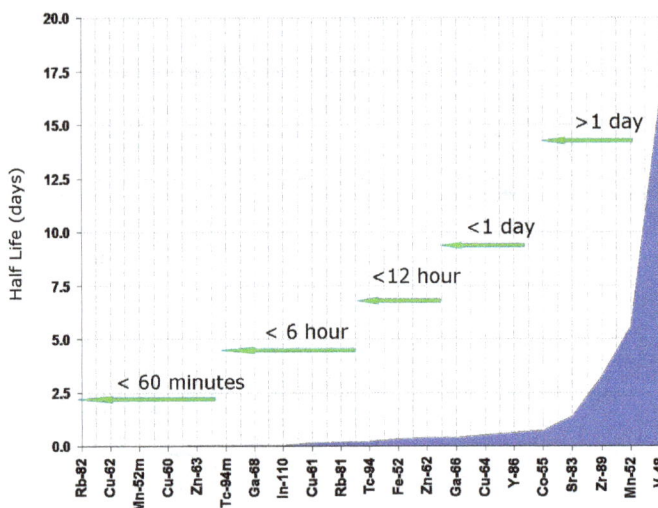

Fig. 3. Classification of positron emitting radioisotopes.

^{52}Fe ^{82}Sr ^{62}Zn

t $_{1/2}$ = 8.28 h   t $_{1/2}$ = 25.5 d   t $_{1/2}$ = 9.19 hr

^{52m}Mn ^{82}Rb ^{62}Cu

t $_{1/2}$ = 21.1 m   t $_{1/2}$ = 21.1 m   t $_{1/2}$ = 1.27 m   t $_{1/2}$ = 9.74 m

^{52}Cr $\xleftarrow{\text{t}_{1/2} = 5.6\,\text{d}}$ ^{52}Mn ^{82}Kr ^{62}Ni

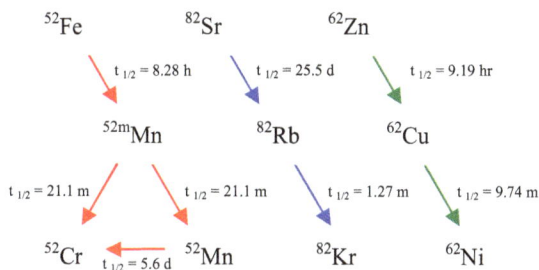

Fig. 4. Decay scheme for generator produced ^{52m}Mn, ^{82}Rb and ^{62}Cu

For example, ^{82}Rb$^+$ ($t_{1/2}$= 1.3 min) is used in myocardial perfusion studies; it acts as a surrogate for K$^+$. The very short half life of ^{82}Rb makes whole body imaging a challenge and precludes imaging after 10 mins. As a result activities of up to 1.5 GBq are administered to patients. Heart, kidney, lungs and pancreas receive an absorbed radiation dose of 5.1, 5.0, 2.8 and 2.4 mGy per MBq, respectively. This is significantly less than both ^{201}Tl and ^{99m}Tc sestamibi SPECT agents (Senthamizhchelvan, S. et al., 2011). The long half life of the parent ^{82}Sr ($t_{1/2}$= 25.55 days) permits its transport throughout the world.

The radioisotopes ^{60}Cu ($t_{1/2}$= 23.7 min) and ^{62}Cu ($t_{1/2}$= 9.6 min) have identical chemistry and can be interchanged depending on the carrier agent and the biological processes under investigation. The ^{62}Cu is produced from the parent ^{62}Zn ($T_{1/2}$ = 9.3 hour); its short half life requires the generator to be prepared daily.

The story with ^{52m}Mn is more complex. It is produced from the positron-emitting parent ^{52}Fe ($t_{1/2}$=8.3 hours) and then decays by positron emission to the long-lived daughter ^{52}Mn radioisotope ($t_{1/2}$=5.591 days). While the ^{52}Mn emits a low energy positron its extraordinarily high energy gamma detracts from its use in PET imaging. A recent study, investigating the effect of the daughter ^{52m}Mn on the ^{52}Fe PET image quality showed there was minor degradation, compared to ^{18}F.

The positron energies for ^{60}Cu and ^{63}Zn are of similar intensity, the latter having comparatively less gamma emissions. It is surprising ^{63}Zn has not been reported in the literature more often. As Cu(II) and Zn(II) complex with similar chelating agents, there is potential to substitute ^{63}Zn for ^{60}Cu in emerging blood flow or hypoxia markers.

Technetium-94m also has a short half life with a relatively high intensity positron emission. It decays to the positron emitter ^{94}Tc, which further decays giving off a large proportion of high energy gammas. The ^{94m}Tc could potentially act as a surrogate for the SPECT ^{99m}Tc agents but negligible improvement in imaging resolution is unlikely to compensate for the increased dose both to operator and subject.

The application of radioisotopes ranging in half life from 1 hour to less than a day is varied. They are readily incorporated into chelators or carrier agents but their stability will be dependent on their respective coordination chemistry and redox chemistry. Generator produced ^{68}Ga ($t_{1/2}$=1.1 hours) has been used to radiolabel a range of peptides including derivatives of octreotide (Gabriel, M. et al., 2007; Henze, M. et al., 2001) and arginine-glycine-aspartate (RGD) (Li, Z.B. et al., 2008; Liu, Z.F. et al., 2009). However, the short half life of ^{68}Ga can create challenges for the synthesis of the radiopharmaceutical and its imaging. The high energy positrons given off by ^{110}In and ^{66}Ga would prohibit their use as surrogates for ^{111}In and ^{67}Ga SPECT agents, while their high intensity gammas (~1 to 4 MeV) require considerably more lead shielding for handling than classical PET agents.

All PET isotopes with half lives greater than 12 hours have greater flexibility in PET imaging. Of the radioisotopes selected ^{64}Cu, [89]Zr and [48]V have the most favourable positron energies for imaging but only ^{64}Cu does not have significant gammas to compromise its application.

5.2 Gamma emissions - Safety

As mentioned previously, some gamma emissions from the PET isotopes can affect PET image quality, choice of shielding in production facilities and cause unnecessary absorbed dose to the subject. Shielding suitable for ^{18}F may not be sufficient to shield gammas with energies significantly greater than 500 keV (Madsen, M.T. et al., 2006). Figure 5 illustrates a simple comparison of the relative energy and the intensities of gammas emitted from some of the radioisotopes given in Table 3. The data clearly show that gammas generated from the decay of [60]Cu, [86]Y and [89]Zr are of significantly higher energies than that of ^{18}F; therefore these radioisotopes will require substantially more lead shielding for processing compared to that used for ^{18}F productions. Gammas in the range of 350 – 700 keV (shaded green on image) from decay of [62]Zn and [86]Y will interfere with PET image quality, resulting in a reduced signal to noise ratio. In stark contrast, [62]Cu and ^{64}Cu have no significant gammas that will interfere with camera quality, nor do they require significant changes to current lead shielding used in PET facilities.

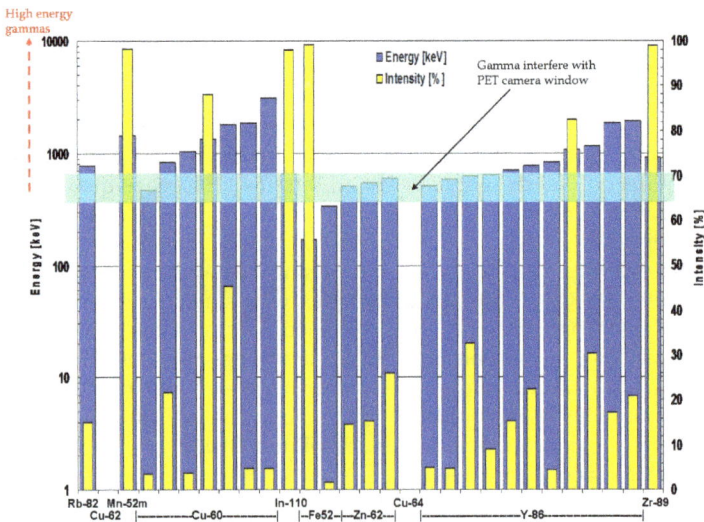

Fig. 5. Compares relative gamma emissions (blue) and their intensities (yellow) for selected radioisotopes from Table 3.

6. Radiation dose

As the energy and the probability of gamma emissions is specific for each radioisotope, they can be used to calculate the energy released per decay in the form of photons. The energy released per decay may be represented as MeV (Bq s)$^{-1}$, where 1 Bq = 1 decay per second.

Radiation dose received by an operator or patient is measured in Sievert, Sv. One Sv is equivalent to the energy absorbed per kilogram of mass (J/kg). In the case of the operator who is positioned at some distance from the radiation source, the dose received in Sv is a function of the intensity of the radiation, type of radiation, distance from the source and the time exposed to the source. It is convenient to discuss the dose rate in Sv/hr, so that the operator can determine the dose received for a specific time period.

Figure 6 illustrates the dose rate per unit of radioactivity received by an operator at one metre distance from each unshielded PET radioisotope source. It shows the effect of high energy gammas can vary by 10-fold across the series of radioisotopes listed in Table 3. These doses also need to be considered when dispensing the radiopharmaceutical and when injecting the subject (i.e. patient or animal) as they effectively become a radioactive source and shielding under these circumstances is not always practical.

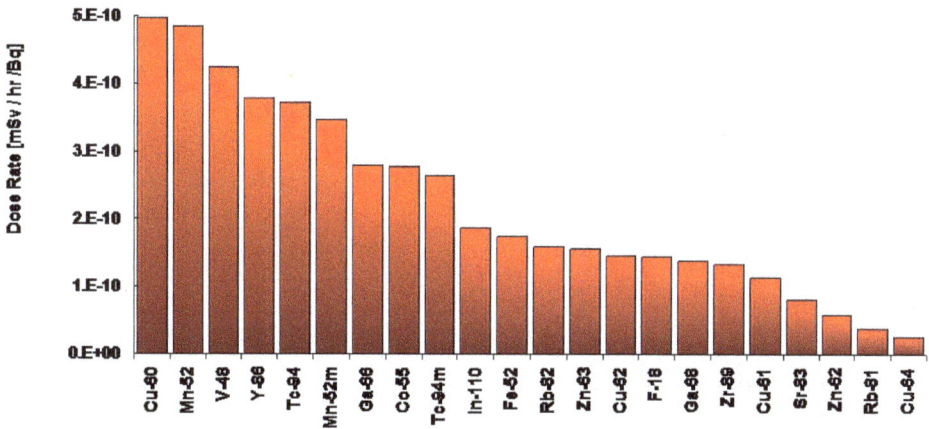

Fig. 6. The dose rate per unit of radioactivity received by an operator at one metre from an unshielded source.

6.1 Shielding

The dose rate values in Figure 6 can be used to determine the shielding for the production, handling and storage of the PET isotopes. The shielding can then be adjusted to ensure the radiation exposure to the operator is kept within an acceptable level. The following relationship can be used to determine the shielding required:

$$I = I_0 \, e^{-\mu t} \tag{1}$$

where:

I = intensity of the radiation after shielding,

I_0 = intensity before shielding,

t = thickness [cm] of shielding material (in this case lead), and

μ = linear attenuation coefficient [cm^{-1}].

The linear attenuation coefficient takes into account the effect of photoelectric absorption, Compton collision, pair production and Rayleigh scattering. The relationship between linear attenuation and the energy coefficients for a range of photon energies is illustrated in Figure 7 (Hubbell, J.H., 1969). The curve shows that for higher energy gammas travelling through lead, the attenuation is lower per unit of distance travelled. As a result, for the same intensity of radiation, a greater thickness of lead is required to shield an operator from higher energy photons than for lower energy photons.

Photoelectric effect, Compton scattering and pair production can result in secondary photons. These photons have a finite probability of contributing to the total dose received by the operator. The extent to which these secondary photons contribute is described through the use of a build-up factor. For simplification this will be ignored here, but it should be noted that it is a multiplicative factor, that is, always greater than one. Further the multiplication factor is dependent on the energy of the incident photons, the atomic number of the shielding material and the thickness of the shielding. For further information regarding shielding calculations refer to Principles of Radiation Shielding (Chilton, A.B. et al., 1984). To calculate the thickness of lead required to shield an operator for a quantity of radioactivity, the radiation is assumed to be at the highest energy gamma associated with the radioisotope decay. Further if we set the maximum permitted operator dose at 12 mSv per working year, the maximum dose rate for operator exposure is ~0.00625 mSv/hr where one year is equivalent to 48 wks at 40 hours/wk.

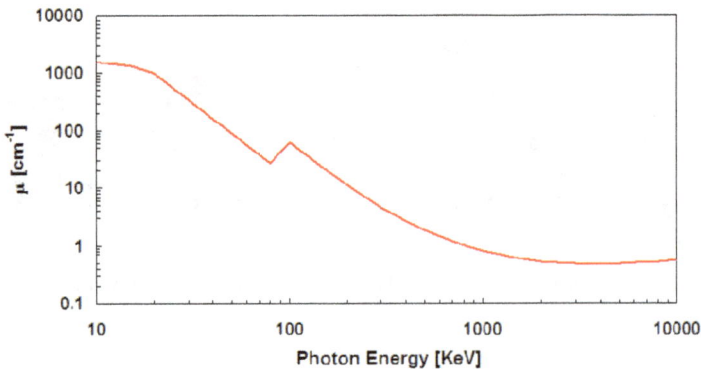

Fig. 7. Log-Log plot of linear attenuation coefficients for lead as a function of energy.

To obtain the thickness of lead required to reduce the original intensity I_0 to the desired value I, equation (1) is rearranged to:

$$t = -(\mu)^{-1} \ln(I/I_0) \text{ [cm]} \tag{2}$$

Assuming the source has 1Ci (37 GBq) of a PET radioisotope, the thickness of lead required to reduce the operator dose rate to 0.00625 mSv/hr may be calculated using (2). It is useful to combine the data of Figure 6 with Figure 7 and compare the dose rate per activity plotted with the relative thickness of required lead shielding to reduce the dose to specified levels in Figure 8. In other words, the data show the relative thickness of shielding for the same amount of radioactivity for each PET isotope. For ^{60}Cu, a 5 fold greater thickness of lead shielding is required compared to an equivalent amount of ^{64}Cu radioactivity.

It is important however to note that the relative order of shielding required for each PET radioisotope in Figure 8 does not always correspond directly to the total dose rate discussed above. For example, shielding for ^{86}Y, ^{66}Ga, ^{55}Co, ^{94}mTc, ^{63}Zn, ^{68}Ga, ^{89}Zr, ^{61}Cu, ^{83}Sr, ^{62}Zn, ^{81}Rb and ^{64}Cu is significantly higher than one might have anticipated when examining only dose. In contrast, the shielding required for ^{52}Mn, ^{48}V, ^{94}Tc, ^{110}In, ^{52}Fe ^{82}Rb, ^{18}F , ^{62}Cu and ^{68}Ga, are all lower. This emphasises the importance of considering dose and shielding together, even when the high energy gammas are of low intensity, when handling and administering the PET radiopharmaceuticals. These results concur with recently reported findings (Holland, J.P. et al., 2011) where the occupational exposure of staff increased when handling low amounts of activity (~1-25 mCi) of ^{86}Y and ^{89}Zr compared to significantly higher quantities of ^{18}F activity (~400-500 mCi).

Furthermore, transport of these types of PET isotopes requires modification of shipment containers or significant reduction in the amount of activities to be shipped. The emission of high energy gammas from ^{86}Y and ^{89}Zr allow only ~1/162 and ~1/44 respectively of the maximum activity of ^{18}F to be transported in approved PET radioisotope shipment containers in the USA (Holland, J.P. et al., 2011). Negligible change in shielding would be required for ^{64}Cu.

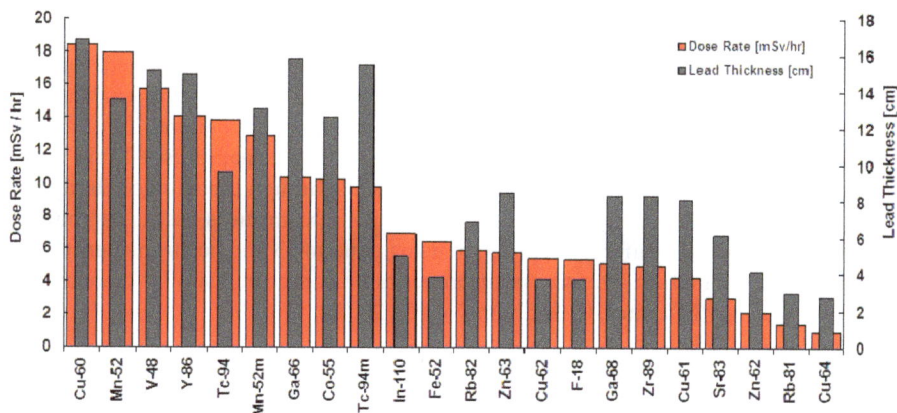

Fig. 8. Dose rate per unit activity and the required lead shielding for each PET radioisotope.

7. Positron energy - Image quality

It is important to consider the energies of the positrons emitted from the PET radioisotopes when determining their application. Because the positron has to lose energy before it annihilates, it travels and interacts with tissue over a significant range (i.e. millimetres). As the density of the tissue can affect the rate at which the energy is lost from the positron, one may also expect the positron to travel further in lung tissue (because it is predominantly air). PET images in this region will be "blurred" and the spatial resolution will be poorer compared to other parts of the body where the density of tissue is higher. In contrast, for bone, which is considered electron rich and of dense composition, the PET image appears comparatively sharper.

For small animal PET cameras, PET radioisotopes emitting high energy positrons have a detrimental effect on image quality (Laforest, R. & Liu, X., 2008; Partridge, M. et al., 2006). Laforest and Lui 2008, reported that both the range the positron travels and the presence of cascades of gamma rays contributed to poor signal to noise ratios. In the same work, they compare phantoms containing either ^{60}Cu, ^{61}Cu, ^{64}Cu, ^{94m}Tc or ^{86}Y and demonstrate how the positron energy can affect the image resolution.

Improvement in spatial resolution of animal PET cameras using these radioisotopes can be achieved by an increase in imaging time and a reduction in the coincidence window limits to prevent contamination from other decay modes (Vandenberghe, S., 2006). To do this the subject either needs to be injected with more radioactivity or kept in the field of view of the camera for longer periods of time. A more sophisticated approach to improve image quality is to develop image reconstruction algorithms that consider the nuclear decay and tomography response (Laforest, R. & Liu, X., 2008; Ruangma, A. et al., 2006).

Though the development of cyclotrons continues to flourish and in parallel the range of PET isotopes, understanding the track structure of positrons emitted is largely unresolved. Attempts to use Monte Carlo modelling to illustrate the tracking of positrons in tissues have been made by a number of groups. Work by Sanchez-Crespo et al. (Sanchez-Crespo, A. et al., 2004) has shown how the density of tissue can change the resolution of an image. But such modelling is unlikely to comprehensively predict the positron behaviour or its microdosimetry, until accurate measurements of the cross sections for the interaction of positrons and electrons with various biological molecules have been attained. Such data have the potential to improve the analysis of PET data and therefore the quality of PET images, as well as improve accuracy of absorbed radiation dose calculations.

A closer examination of the PET radioisotopes considered in this chapter (see Table 3) shows that many of the isotopes can emit a number of positrons through the allowed decay paths. Understanding the kinetic energy spectra for each of these positrons is useful, as they affect the quality of the image to different degrees. In using equation 3 the positron energy spectra are calculated for the highest intensity positron per radioisotope in a similar manner to that reported by Le Loirec (Le Loirec, C. & Champion, C., 2007a).

$$N(E) = C\ F(Z,E)\ p\ E\ (E_{max} - E)^2 \tag{3}$$

$$F(Z,E) = 2\ \pi\ \eta\ (\ 1 - e^{-2\pi\eta})^{-1}$$

$$\eta = -Z\ \alpha\ E/p$$

Where:
E = total energy of the positron;
Z = atomic number;
α = fine structure constant;
p = momentum of positron;
E_{max} = maximum allowed energy of positron;
C = normalisation constant.

The spectrum of the dominant positron emission for each radioisotope is illustrated in Figure 9. The data demonstrate the range of positron energies per isotope and therefore the relative distance they are likely to travel *in vivo*. Of course, these values do not take into account the tissue density, but they can be used to give an indication of the relative quality of images to be obtained for each PET radioisotope. The green shaded area represents the range of positron energy spectrum for ^{18}F for comparison. Clearly positron spectra of [52]Mn, [62]Zn, ^{64}Cu and [48]V correlate best with ^{18}F and potentially could give similar image quality. However the interfering gammas for all of these radioisotopes, except ^{64}Cu, may compromise image quality.

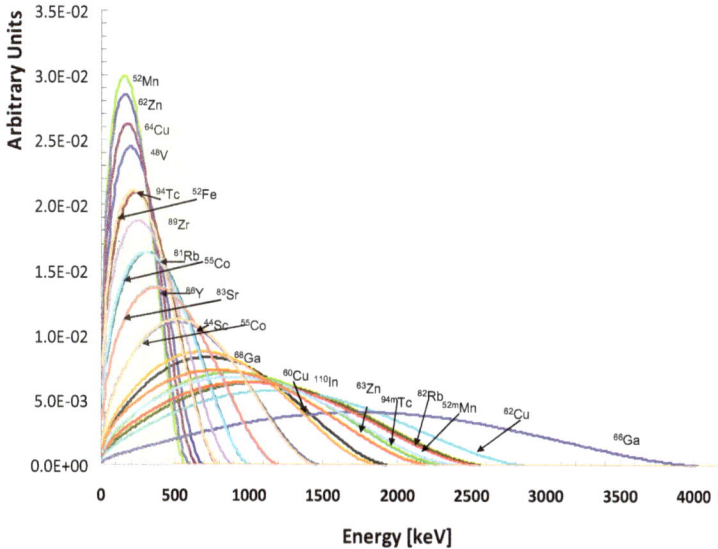

Fig. 9. The spectrum of the dominant positron emission for each radioisotope

Figure 10 (a) and (b) illustrate the relative intensity of kinetic energy spectra for β+ decays (in colour dotted curves) of [61]Cu and [66]Ga respectively. These spectra do not take into account the relative intensity of each positron. Conversely the black curve is the result of the accumulated spectra taking into consideration the relative intensities of each. Figure 10a shows how a low energy positron (559 keV) can become insignificant when a high intensity of the higher energy positron (1215 keV) is present. A similar effect is evident for [66]Ga however the shape is significantly different and suggests the signal to noise ratio would be poorer.

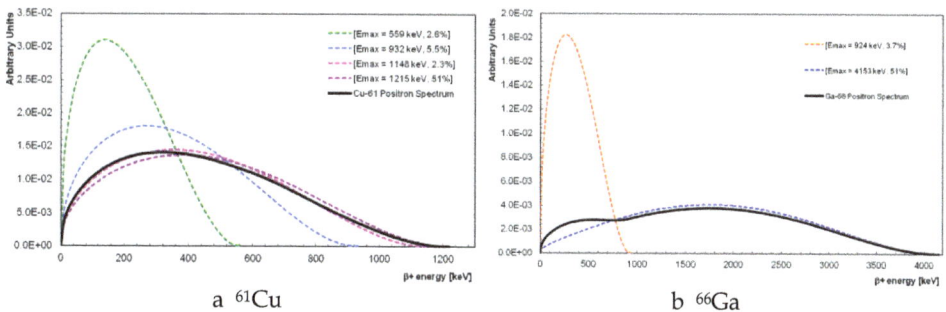

a [61]Cu

b [66]Ga

Fig. 10. The relative kinetic energy spectra of the positrons emitted by [61]Cu and [66]Ga (Le Loirec, C. & Champion, C., 2007a; 2007b)

8. Production of metal positron emitting isotopes

There are many radioisotopes that can be produced with a cyclotron. The challenge in each case is to develop targetry (target material and loading system) that allows it to be kept stable in the beam of the accelerated particles and to remove the desired radioisotope quickly and efficiently. Unlike the well-known PET isotopes ^{11}C, ^{18}F and ^{15}N which are produced using gas or liquid based targetry, the metal radioisotopes listed in Table 3 are mostly generated from solid targets of metal based materials (e.g. metal element or metal oxide). The target can be placed within the cyclotron (under vacuum) or external to the cyclotron. In each case the target system needs to tolerate heat generated from the particle reactions. Depending on the energy and current of the beam (μA or mA) the cooling systems can be quite elaborate.

As the number of nuclear reactions is potentially large, one needs to design the target so that the preferred nuclear reaction dominates. This may be achieved in a number of ways:

a. choice of beam particle [i.e. p, d, or α]

b. energy of the beam

c. target material [i.e. natural or enriched isotope of target element].

The use of enriched isotopes for these types of reactions can add considerable cost to the production of a radioisotope. It is wise to treat the material as capital and recycle it for re-use. By using an enriched isotope for the target, one can also prevent or reduce the production of contaminating radioisotopes. If these contaminating radioisotopes have high gamma emissions or different half lives from the desired radioisotope, they can affect the shielding and timing of a production process, as well as delay the recycling of enriched target material. For example, ^{67}Ga is a SPECT agent produced by proton bombardment of enriched ^{68}Zn. By-products of this reaction are ^{66}Ga ($t_{1/2}$=9.5 hr) and ^{65}Zn ($t_{1/2}$=244 days), both of which produce high energy gammas. The former is often left to decay before the ^{67}Ga can be processed. Furthermore shielding of hot cells needs to be increased to reduce the dose to operators. The presence of the latter, ^{65}Zn, prevents recycling of the ^{68}Zn target material. Therefore, large inventories (~100g costing ~$250,000 US) of enriched ^{68}Zn is maintained so irradiated samples can be left to decay before recycling. The presence of a long lived radioisotope contaminant in the desired PET product can impact on its expiry time and use. Good target design is essential to ensure cost effectiveness and high quality product.

8.1 Generator produced PET isotopes

There is added complexity to the production and separation of short lived radioisotopes. For example, ^{82}Rb ($T_{1/2}$ = 1.3 min), ^{68}Ga ($T_{1/2}$ = 67.6 min) and ^{62}Cu ($T_{1/2}$ = 7.6 min) are all produced from their respective parent isotopes in a generator. These generators are then supplied to the PET facilities and milked periodically to resource the desired isotopes.

8.2 Nuclear reactions for PET radioisotopes

The type of nuclear reactions to produce a PET radioisotope can be quite varied. Table 4 summarises some of the more successful production routes for the PET radioisotopes, their nuclear reactions, useable energy range and the maximum cross section from the excitation function. The abundance of target element used for each reaction is also listed.

Nuclide	Nuclear Reaction	Energy Range [MeV]	Natural Abundance [%]	Energy at Maximum [MeV]	Maximum Cross-Section [mb]
^{44m}Sc	$^{nat}Ti(p, x)^{44m}Sc$	28 - 48	Natural	34	17
^{48}V	$^{48}Ti(p, n)^{48}V$	6 - 50	73.8	12	380
^{52}Mn	$^{nat}Fe(d, x)^{52}Mn$	28 - 49	Natural	38	166
$^{52}Fe^{**}$	$^{50}Cr(\alpha, 2n)^{52}Fe$	20 - 42	4.4	29 , 25	21 , 60
^{52}Fe	$^{55}Mn(p, 4n)^{52}Fe$	40 – 60	100.0	54	100
^{52}Fe	$^{52}Cr(^{3}He, 3n)^{52}Fe$	25 – 40	83.8	19	550
^{55}Co	$^{54}Fe(d, n)^{55}Co$	5 - 10	5.8	7	84
^{55}Co	$^{56}Fe(p, 2n)^{55}Co$	20 – 30	91.7	25	70
^{55}Co	$^{58}Ni(p, \alpha)^{55}Co$	10 – 24	68.3	17	40
^{55}Co	$^{nat}Ni(p, x)^{55}Co$	11 – 22	Natural	17	24
^{55}Co	$^{nat}Fe(p, x)^{55}Co$	16 – 42	Natural	25	60
^{55}Co	$^{55}Mn(^{3}He,3n)^{55}Co$	15 – 25	100.0	19	550
^{61}Cu	$^{61}Ni(p, n)^{61}Cu$	3 – 19	1.1	10	480
^{61}Cu	$^{nat}Ni(\alpha, p)^{61}Cu$	15 – 25	Natural	27	450
^{61}Cu	$^{59}Co(a, 2n)^{61}Cu$	18 – 60	100.0	28	470
^{61}Cu	$^{61}Ni(p, n)^{61}Cu$	4 – 19	1.1	10	500
^{62}Zn	$^{nat}Cu(p, x)^{62}Zn$	14 - 61	Natural	23	75
^{62}Zn	$^{nat}Zn(p, x)^{62}Zn$	28 - 70	Natural	37	45
^{63}Zn	$^{nat}Cu(d, x)^{63}Zn$	6 – 40	Natural	17	420
^{63}Zn	$^{nat}Cu(p, x)^{63}Zn$	6 – 50	Natural	12	350
^{63}Zn	$^{63}Cu(p, n)^{63}Zn$	4 – 40	69.2	12	460
^{63}Zn	$^{60}Ni(\alpha\, n)^{63}Zn$	8 – 25	26.1	19	550
^{64}Cu	$^{61}Ni(\alpha, p)^{64}Cu$	5 – 15	0.9	10	148
^{64}Cu	$^{64}Ni(p, n)^{64}Cu$	3 - 40	0.9	11	700
^{64}Cu	$^{64}Ni(d, 2n)^{64}Cu$	5 - 50	0.9	15	800
^{64}Cu	$^{66}Zn(d, \alpha)^{64}Cu$	8 - 14	27.9	11	27
^{64}Cu	$^{nat}Zn(p, x)^{64}Cu$	20 – 30 / 30 - 65	Natural	23 / 40	15 / 25
^{64}Cu	$^{68}Zn(p, \alpha n)^{64}Cu$	16 – 45 / 45-100	18.8	25 / 70	75 / 60
^{66}Ga	$^{66}Zn(p,n)^{66}Ga$	7 - 30	27.8	12.5	650
^{66}Ga	$^{nat}Zn(p\ x)^{66}Ga$	5 - 70	Natural	15	150
^{66}Ga	$^{63}Cu(\alpha, n)^{66}Ga$	10 - 26	69.2	16	690
^{68}Ga	$^{68}Zn(p,n)^{68}Ga$	5 - 35	18.8	11	830
^{68}Ga	$^{65}Cu(\alpha,n)^{68}Ga$	11- 22	30.83	17	850
^{68}Ge	$^{nat}Zn(\alpha, x)^{68}Ge$	15 - 100	Natural	30	155
^{68}Ge	$^{66}Zn(\alpha, 2n)^{68}Ge$	20 - 40	27.8	30	550
^{68}Ge	$^{69}Ga(p, 2n)^{68}Ge$	13 – 36	60.1	20	558

^{68}Ge	^{nat}Ga(p, x)^{68}Ge	13 - 56	Natural	21	325
^{81}Rb	^{81}Kr(p, x)^{81}Rb	15 - 35 / 35 - 100	Natural	25 / 50	100 / 190
^{82}Rb	^{85}Rb(p, 4n)^{82}Sr	38 - 100	72.2	52	150
^{82}Rb	^{nat}Rb(p, x)^{82}Sr	45 - 60	72.2	52	100
^{82}Rb	^{82}Kr(α, 3n)^{83}Sr	25 - 40	11.6	30	280
^{82}Sr	^{82}Kr(^{3}He, 3n)^{82}Sr	19 - 35	11.6	35	275
^{86}Y	^{86}Sr(p, n)^{86}Y	7 - 50	9.9	13	950
^{86}Y	^{nat}Zr(d, x)^{86}Y	31 - 50	Natural	43	39
^{86}Y	^{86}Sr(d, 2n)^{86}Y	5 - 30	9.9	19	691
^{86}Y	^{88}Sr(p, 3n)^{86}Y	30 - 85	82.6	40	470
^{86}Y	^{85}Rb(α, 3n)^{86}Y	29 - 50	72.2	41	800
^{89}Zr	^{89}Y(p, n)^{89}Zr	5 - 50	100.0	14	780
^{89}Zr**	^{89}Y(d, 2n)^{89}Zr	5 - 40	100.0	16 , 20	700 , 1100
^{94m}Tc	^{94}Mo(p, n)^{94m}Tc	7 - 18	9.3	12	460

** Two references quoting different values.

Table 4. Common production routes of PET radioisotopes (Al-Abyad, M. et al., 2009; Aydin, A. et al., 2007; Brodzinski, R.L. et al., 1971; Fulmer, C.B. & Williams, I.R., 1970; Hermanne, A. et al., 2000; Hille, M. et al., 1972; Ido, T. et al., 2002; International Atomic Energy Agency, 2009; Khandaker, M.U. et al., 2009; Sadeghi, M. et al., 2010; Sadeghi, M. et al., 2009; Szelecsényi, F. et al., 2001; Szelecsenyi, F. et al., 2006; Takacs, S. et al., 2003; Takacs, S. et al., 2007; Tarkanyi, F. et al., 2005)

The cross sections for the production of ^{68}Ga via p and α bombardment of ^{68}Zn and ^{65}Cu targets respectively, are similar. The former production route would be preferred due to the wider availability of p producing cyclotrons. However it is important to note that ^{66}Ga which has a longer half life, can be produced via p bombardment of ^{66}Zn in the same energy region. As the natural abundances for ^{64}Zn, ^{66}Zn and ^{68}Zn are 48.6%, 27.9 % and 18.8 %, respectively, even small amounts of ^{64}Zn and ^{66}Zn present in enriched (>95%) ^{68}Zn will contribute to the production of contaminating isotopes.

Generators for the production of ^{82}Rb ($T_{1/2}$ = 1.3 min) and ^{68}Ga ($T_{1/2}$ = 67.6 min) are commercially available. The long half lives of their respective parents, ^{82}Sr ($T_{1/2}$ = 25.3 days) and ^{68}Ge ($T_{1/2}$ = 270.8 days) means the generators have a life-time of months to almost a year. However the challenge in the production of these generators is that they must stand up to repeated milking over at least twice the lifetime of the generator to demonstrate that there is no breakdown of the column material or any leaching of contaminating long lived radioisotopes. The parent of ^{62}Cu, ^{62}Zn has a short half-life and needs to be prepared daily. This generator can be milked multiple times however the short half-life of the ^{62}Zn limits the lifetime of its use to less than a day.

A comparison of the maximum cross section and the energy range suitable for the production of each PET isotope on a 30 MeV cyclotron is illustrated in Figure 11. The green and blue icons represent p and d reactions at the maximum cross section. The associated error bar reflects the useable energy ranges (cross section >10 mb) for the respective reactions.

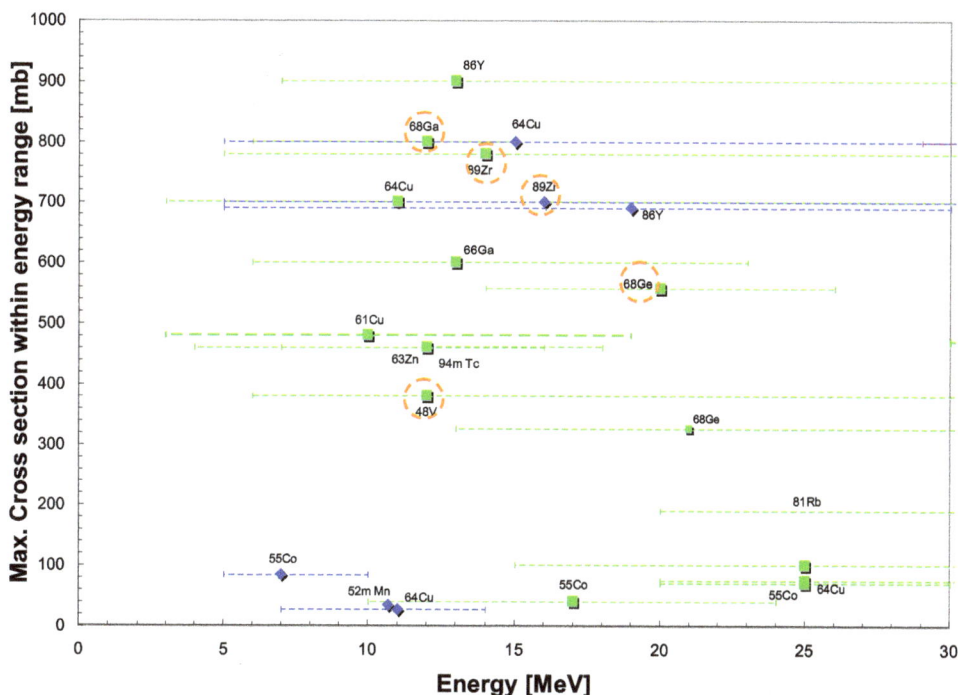

Fig. 11. Radioisotopes produced using protons and deuteron in the energy range 0-30 MeV.

In general, the production of radioisotopes via proton bombardment are more common across the energy ranges, as their high cross sections provide for higher yields. In contrast, nuclear reactions with deuterons are less common. This is not surprising as maintaining a stable beam of these particles at a high current is more difficult.

Reactions for ^{86}Y, ^{68}Ga, ^{64}Cu, ^{89}Zr, ^{66}Ga, ^{68}Ge, ^{61}Cu , ^{63}Zn, ^{94m}Tc and ^{48}V in Figure 11 have a very high probability. However the cost effectiveness of their production relies on the enrichment of their target material. The targets used to produce ^{89}Zr, ^{48}V, ^{63}Zn , ^{68}Ga and ^{68}Ge are of high (>60%) natural abundance (identified by the orange circles in Figure 11) and therefore will yield high quantities of these radioisotopes at comparatively reasonable cost. Once the preferred reaction route is identified then a separation method for the radioisotope needs to be developed. The separation methods need to be robust, reliable and cost effective in order to maintain a routine supply of the desired radioisotope.

9. Conclusions

Designing a PET radiopharmaceutical relies on careful consideration of a number of factors. They include the physical characteristics of the radioisotope, its chemistry and half life. For the former the critical aspects are the positron and gamma energies. A radioisotope with a low maximum positron energy is always preferred. The presence of any gammas contribute to radiation dose. High energy gammas require additional lead shielding and this significantly increases the cost of infrastructure. Gammas in the 350-700 keV range decrease PET image resolution and will require longer imaging times. The application of the radiopharmaceuticals in PET imaging will depend on the radioisotope's chemistry and biological half-life. The former requires the separation and radiolabelling to take place rapidly without extensive infrastructure and technical skills. Advances in the ability of cyclotrons to produce p, d and α particles have substantially increased the range of metal PET radioisotopes that can be produced. However the yield and the cost effectiveness of the radioisotope production are ultimately dependent on the choice of nuclear reaction and the natural abundance of the target material used. Metal PET radioisotopes are ideal for labelling small molecules, peptides, proteins and particles via chelators and clearly have a role in the diagnosis of disease and risk assessment of new drugs. Of the 21 metal PET radioisotopes evaluated, ^{61}Cu and ^{64}Cu best meet the criteria for human use.

10. References

Al-Abyad, M., Comsan, M.N.H. & Qaim, S.M. (2009). Excitation functions of proton-induced reactions on Fe-nat and enriched Fe-57 with particular reference to the production of Co-57. *Applied Radiation and Isotopes*, Vol. 67, 1, (Jan, 2009), pp. (122-128), 0969-8043

Aydin, A., Sarer, B. & Tel, E. (2007). New calculation of excitation functions of proton-induced reactions in some medical isotopes of Cu, Zn and Ga. *Applied Radiation and Isotopes*, Vol. 65, 3, (2007), pp. (365-370), 0969-8043

Baker, M. (2010). The whole picture. *Nature*, Vol. 463, 7283, (2010), pp. (977-980), 0028-0836

Bockisch, A., Freudenberg, L.S., Schmidt, D. & Kuwert, T. (2009). Hybrid imaging by SPECT/CT and PET/CT: Proven outcomes in cancer imaging. *Seminars in Nuclear Medicine*, Vol. 39, 4, (Jul, 2009), pp. (276-289), 0001-2998

Brodzinski, R.L., Rancitelli, L.A., Cooper, J.A. & Wogman, N.A. (1971). High-energy proton spallation of titanium. *Physical Review C*, Vol. 4, 4, (October, 1971), pp. (1250–1257),

Cai, W.B. & Chen, X.Y. (2008). Multimodality molecular imaging of tumor angiogenesis. *Journal of Nuclear Medicine*, Vol. 49, Suppl. 2, (June, 2008), pp. (113S-128S), 0161-5505

Cañadas, M.E., M. Lage, E. Desco, M. Vaquero, J.J. Pérez, J.M. (2008). Nema nu 4-2008 performance measurements of two commercial small-animal PET scanners: Clear PET RPET-1 *Nuclear Science, IEEE Transactions on* Vol. 58, 1 2008), pp. (4773-4779), 0018-9499

Cropley, V.L., Fujita, M., Innis, R.B. & Nathan, P.J. (2006). Molecular imaging of the dopaminergic system and its association with human cognitive function. *Biological Psychiatry*, Vol. 59, 10, (May, 2006), pp. (898-907), 0006-3223

Frost & Sullivan (2010). Emerging technology developments in hybrid imaging. Frost and Sullivan Technical Insights Reports.

Fulmer, C.B. & Williams, I.R. (1970). Excitation functions for radioactive nucleides produced by deuteron-induced reactions in copper. *Nuclear Physics* Vol. A155, (June, 1970), pp. (40-48),

Gabriel, M., Decristoforo, C., Kendler, D., Dobrozemsky, G., Heute, D., Uprimny, C., Kovacs, P., Von Guggenberg, E., Bale, R. & Virgolini, I.J. (2007). Ga-68-dota-tyr(3)-octreotide PET in neuroendocrine tumors: Comparison with somatostatin receptor scintigraphy and ct. *Journal of Nuclear Medicine*, Vol. 48, 4, (Apr, 2007), pp. (508-518), 0161-5505

Gomes, C.M.F., Abrunhosa, A.J. & Pauwels, E.K.J. (2011). Molecular imaging with SPECT and PET in exploratory investigational new drug studies. *Drugs of the Future*, Vol. 36, 1, (Jan, 2011), pp. (69-77), 0377-8282

Gregoire, V., Haustermans, K., Geets, X., Roels, S. & Lonneux, M. (2007). PET-based treatment planning in radiotherapy: A new standard? *Journal of Nuclear Medicine*, Vol. 48, (Jan, 2007), pp. (68S-77S), 0161-5505

Haddad, F., Ferrer, L., Guertin, A., Carlier, T., Michel, N., Barbet, J. & Chatal, J.F. (2008). Arronax, a high-energy and high-intensity cyclotron for nuclear medicine. *European Journal of Nuclear Medicine and Molecular Imaging*, Vol. 35, 7, (Jul, 2008), pp. (1377-1387), 1619-7070

Henze, M., Schuhmacher, J., Hipp, P., Kowalski, J., Becker, D.W., Doll, J., Macke, H.R., Hofmann, M., Debus, J. & Haberkorn, U. (2001). PET imaging of somatostatin receptors using (68)Ga dota-d-phe(1)-tyr(3)-octreotide: First results in patients with meningiomas. *Journal of Nuclear Medicine*, Vol. 42, 7, (Jul, 2001), pp. (1053-1056), 0161-5505

Hermanne, A., Sonck, M., Takacs, S. & Tarkanyi, F. (2000). Experimental study of excitation functions for some reactions induced by deuterons (10-50 MeV on natural Fe and Ti. *Nuclear Instruments & Methods in Physics Research Section B-Beam Interactions with Materials and Atoms*, Vol. 161, (Mar, 2000), pp. (178-185), 0168-583X

Hille, M., Hille, P., Uhl, M. & Weisz, W. (1972). Excitation functions of (p, n) and (a ,n) reactions on Ni, Cu and Zn. *Nuclear Physics*, Vol. A198, (1972), pp. (625-640),

Holland, J.P., Williamson, M.J. & Lewis, J.S. (2010) Unconventional nuclides for radiopharmaceuticals. *Molecular Imaging*, Vol. 9, 1, (Jan-Feb 2010), pp. (1-20), 1535-3508 http://www.nndc.bnl.gov/. Retrieved 21 May 2011.

Hubbell, J.H. (1969). Photon cross sections, attenuation coefficients, and energy absorption coefficients from 10 keV to 100 GeV National Standard Reference Data System.

Ido, T., Hermanne, A., Ditroi, F., Szucs, Z., Mahunka, I. & Tarkanyi, F. (2002). Excitation functions of proton induced nuclear reactions on Rb-nat from 30 to 70 MeV. Implication for the production of Sr-82 and other medically important rb and sr radioisotopes. *Nuclear Instruments & Methods in Physics Research Section B-Beam Interactions with Materials and Atoms*, Vol. 194, 4, (Oct, 2002), pp. (369-388), 0168-583X

International Atomic Energy Agency (2006). Directory of cyclotrons used for radionuclide production in member states-2006 update. Vienna, Austria.

International Atomic Energy Agency (2009). Cyclotron produced radionuclides : Physical characteristics and production methods. Technical reports series No 468. Vienna, Austria.

Khandaker, M.U., Kim, K., Lee, M.W., Kim, K.S., Kim, G.N., Cho, Y.S. & Lee, Y.O. (2009). Investigations of the Ti-nat(p,x)Sc-43,Sc-44m,Sc-44g,Sc-46,Sc-47,Sc-48,V-48 nuclear processes up to 40 MeV. *Applied Radiation and Isotopes*, Vol. 67, 7-8, (Jul-Aug, 2009), pp. (1348-1354), 0969-8043

Kitajima, K., Murakami, K., Sakamoto, S., Kaji, Y. & Sugimura, K. (2011). Present and future of FDG-PET/CT in ovarian cancer. *Annals of Nuclear Medicine*, Vol. 25, 3, (Apr, 2011), pp. (155-164), 0914-7187

Laforest, R. & Liu, X. (2008). Image quality with non-standard nuclides in PET. *Quarterly Journal of Nuclear Medicine and Molecular Imaging*, Vol. 52, 2, (Jun, 2008), pp. (151-158), 1824-4661

Lancelot, S. & Zimmer, L. (2010) Small-animal positron emission tomography as a tool for neuropharmacology. *Trends in Pharmacological Sciences*, Vol. 31, 9, (Sep 2010), pp. (411-417), 0165-6147

Langer, A. (2010) A systematic review of PET and PET/CT in oncology: A way to personalize cancer treatment in a cost-effective manner? *MC Health Services Research*, Vol. 10, (Oct), pp. (16), 1472-6963

Le Loirec, C. & Champion, C. (2007a). Track structure simulation for positron emitters of medical interest. Part I: The case of the allowed decay isotopes. *Nuclear Instruments & Methods in Physics Research Section A-Accelerators Spectrometers Detectors and Associated Equipment*, Vol. 582, 2, (Nov, 2007a), pp. (644-653), 0168-9002

Le Loirec, C. & Champion, C. (2007b). Track structure simulation for positron emitters of physical interest. Part II: The case of the radiometals. *Nuclear Instruments & Methods in Physics Research Section A-Accelerators Spectrometers Detectors and Associated Equipment*, Vol. 582, 2, (Nov, 2007b), pp. (654-664), 0168-9002

Lee , J.S. (2010). Technical advances in current PET and hybrid imaging systems. *The Open Nuclear Medicine Journal*, Vol. 2,(2010), pp. (192-208),

Lewellen, T.K. (2008). Recent developments in PET detector technology. *Physics in Medicine and Biology*, Vol. 53, 17, (Sep, 2008), pp. (R287-R317), 0031-9155

Li, Z.B., Chen, K. & Chen, X. (2008). Ga-68-labeled multimeric RGD peptides for microPET imaging of integrin alpha(v)beta(3) expression. *European Journal of Nuclear Medicine and Molecular Imaging*, Vol. 35, 6, (Jun, 2008), pp. (1100-1108), 1619-7070

Liu, Z.F., Yan, Y.J., Liu, S.L., Wang, F. & Chen, X.Y. (2009). F-18, Cu-64, and Ga-68 labeled rgd-bombesin heterodimeric peptides for PET imaging of breast cancer. *Bioconjugate Chemistry*, Vol. 20, 5, (May, 2009), pp. (1016-1025), 1043-1802

Lonsdale, M.N. & Beyer, T. (2010)Dual-modality PET/CT instrumentation -today and tomorrow. *European Journal of Radiology*, Vol. 73, 3, (Mar 2010), pp. (452-460), 0720-048X

Madsen, M.T., Anderson, J.A., Halama, J.R., Kleck, J., Simpkin, D.J., Votaw, J.R., Wendt, R.E., Williams, L.E. & Yester, M.V. (2006). Aapm task group 108: PET and PET/CT shielding requirements. *Medical Physics*, Vol. 33, 1, (Jan, 2006), pp. (4-15), 0094-2405

Moses, W.W. (2007). Recent advances and future advances in time-of-flight PET. *Nuclear Instruments & Methods in Physics Research Section A-Accelerators Spectrometers Detectors and Associated Equipment*, Vol. 580, 2, (Oct, 2007), pp. (919-924), 0168-9002

Moses, W.W., Virador, P.R.G., Derenzo, S.E., Huesman, R.H. & Budinger, T.F. (1997). Design of a high-resolution, high-sensitivity PET camera for human brains and small animals. *IEEE Transactions on Nuclear Science*, Vol. 44, 4, (Aug, 1997), pp. (1487-1491), 0018-9499

Partridge, M., Spinelli, A., Ryder, W. & Hindorf, C. (2006). The effect of positron energy on performance of a small animal PET camera. *Nuclear Instruments and Methods in Physics Research A*, Vol. 568, (2006), pp. (933-936),

Petrusenko, Y.T., Lymar, A.G., Nikolaichuk, L.I., Tutubalin, A.I., Shepelev, A.G., Ponomarenko, T.A. & Nemashkalo, O.V. (2009). Analysis of data arrays on cyclotron production of medical radioisotopes. *Problems of Atomic Science and Technology*, Vol., 3,(2009), pp. (82-88), 1562-6016

Pimlott, S.L. & Sutherland, A.(2011) Molecular tracers for the pet and spect imaging of disease. *Chemical Society Reviews*, Vol. 40, 1), pp. (149-162), 0306-0012

Roach, M., Alberini, J.L., Pecking, A.P., Testori, A., Verrecchia, F., Soteldo, J., Ganswindt, U., Joyal, J.L., Babich, J.W., Witte, R.S., Unger, E. & Gottlieb, R. (2011). Diagnostic and therapeutic imaging for cancer: Therapeutic considerations and future directions. *Journal of Surgical Oncology*, Vol. 103, 6, (May, 2011), pp. (587-601), 0022-4790

Ruangma, A., Bai, B., Lewis, J.S., Sun, X.K., Welch, M.J., Leahy, R. & Laforest, R. (2006). Three-dimensional maximum a posteriori (map) imaging with radiopharmaceuticals labeled with three Cu radionuclides. *Nuclear Medicine and Biology*, Vol. 33, 2, (2006), pp. (217-226), 0969-8051

Sadeghi, M., Enferadi, M., Aref, M. & Jafari, H. (2010). Nuclear data for the cyclotron production of Ga-66, Y-86, Br-76, Cu-64 and Sc-43 in PET imaging. *Nukleonika*, Vol. 55, 3, 2010), pp. (293-302), 0029-5922

Sadeghi, M., Zali, A. & Mokhtari, L. (2009). Calculations of Y-86 production via various nuclear reactions. *Nuclear Science and Techniques*, Vol. 20, 6, (Dec, 2009), pp. (369-372), 1001-8042

Sanchez-Crespo, A., Andreo, P. & Larsson, S.A. (2004). Positron flight in human tissues and its influence on pet image spatial resolution. *European Journal of Nuclear Medicine and Molecular Imaging*, Vol. 31, 1,(2004), pp. (44-51), 1619-7070

Schmor, P.W. (2010). Review of cyclotrons used in the production of radioisotopes for biomedical applications, *Proceedings of Cyclotrons 2010*, Lanzhou, China, September, 2010

Seam, P., Juweid, M.E. & Cheson, B.D. (2007). The role of FDG-PET scans in patients with lymphoma. *Blood*, Vol. 110, 10, (Nov, 2007), pp. (3507-3516), 0006-4971

Senthamizhchelvan, S., Bravo, P.E., Lodge, M.A., Merrill, J., Bengel, F.M. & Sgouros, G. (2011). Radiation dosimetry of Rb-82 in humans under pharmacologic stress. *Journal of Nuclear Medicine*, Vol. 52, 3, (Mar, 2011), pp. (485-491), 0161-5505

Smith, S. V. (2004). Molecular imaging with copper-64. J. Inorg. Biochem., Vol. 98, 11, (2004), pp. (1874-1901), 0162-0134.

Smith, S.V. (2005). Challenges and opportunities for positron-emission tomography in personalized medicine. *Idrugs*, Vol. 8, 10, (Oct, 2005), pp. (827-833), 1369-7056

Smith, S.V. (2007). Molecular imaging with copper-64 in the drug discovery and development arena. *Expert Opinion on Drug Discovery*, Vol. 2, 5, (May, 2007), pp. (659-672), 1746-0441

Szelecsényi, F., Kovács, Z., Suzuki, K., Takei, M. & Okada, K. (2001). Investigation of direct production of ^{62}cu radioisotope at low energy multiparticle accelerator for PET studies, *Proceedings of International Symposium on Utilization of Accelerators*, Sao Paulo, Brazil, November, 2001

Szelecsenyi, F., Steyn, G.F., Kovacs, Z., van der Walt, T.N. & Suzuki, K. (2006). Comments on the feasibilty of Cu-61 production by proton irradiation of (natZn) on a medical cyclotron. *Applied Radiation and Isotopes*, Vol. 64, 7, (Jul, 2006), pp. (789-791), 0969-8043

Takacs, S., Tarkanyi, F., Hermanne, A. & de Corcuera, R.P. (2003). Validation and upgrading of the recommended cross section data of charged particle reactions used for production of pet radioisotopes. *Nuclear Instruments & Methods in Physics Research Section B-Beam Interactions with Materials and Atoms*, Vol. 211, 2, (Oct, 2003), pp. (169-189), 0168-583X

Takacs, S., Tarkanyi, F., Kiraly, B., Hermanne, A. & Sonck, M. (2007). Evaluated activation cross sections of longer-lived radionuclides produced by deuteron induced reactions on natural nickel. *Nuclear Instruments & Methods in Physics Research Section B-Beam Interactions with Materials and Atoms*, Vol. 260, 2, (Jul, 2007), pp. (495-507), 0168-583X

Tarkanyi, F., Ditroi, F., Csikai, J., Takacs, S., Uddin, M.S., Hagiwara, M., Baba, M., Shubin, Y.N. & Dityuk, A.I. (2005). Activation cross-sections of long-lived products of proton-induced nuclear reactions on zinc. *Applied Radiation and Isotopes*, Vol. 62, 1, (Jan, 2005), pp. (73-81), 0969-8043

Truong, M.T., Viswanathan, C. & Erasmus, J.J. (2011). Positron emission tomography/computed tomography in lung cancer staging, prognosis, and assessment of therapeutic response. *Journal of Thoracic Imaging*, Vol. 26, 2, (May, 2011), pp. (132-146), 0883-5993

Vandenberghe, S. (2006). Three-dimensional positron emission tomography imaging with I-124 and Y-86. *Nuclear Medicine Communications*, Vol. 27, 3, (Mar, 2006), pp. (237-245), 0143-3636

Vyas, N.S., Patel, N.H., Nijran, K.S., Al-Nahhas, A. & Puri, B.K. (2011). The use of pet imaging in studying cognition, genetics and pharmacotherapeutic interventions in schizophrenia. *Expert Review of Neurotherapeutics*, Vol. 11, 1, (Jan, 2011), pp. (37-51), 1473-7175

Wahl, R.L., Jacene, H., Kasamon, Y. & Lodge, M.A. (2009). From recist to percist: Evolving considerations for PET response criteria in solid tumors. *Journal of Nuclear Medicine*, Vol. 50, (May, 2009), pp. (122S-150S), 0161-5505

Wang, C., Li, H.D., Ramirez, R.A., Zhang, Y.X., Baghaei, H., Liu, S.T., An, S.H. & Wong, W.H.(2010) A real time coincidence system for high count-rate TOF or non-TOF PET cameras using hybrid method combining and-logic and time-mark technology. *IEEE Transactions on Nuclear Science*, Vol. 57, 2, (Apr-2010), pp. (708-714), 0018-9499

Winant, C.D., Kim, S. & Seo, Y.(2010) Functional imaging combined with multi-detector CT: A radionuclide imaging perspective. *Current Medical Imaging Reviews*, Vol. 6, 2, (May), pp. (100-111), 1573-4056

Wolf, W. (2011). The unique potential for noninvasive imaging in modernizing drug development and in transforming therapeutics: PET/MRI/MRS. *Pharmaceutical Research*, Vol. 28, 3, (Mar, 2011), pp. (490-493), 0724-8741

Nuclear Medicine in the Imaging and Management of Breast Cancer

Luciano Izzo[1], Sara Savelli[1], Andrea Stagnitti[2] and Mario Marini[2]
Sapienza University of Rome, Umberto I hospital
[1]Department of Surgery "P. Valdoni"
[2]Department of Radiology
Italy

1. Introduction

Axillary nodal status is the most important prognostic factor for patients with breast cancer. Clinical assessment and imaging modalities are not always reliable.

Surgical removal and histopathological examination of axillary lymph nodes remain essential methods of staging the axilla and planning breast cancer therapy (Chu et al, 2010; Perry, 2001) But whether axillary lymph node dissection improves survival remains controversial.

Historically, nodal involvement was determined by conventional axillary dissection. (Gill, 2009)

Till 1990s conventional axillary dissection was performed in all women with breast cancer with great incidence of morbidities such as pain, numbness, shoulder joint stiffness, scarring, infection and long term lymphoedema. (Mansel et al, 2006)

Sentinel lymph node biopsy has recently been introduced in the treatment of women with breast cancer.

The American Society of Clinical Oncology in 2005 (Lyman, 2005), and more recently the British Association of Surgical Oncology (Association of Breast Surgery, 2009) endorsed sentinel lymph node biopsy (SLNB) as the recommended method of staging early breast cancer in clinically node negative patients.

The sentinel lymph node hypothesis is that the sentinel lymph nodes are the first nodes draining a tumor and that focused histological analysis of sentinel lymph nodes is predictive of the status of the regional nodes thus accurately staging the breast disease with lower incidence of complications. (Krag et al, 1998; Morton et al, 1992)

Giuliano et al (Giuliano et al, 1994) demonstrated that the status of sentinel nodes, with respect to the presence of metastases, accurately reflected the metastatic content of the local nodal basin.

Early works support the feasibility and accuracy of sentinel lymph node biopsy.

However, the optimal management of the axilla remains uncertain as axillary lymph node dissection, sentinel lymph node biopsy and lymphoscintigraphy have been described as reliable alternative procedures.

By combining preoperative lymphatic mapping with intraoperative gamma probe detection nuclear medicine procedures are increasingly used to identify and detect the sentinel node in breast cancer, and in other malignancies.

Some surgeons have begun to offer sentinel lymph node biopsy as an alternative to axillary dissection for women with small infiltrating breast cancer; axillary dissection is performed only if tumor is found in the sentinel lymph node.

Moreover in the recent years percutaneous imaging guided localization has been increasingly used to diagnose carcinoma previously detected at screening mammography.

In this setting sentinel lymph node biopsy is considered the gold standard for axillary staging according to the low rate of node metastasis in nonpalpable infiltrating breast cancer.

2. Sentinel lymph node biopsy

In contrast to the procedure of lymphoscintigraphy for melanoma, in breast cancer there has not yet been reached a consensus for many topics regarding sentinel lymph node biopsy methodology such as tracer characteristics, injection volume, and principally the site of administration.

2.1 Tracer characteristics

Sentinel lymph node biopsy is commonly performed using radioisotopes and/or blue dye that facilitate sentinel lymph node mapping. (Jakub et al, 2010)

However, it is still undefined which reagent is more suitable for lymphoscintigraphy. Several data from literature suggest that radioisotopes based on technetium-99m colloid are superior to blue dye in detecting SLN in breast cancer. (Somasundaram et al, 2007)

Historically, the colorimetric detection of lymph nodes by the trypan blue (or equivalent) preceded the use of radioisotopes. For its rapid migration kinetics, the dye is injected in 1-4 ml intra-operatively either into or around the tumour, or superficially into the section of areolar tissue that correlates with the index quadrant, 5 to 10 minutes before surgery. Patent blue dyes bind to interstitial albumin and are taken up by local lymphatic tissue. The efficiency with which the lymphatics are converted to bright blue channels by the vital dyes reflects their small hydrodynamic diameter, their ability to disperse quickly and even their capacity to readily progress through and beyond the nodes. The dissection is guided by the path of the blue lymphatic channels that lead to one or more nodes, more or less intensely colored. The choice of vital dye varies between operating surgeons. In the United States, Isosulfan blue, as a 1% sterile solution is the vital dye approved for use in humans by the Food and Drug Administration. Isosulfan blue is a monosodium isomer of the 2,5 disulfonated triphenylmethanes which are also known as the patent blue dyes (Leong et al, 2000). One of these patent blue dyes, Patent Blue Violet (or Patent Blue V), is preferentially used in Australia and Europe. Another vital dye, Indocyanine green has been used in Japan. All the vital dyes are selectively absorbed into lymphatic tissue and therefore facilitate localisation of the lymphatic channels and nodes. (White et al, 2011)

After the introduction of the isotope method, some teams have remained faithful to the colorimetric method while others are converted to the exclusive use of the radiocolloides; but the majority of teams use the combination of the two techniques (Tafra et al, 2002), the results are usually best with only one, like a radiocolloide. The advantage of vital dyes is their "real time" efficiency in guiding the gamma probe to the sentinel node. This reduces exploratory dissection and tissue plane disruption and leads to a surgical exploration much less aggressive than with isotope technique, which has the advantage of transparency of the tissues to the gamma photons. Moreover the injection of blue presents a number of disadvantages, the most serious is the risk of allergic shock, whose incidence is approximately 1%; review of the data suggests otherwise that the risk of severe anaphylaxis (grade 3) can be as low as 0.06%, and up to 0.4% for patients undergoing SLNB when data is analysed from large trials or databases. (Barthelmes et al, 2010; Hunting et al, 2010) Methylene blue has been used as an alternative to the other vital dyes. Its attraction as a mapping agent is its low risk profile. The risk of anaphylaxis from patent blue dyes may also be reduced by pre-operative prophylaxis with corticosteroids, antihistamines and H2-receptor antagonists.

In a study published recently Derossis and al. (Derossis et al, 2001) show that the rate of identification of the sentinel node by dye ganglion has remained stable over time (89% in the first 500 patients and 90% in the last 500 patients) while it has increased with the isotopic technique (respectively from 88% to 98%)(p < 0.0015). Moreover blue dyes also contribute to increased sensitivity of the SLNB procedure when used in conjunction with radioactive nanocolloids.

Sentinel node biopsy (SNB) by radioisotopes is a widely accepted and reliable surgical method for staging breast cancer in patients with unknown positive axillary lymph nodes involvement.

The reproducibility of lymphoscintigraphy for sentinel node detection varies from 85% to 88% and the method appears to have a high interobserver agreement.

Injection of the radio-isotope occurs between 2 and 24 h before surgery. Radioactive nanocolloid is injected preoperatively, with the assistance of radiological imaging either around the tumour, or into the overlying skin. It is thought that nanocolloids become entrapped within the sentinel lymph nodes either through a function of their particulate size (the larger hydrodynamic diameter of 50-100 nm for nanocolloid requires a transit time of usually more than 1 h and 15 min) or because of phagocytosis by leukocytes which migrate to and are retained within the draining lymph nodes. These entrapment processes are unlikely to be mutually exclusive, and other mechanisms may also exist, but the end result is localisation of the nanocolloid within the sentinel nodes rather than its diffuse spread to secondary nodes (Cody et al, 2001).

2.2 Site of administration

The exact site of injection of the mapping agent varies between studies, but can be broadly divided into "superficial" or "deep". (Nieweg et al, 2004; Celliers et al, 2003)

Lymphoscintigraphy by subdermal tracer administration is able to detect axillary lymph nodes in 98% of the cases but the method is accompanied by a low visualization incidence of drainage outside the lower axilla such as the internal mammary chain. This latter aspect appears to occur in 16% to 35% in the series using peri- or intratumoural administration with an axillary rate of visualization of 75% to 98%. Although peritumoural administration is predominantly associated with late lymph node detection, the early appearance observed after subdermal and intratumoural tracer injection justifies the obtention of early gamma camera images (McMasters et al, 2001; Boolbol et al, 2001). Intradermal injection delivers macromolecules to loco-regional lymph nodes faster than subcutaneous injection, suggesting easier lymphatic vessel access. Although the majority of patients has different pathways of lymphatic drainage from the ipsilateral breast and upper limb, in a small minority of patients the drainage pathway is through a common SLN. Such patients may be at increased risk of developing upper limb breast cancer-related lymphoedema after SLN biopsy (Cody et al, 1999).

The best injection site remains a hotly debated topic. The options are injected intratumor, peritumoral, subareolar, periareolar, subcutaneous and intradermal. The number of injection sites is variable, ranging between one and six.

Most authors agree intratumoral site. Other methods of injection have all excellent results, without any real objective comparisons have been made, but now the simplest and most favorable for the intraoperative detection, is the intradermal periareolar apprauch.

2.3 Injection volume

The choice of an intradermal injection allows only a small volume (0.2 mL) per injection. When the intradermal injection is not possible there is a disagreement between American and European teams.

The first team is for a large volume (up to 6 mL) to dilate the lymphatic channels and force the passage of the suspension of radiocolloid to the sentinel node. In contrast, the European teams, particularly under the impetus of the European Institute of Oncology in Milan, defend the injection of a small volume (0.1 to 0.4 mL) according to best physiological conditions.

The debate is difficult because the reported results are excellent in both cases.

Moreover, some prominent American teams also began to advocate a volume of 0.1 mL.

2.4 Histopathological analysis

The histopathological analytical methods for excised sentinel lymph nodes have been improved in order to increase staging accuracy and reduce false-negative rates.

The dissected tissue containing the sentinel lymph node is placed in 4% formalin solution. Frozen slide analysis is not performed. The sentinel lymph nodes are separated from surrounding adipose tissue, bisected, lamellated and embedded. Subsequent serial sections are made.

The relevant sentinel lymph node pathology must include multisection staining with hematoxylin and eosin.

Immunohistochemical analysis with antibodies to cytokeratin is also commonly used for hematoxylin and eosin-negative cases and may facilitate the detection of small deposits, thereby enhancing the screening of sentinel lymph nodes sections.

The majority of sentinel lymph node cases can be classified as positive or negative based on the presence or absence of macrometastasis. The 2003 edition of the TNM classification uses 2.0 mm as the cutoff size that distinguishes between micro- and macrometastasis.

The cutoff value for isolated tumor cells or so-called submicrometastasis is 0.2 mm.

The significance and practical relevance of micrometastases, isolated tumor cells and the value of immunohistochemical analysis also require further definition.

2.5 Axillary excision complications

Lymphedema is one of the most feared complication of breast cancer surgery.

Sentinel lymph node biopsy has markedly reduced its incidence but has not eliminated it. The incidence of measured lymphedema after sentinel lymph node biopsy has been reported to be 0 to 22%; most studies report an incidence of 3 to 7%.

The prevalence of subjectively reported lymphedema ranges from 0 to 15%. (Veronesi et al, 2003; Leidenius et al, 2005; Langer et al, 2007; McLaughlin et al, 2008; Armer J et al, 2004; Lucci A et al, 2007; Crane-Okada et al, 2008; Helyer LK et al, 2009; Schrenk et al, 2000; Kwan et al, 2010)

The variation in incidence rates is influenced by the length of follow-up after the axillary surgery, by the absence of a standard definition of both measured and perceived lymphedema, and by the lack of agreement between patient perceptions and objective measures.

Many authors have shown that in women treated for breast cancer, there is a large variation in the prevalence of lymphedema depending on the definition used. In their population of 211 women treated for breast cancer, they found the incidence of lymphedema at 30 months post-treatment varied widely depending on the measure and definition used: 91% using the definition of a 2-cm difference in arm circumference, 67% using the definition of a 200-ml increase in volume by perometry, 45% using the definition of a 10% increase in volume by perometry, 41% using self-report of limb heaviness or swelling (Armer et al 2009).

Dayes et al (Dayes et al, 2006, 2008) found that more than 20% of women presenting (presumably with the perception of arm swelling) to lymphedema clinics at large regional cancer centers had no measurable lymphedema (5% excess arm volume compared with the contralateral arm). Similarly, Querci della Rovere et al (Querci della Rovere et al, 2003) reported on a series of 199 patients, 10% (20 of 199) of whom perceived current arm swelling. Half of these patients with perceived swelling were not found to have objectively measured lymphedema by circumferential measurements.

Women who had surgery on their nondominant axilla less frequently reported perceived arm swelling in the absence of measured arm swelling, and less frequently perceived arm swelling even when there was objectively measured lymphedema. This suggests that patient perceptions of lymphedema are less influenced by sensory changes in the nondominant arm compared with the dominant arm.

It is possible that patient perceptions of lymphedema are influenced by patient knowledge of the number of nodes that was removed at the time of the sentinel lymph node biopsy.

2.6 Limitations

Several factors affect the technical success of sentinel lymph node. First of all the success rate is higher with experience.

The reported technical success rates for sentinel lymph node biopsy are 69%-99% for radioisotope alone, 67%-93% for blue dye alone and 90%-100% for the combined use. Radioisotopes injection allows the surgeon to identify the sentinel nodes prior to making any incision, blue dye allows the surgeon to visually identify the sentinel nodes; the combined method may shorten the learning curve and increase the likelihood of technical success.

The main limitation of this method is due to the appearance of false negatives that may be caused by tumor lymph node blockage of the sentinel lymph node and uptake in the neighboring lymph nodes. Infiltered sentinel nodes are generally increased in size and firm. Thus, they can be detected by intraoperative palpation, even when there is no uptake by the radiotracer. Intraoperative axillary palpation, once the sentinel lymph node is done, reduces the false negative rate.

Moreover identification of sentinel lymph node is dependent on uptake of the labeling agent by the lymphatic pathway that drain the tumor. Previous surgery or surgical biopsy may disrupt this lymphatic pathway thus lowering the success rate and the accuracy of sentinel lymph node biopsy. Krag et al (Krag et al, 1998) reported a significantly higher frequency of failure to identify a hot spot in women who had undergone prior surgical biopsy when compared to women that have undergone percutaneous breast biopsy or no surgical biopsy at all. Other authors did not find a significant difference between the groups.

A reduced identification rate in older patients is consistently reported. The increased fatty tissue in the breast in elderly patients might cause a decreased lymphatic flow. It is also suggested that the replacement of lymph nodes by fatty tissue decreases the capacity of lymph nodes to retain the radioactive colloid.

Tumor size is also associated with the detection rate, showing a low identification rate in tumors >5 cm in size. Patients with large tumors have in fact a greater risk of extensive axillary tumor burden, which decreases the lymphoscintigraphic visualization.

3. Radioguided occult lesion localization

With increasing awareness of early breast cancer detection among the population, the use of mammography and breast ultrasound in screening has gained prominence in recent years.

As a result, small non-palpable occult breast lesions are detected with increasing frequency. These lesions are initially characterised by percutaneous biopsy. If these turn out to be malignant, in terms of being invasive or in situ, or heterogeneous (such as atypical ductal hyperplasia, radial scar, papillary lesions, or those with imaging-pathologic discordance), diagnostic or therapeutic surgical excision becomes warranted.

It is then necessary to accurately localize such occult breast lesions, with the aim to excise the smallest amount of breast tissue and yet remove the entire lesion whilst achieving adequate clear margins.

A variety of preoperative localisation techniques for nonpalpable breast lesions has been described in the literature: hypodermic needles, wires (the most widely accepted technique), carbon solution, methylene blue dye. For the past 10 years, the radioguided occult lesion localisation (ROLL) technique has also been proposed either by means of radiotracer injection or by intratumoral deposits of radioactive seeds, and its role has been increasing in recent years.

3.1 Hookwire localization

For years, hookwire localization has been the traditional localization procedure in patients undergoing lumpectomy or wide local excision of clinically occult breast lesions (Chen et al, 2005).

This technique has several well described disadvantages such as incidental migration, kinking or fracture of the wire, and difficult logistics between the radiology, surgical and nuclear medicine departments.

Moreover hookwire localization has a number of other drawbacks.

Negotiating the wire through a very dense breast can be technically difficult for radiologists. The hookwire may be imprecisely placed or even displaced in fatty breasts, which is of particular concern when a patient has to move from one centre to another for surgery. The pathologist may find the wire difficult to dissect, with possible subsequent damage to the specimen.

3.2 ROLL radioguided occult lesion localization

Radioguided occult lesion localization is a new technique which allows identification of non palpable breast lesion in breast cancer using, on the model of sentinel node procedure, injection of a radiotracer over the tumour lesion. With a gamma detection probe, it is then possible during surgery to identify the lesion. (Belloni et al, 2011)

Radioguided occult lesion localization was pioneered in 1996 at the European Institute of Oncology, Milan. Inspired by the rationale for sentinel node biopsy, radioguided occult lesion localization involved the inoculation of macroaggregates of human serum albumin labelled with radioactive technetium99 (99mTc) directly into the site of a nonpalpable breast lesion, with mammography or ultrasound guidance. After radiotracer injection, a gamma-detecting probe was used to locate the lesion during surgery, allowing the surgeon to evaluate its skin projection and decide the best approach with an acceptable cosmetic outcome (Luini et al, 1998).

After the first publications on ROLL it was proposed that human serum albumin should be used as a radiotracer, with large particles ranging from 10 to 150 µm in diameter. For radioguided localisation of non-palpable breast lesions it is preferable that radiotracers do not migrate rapidly by lymphatic channels, in order to allow lesion identification by the surgeon using the probe. For this reason, it is recommended that larger 99mTc-labelled colloid particles of albumin should be used than are employed for sentinel lymph node identification, for which small particles from 200 to 1,000 nm are preferred. Other small particles, such as sulphur colloid, which are used in most services in the United States, flow quickly to the axillary lymph nodes and are often trapped in several nodes.

99mTc-labelled dextran was first used in lymphoscintigraphy by Henze et al in 1982 (Henze et al, 1982) that demonstrated its usefulness in SLN mapping in breast carcinoma, achieving successful identification of the SLN in 98% of cases. The sensitivity of SLN biopsy in predicting axillary node status was 100%, with no false-negative results. Another argument in favour of dextran is its low cost: it is much less expensive than human albumin.

Radiation safety to patient and staff in radioguided occult lesion localisation procedures has been secured and radioguided occult lesion localization is considered a reliable and safe alternative to wire localization.

Technetium-99m has a short half-life of 6 hours and low-dose gamma radiation is used. Cremonesi et al (Cremonesi et al, 1999) calculated the effective dose for patient, which is 9.25 µSv, less than half the dose of a chest X-ray (0.02 mSv).

Radiation due to additional mammograms needed for hookwire localization (1-2 mSv) may exceed that involved for 100 to 200 radioguided occult lesion localisation procedures.

Finger doses to breast surgeons and radiologists are also minimal, as stated by Rampaul et al (Rampaul et al, 2003) and amount to 9.3±3.3 mSv and 0.5±0.1 mSv, respectively.

The advantages of this technique can be summarised as: precise localisation and accurate surgical removal; reduction of tissue damage in the final pathological specimen (as the wire is not present in the specimen); accurate frozen section (if indicated); improved rate of clear margins; reduced size of the excised specimen; increased patient comfort; decreased operative time; and reduced number of reoperations (cost effectiveness).

One of the important drawbacks of such a technique is radiotracer spillage within the mammary gland that makes the precise lesion resection difficult, and this requires the use of a hook-wire collocation to reach the lesion. The possibility of obtaining an intraoperative image of the specimen could help to confirm whether the lesion is correctly removed. Some types of portable gamma cameras have been designed, but up to now, intraoperative use has been confined to surgery of parathyroid adenomas and sentinel lymph node location.

The intraoperative acquisition of such images can predict the involvement of surgical margins, avoiding future surgical procedures.

Reported advantages of radioguided occult lesion localization are (van der Ploeg et al, 2008): precise localisation and accurate surgical removal, reduction in the tissue damage within the final pathological specimen, accurate frozen section (in the case a differential diagnosis is required between DCIS and invasive cancer), improved rate of clear margins avoiding the emotional trauma of another operation, reduced size of the excised specimen, better concentricity of the lesion, increased patient comfort, decreased operative time, and reduced rate in re-do surgery reducing costs. Moreover no control mammogram is needed after localisation, allowing reduction in both costs and radiation dose.

When comparing the radioguided occult lesion localization technique with the hookwire localization in the radioguided occult lesion localization group between 69% and 84% of the lesions were radically excised, compared with 44-60% of the lesions in the hookwire localization group.

The studies that combined radioguided occult lesion localization and the sentinel node procedure mentioned even higher percentages of radically excised specimens ranging from 90% to 95% and an identification rate of sentinel nodes up to 100%, resulting in minimisation of the surgical intervention and a decrease in postoperative morbidity.

One randomised controlled trial mentioned that the radioguided occult lesion localization procedure had a faster localisation time, was easier to perform, was less painful and gave a better cosmetic result than hookwire localization.

Many studies have reported good margin clearance with radioguided occult lesion localization, which ranged from 75 to 100%. In a recent publication, the free margin clearance rate was 84% after radioguided occult lesion localization compared with 60% after hookwire localization.

An important advantage of the ROLL technique is that it takes advantage of the radioactive tracer that is used anyway to detect the sentinel node. Therefore, it is not necessary to use a separate injection for the SN detection as was used in some studies. Feggi et al (Feggi et al, 2001) demonstrated successfully, that it is possible to perform ROLL and sentinel lymph node mapping with a single injection of Technetium-labelled Nanocolloid.

Ductal migration of isotope is an uncommon problem with radioguided occult lesion localization.

Another constraint is the timing of surgery, which has to be performed within 4 hours of radioactive occult breast lesion localization completion, lest the tracer count is not high enough to enable accurate gamma probe–guided excision.

3.3 SNOLL sentinel node and occult lesion localisation

Sentinel node localization can be performed together with radioguided occult lesion localization. (Strnad et al, 2006)

How to execute lymphoscintigraphy in patients with non-palpable lesions scheduled to undergo radioguided surgery for the primary lesion is still the subject of debate; it is not clear whether it is better to employ a single tracer and single injection method, as recently proposed, or a dual tracer method with injection at different sites. (Patel et al, 2004)

Filtered Tc-99m labelled sulphur colloid with a particle size of smaller than 100 nanometers is injected. Tracers flow via the lymphatics and accumulate in the sentinel nodes. Subsequent scintigraphy can be performed as usual. The application of a radioisotope in sentinel node and occult lesion localization has revolutionized the original two-step procedure into one with very promising results.

Otherwise some authors report that a dual administration of radiopharmaceuticals, i.e. the radioguided occult lesion localization procedure with intralesional injection of macroaggregates of 99mTc-labelled human serum albumin followed by lymphoscintigraphy with subdermal injection of nanocolloids, represents the best approach to localise both non-palpable lesions of the breast and axillary sentinel nodes on the same day.

Also the Italian group of the European Institute of Oncology (De Cicco et al, 2004, 2002) confirmed in a recent paper that the combination of the ROLL procedure with direct injection of human serum albumin into the lesion and lymphoscintigraphy performed with subdermal injection of radiocolloids represents the method of choice for accurate localisation of both non-palpable lesions and sentinel nodes.

3.4 RSD radioactive seed localization

There are a few reports describing alternative techniques to localise non-palpable breast lesions. In 2001, Gray et al (Gray et al, 2001) reported on a refinement of the technique of radioguided surgery. In this technique, known as "radioactive seed localization" (RSL), a radio-opaque titanium seed containing Iodine-125 is inserted into the tumor under stereotactic or ultrasound guidance, and the gamma probe is again used to guide surgical resection.

Use of radioactive seed localization such as an iodine-125-radiolabelled (I-125) seed for localization of non-palpable breast tumours could potentially prevent the problems related to hookwire localization and seems a promising approach for the resection of nonpalpable breast lesions. (Barros et al, 2002)

Radioactive seed localization is at least equivalent compared with wire localization in terms of the ease of the procedure, removing the target lesion, the volume of breast tissue excised, obtaining negative margins, avoiding a second operative intervention, and allowing for simultaneous axillary staging.

Radioactive seed localization is safe, effective, and compared to wire localization, reduces the rates of intraoperative re-excision and reoperation for positive margins by 68%. Patient satisfaction is improved with radioactive seed localization.

RSL has been reviewed by Jakub et al (Jakub et al, 2010). Subsequent randomized and non-randomized trials have shown improvements in positive margin rates compared to standard hookwire guided localization. In studies of both radioguided occult lesion localization and radioguided seed localization, there was rapid adoption by surgeons and radiologists, and satisfaction and acceptance by patients.

3.5 Probes

Currently, tumoral harvesting is aided by using a hand-held gamma probe. This device locates the highest activity point and the point where activity fades, thus indicating the surgical resection margins. Once the tumour has been removed, handheld scanning of the surgical area will confirm that activity is only located within the tumoral specimen and that there is no residual activity requiring further excision.

Intraoperative probes are capable of detecting a very low flow of gamma photons. The sensitive part is at the end of a cylindrical tube of 1 to 2 cm in diameter. To ensure sterility, the tube is slipped into a long sleeve sterile flexible plastic disposable. It is connected by cable to an electronic unit which analyzes the signal, displays the count rate and beeps modulated by the level of radioactivity detected. The detection of gamma photons is ensured either by absorption in a scintillating crystal type sodium iodide or cesium iodide activated with thallium (NaI (Tl) or CsI (Tl)) or in a semiconductor-type cadmium-tellurium (Cd-Te) and cadmium-zinc-telluride (Cd-Zn-Te, sometimes abbreviated as CZT). In the first case, the signal emitted by the crystal must be amplified by a photomultiplier vacuum tube powered by a voltage of about 1000 volts, whereas with the semiconductor signal is processed immediately in low voltage and the system does not include a photomultiplier. However, detection volume identical stopping power and hence the detection sensitivity of sparkling crystals are higher than those of semiconductors.

The intraoperative detection probes currently available have a very high sensitivity to gamma photons emitted by technetium-99m. They are also capable, with a setting often already pre-programmed to detect the emission of other isotopes commonly used in nuclear medicine (indium 111, iodine 131), even as the signal from the positron emitters such as 18-fluoro- deoxyglucose (FDG). (Paredes et al, 2008)

Many studies showed that the use of a portable c-camera equipped for simultaneous 99mTc signal and Gd-153 pointer signal display can improve sentinel lymph node excision identification in breast carcinoma, prostate cancer, melanoma and other cutaneous malignancies.

Furthermore some authors adfirm that intraoperative realtime imaging by a c-camera can be used as an additional technique, besides preoperative lymphoscintigraphy, preoperative SPECT/CT and the intraoperative application of a hand-held c-probe (Cho et al 2009).

4. Radioguided localization after neo-adjuvant chemotherapy

The use of neo-adjuvant chemotherapy has increased in the treatment of loco-regionally advanced primarily operable breast cancer. As a result of improved neo-adjuvant chemotherapy regimens the number of clinical as well as radiological responses have increased. In case of a complete response it is difficult to identify residual disease and to perform an adequate radical breast-conserving surgery. Therefore localization of the original tumour bed is mandatory. The (125)I has a half-time of 60 days and is therefore still recognisable with a gamma probe after admittance of several courses of neo-adjuvant chemotherapy. (125)I seed localization is a highly successful technique in localizing the tumour bed in patients who receive neo-adjuvant chemotherapy for breast cancer leading to a high percentage of radical margins in case of breast-conserving surgery. (Straver et al, 2010a, 2010b; van Riet et al, 2010)

Nontheless an important benefit of neoadjuvant chemotherapy is the increased potential for breast-conserving surgery. At present also the response of axillary lymph node metastases to chemotherapy is not easily assessed, rendering axilla-conserving treatment difficult.

The tumour response in the marked lymph node may be used to tailor further axillary treatment, making axilla-conserving surgery a possibility. Moreover the high accuracy in visualizing lymph node metastases and the sufficiently high SUV(max) and tumour to background ratio at baseline suggest that it is feasible to monitor the axillary response with FDG PET/CT, especially in triple-negative tumours. (Choi et al, 2010)

5. Conclusion

Screening mammography is allowing detection of early breast cancer in an increasing number of women with occult, locally invasive or in situ carcinoma. The result is a marked decrease in the mean size of lesions, an increase in the incidence of in situ carcinomas and a reduction in axillary node involvement. Nuclear medicine techniques can help surgeons to identify both the tumor and the sentinel lymph nodes in order to stage the disease.

6. References

Armer J, Fu MR, Wainstock JM, Zagar E K & Jacobs LK. (2004). *Lympheedema following breast cancer treatment, including sentinel lymph node biopsy.* Lymphology; 37:73–91.

Armer JM, Stewart BR & Shook RP. (2009). *30-Month post-breast cancer treatment lymphoedema.* J Lymphoedema; 4:14–8

Ashikaga T, Krag DN, Land SR, Julian TB, Anderson SJ, Brown AM, Skelly JM, Harlow SP, Weaver DL, Mamounas EP, Costantino JP & Wolmark N; National Surgical Adjuvant Breast, Bowel Project. (2010). *Morbidity results from the NSABP B-32 trial comparing sentinel lymph node dissection versus axillary dissection.* J Surg Oncol; 102:111–8.

Association of Breast Surgery. (2009). *Surgical guidelines for the management of breast cancer.* Eur J Surg Oncol; 35(Suppl. 1):1–22.

Barros A, Cardoso MA, Sheng PY, Costa PA & Pelizon C. (2002). *Radioguided localisation of non-palpable breast lesions and simultaneous sentinel lymph node mapping.* Eur J Nucl Med Mol Imaging. 29(12):1561-5.

Barthelmes L, Goyal A, Newcombe RG, McNeill F & Mansel RE. (2010). *Adverse reactions to patent blue V dye - The NEW START and ALMANAC experience.* Eur J Surg Oncol, 36. 4: pp. 399–403.

Belloni E, Canevari C, Panizza P, Marassi A, Rodighiero M, Tacchini S, Zuber V, Sassi I, Gianolli L, Fazio F & Del Maschio A. (2011). *Nonpalpable breast lesions: preoperative radiological guidance in radioguided occult lesion localisation (ROLL).* Radiol Med. 2011 Mar 7. [Epub ahead of print]

Boolbol SK, Fey JV, Borgen PI, Heerdt AS, Montgomery LL, Paglia M, Petrek JA, Cody HS 3rd & Van Zee KJ. (2001). *Intradermal isotope injection : a highly accurate method of lymphatic mapping in breast carcinoma.* Ann Surg Oncol; 8 : 3-6.

Celliers L & Mann GB. (2003). *Alternative sites of injection for sentinel lymph node biopsy in breast cancer.* ANZ J Surg; 73(8):600–4.

Chen C, Chan MC, Hung WK, Lam HS & Yip AW. (2005). *Wire-guided excision of mammographic abnormalities.* Hong Kong Med J; 11:153-7.

Cho N, Moon WK, Han W, Park IA, Cho J & Noh DY. (2009). *Preoperative sonographic classification of axillary lymph nodes in patients with breast cancer: node-to-node correlation with surgical histology and sentinel node biopsy results.* Am J Roentgenol; 193: 1731–1737.

Choi JH, Lim HI, Lee SK, Kim WW, Kim SM, Cho E, Ko EY, Han BK, Park YH, Ahn JS, Im YH, Lee JE, Yang JH & Nam SJ. (2010). *The role of PET CT to evaluate the response to neoadjuvant chemotherapy in advanced breast cancer: comparison with ultrasonography and magnetic resonance imaging.* J Surg Oncol; 102(5):392-7.

Chu TY, Lui CY, Hung WK, Kei SK, Choi CL & Lam HS. (2010). *Localisation of occult breast lesion: a comparative analysis of hookwire and radioguided procedures.* Hong Kong Med J. 16(5):367-72.

Cody HS 3rd, Fey J, Akhurst T, Fazzari M, Mazumdar M, Yeung H, Yeh SD & Borgen PI. (2001). *Complementarity of blue dye and isotope in sentinel node localization for breast cancer : univariate and multivariate analysis in 966 procedures.* Ann Surg Oncol; 8 : 13-9.

Cody HS & Borgen PI. (1999). *State-of-the-art approaches to sentinel node biopsy for breast cancer : study design, patient selection, technique and quality control at Memorial Sloan-Kettering Cancer Center.* Surg Oncol; 8 : 85-91.

Crane-Okada R, Wascher RA, Elashoff D & Giuliano AE. (2005). *Longterm morbidity of sentinel node biopsy versus complete axillary dissection for unilateral breast cancer.* Ann Surg Oncol; 15:1996–2005.

Cremonesi M, Ferrari M, Sacco E, Rossi A, De Cicco C, Leonardi L, Chinol M, Luini A, Galimberti V, Tosi G, Veronesi U & Paganelli G. (1999). *Radiation protection in radioguided surgery of breast cancer.* Nucl Med Commun. 20 (10):919-24)

Dayes IS. (2006). *Current issues in the management of lymphedema in breast cancer patients.* J Support Oncol. 4(8):392-3

Dayes IS, Levine MN, Julian JA, Pritchard KI, D'Souza DP, Kligman L, Reise D, Wiernikowski JA, Bonilla L & Whelan TJ. (2008). *Lymphedema in women with breast cancer: characteristics of patients screened for a randomized trial.* Breast Cancer Res Treat; 110:337–42.

De Cicco C, Pizzamiglio M, Trifirò G, Luini A, Ferrari M, Prisco G, Galimberti V, Cassano E, Viale G, Intra M, Veronesi P & Paganelli G. (2002). *Radioguided occult lesion localization (ROLL) and surgical biopsy in breast cancer. Technical aspects.* Q J Nucl Med.46(2):145-51.

De Cicco C, Trifirò G, Intra M, Marotta G, Ciprian A, Frasson A, Prisco G, Luini A, Viale G & Paganelli G. (2004). *Optimised nuclear medicine method for tumour marking and sentinel node detection in occult primary breast lesions.* Eur J Nucl Med Mol Imaging.31(3):349-54.

Del Bianco P, Zavagno G, Burelli P, Scalco G, Barutta L, Carraro P, Pietrarota P, Meneghini G, Morbin T, Tacchetti G, Pecoraro P, Belardinelli V & De Salvo GL. (2008). *Morbidity comparison of sentinel lymph node biopsy versus conventional axillary lymph node dissection for breast cancer patients: results of the sentinella-GIVOM Italian randomised clinical trial.* Eur J Surg Oncol; 34:508–13.

Derossis AM, Fey J, Yeung H, Yeh SD, Heerdt AS, Petrek J, VanZee KJ, Montgomery LL, Borgen PI & Cody HS 3rd. (2001). *A trend analysis of the relative value of blue dye and isotope localization in 2,000 consecutive cases of sentinel node biopsy for breast cancer.* J Am Coll Surg; 193(5): 473–8.

Feggi L, Basaglia E, Corcione S, Querzoli P, Soliani G, Ascanelli S, Prandini N, Bergossi L & Carcoforo P. (2001). *An original approach in the diagnosis of early breast cancer: use of the same radiopharmaceutical for both non-palpable lesions and sentinel node localisation.* Eur J Nucl Med. 28(11):1589-96

Gill G. (2009). *Sentinel-lymph-node-based management or routine axillary clearance? One-year outcomes of sentinel node biopsy versus axillary clearance (SNAC): a randomized controlled surgical trial.* Ann Surg Oncol; 16(2):266–75.

Giuliano AE, Kirgan DM, Guenther JM & Morton DL. (1994). *Lymphatic mapping and sentinel lymphadenectomy for breast cancer.* Ann Surg; 220(3):391–8. discussion 398-401.

Gray RJ, Giuliano R, Dauway EL, Cox CE & Reintgen DS. (2001). *Radioguidance for nonpalpable primary lesions and sentinel lymph node(s).* Am J Surg. 182(4):404-6.

Helyer LK, Varnic M, Le LW, Leong W & McCready D. (2009). *Obesity is a risk factor for developing postoperative lymphedema in breast cancer patients.* Breast J.; 16:48–54.

Henze E, Schelbert HR, Collins JD, Najafi A, Barrio JR & Bennett LR. (1982). *Lymphoscintigraphy with Tc-99m-labeled dextran.* J Nucl Med. 23(10):923-9.

Hunting AS, Nopp A, Johansson SG, Andersen F, Wilhelmsen V & Guttormsen AB. (2010). *Anaphylaxis to patent blue V. I. Clinical aspects.* Allergy; 65(1):117–23.

Jakub JW, Gray RJ, Degnim AC, Boughey JC, Gardner M & Cox CE. (2010). *Current status of radioactive seed for localization of non palpable breast lesions.* Am J Surg; 199(4):522-8.

Krag D, Weaver D, Ashikaga T, Moffat F, Klimberg VS, Shriver C, Feldman S, Kusminsky R, Gadd M, Kuhn J, Harlow S & Beitsch P. (1998). *The sentinel node in breast cancer: a multicenter validation study.* N Engl J Med;339(14):941–6.

Kwan ML, Darbinian J, Schmitz KH, Citron R, Partee P, Kutner SE & Kushi LH. (2010). *Risk factors for lymphedema in a prospective breast cancer survivorship study: the Pathways Study.* Arch Surg; 145:1055-63.

Leidenius M, Leivonen M, Vironen J & von Smitten K.(2005). *The consequences of long-time arm morbidity in node-negative breast cancer patients with sentinel node biopsy or axillary clearance.* J Surg Oncol; 92:23-31.

Langer I, Guller U, Berclaz G, Koechli OR, Schaer G, Fehr MK, Hess T, Oertli D, Bronz L, Schnarwyler B, Wight E, Uehlinger U, Infanger E, Burger D & Zuber M. (2007). *Morbidity of sentinel lymph node biopsy (SLN) alone versus SLN and completion axillary lymph node dissection after breast cancer surgery: a prospective Swiss multicenter study on 659 patients.* Ann Surg; 245: 452-61.

Leong SP, Donegan E, Heffernon W & Dean S, Katz JA. (2000). *Adverse reactions to isosulfan blue during selective sentinel lymph node dissection in melanoma.* Ann Surg Oncol; 7 : 361-6.

Lucci A, McCall LM, Beitsch PD, Whitworth PW, Reintgen DS, Blumencranz PW, Leitch AM, Saha S, Hunt KK & Giuliano AE. (2007). *Surgical complications associated with sentinel lymph node dissection (SLND) plus axillary lymph node dissection compared with SLND alone in the American College of Surgeons Oncology Group Trial Z0011.* J Clin Oncol.; 25:3657-63.

Luini A, Zurrida S, Galimberti V, Paganelli G. (1998). *Radioguided surgery of occult breast lesions.* Eur J Cancer; 34: 204- 5

Lyman GH, Giuliano AE, Somerfield MR, Benson AB 3rd, Bodurka DC, Burstein HJ, Cochran AJ, Cody HS 3rd, Edge SB, Galper S, Hayman JA, Kim TY, Perkins CL, Podoloff DA, Sivasubramaniam VH, Turner RR, Wahl R, Weaver DL, Wolff AC & Winer EP. (2005). *American Society of Clinical Oncology guideline recommendations for sentinel lymph node biopsy in early-stage breast cancer.* J Clin Oncol; 23(30):7703-20.

Mansel RE, Fallowfield L, Kissin M, Goyal A, Newcombe RG, Dixon JM, Yiangou C, Horgan K, Bundred N, Monypenny I, England D, Sibbering M, Abdullah TI, Barr L, Chetty U, Sinnett DH, Fleissig A, Clarke D & Ell PJ. (2006). *Randomized multicenter trial of sentinel node biopsy versus standard axillary treatment in operable breast cancer: the ALMANAC Trial.* J Natl Cancer Inst 98(9):599-609.

McLaughlin SA, Wright MJ, Morris KT, Giron GL, Sampson MR, Brockway JP, Hurley KE, Riedel ER & Van Zee KJ. (2008). *Prevalence of lymphedema in women with breast cancer 5 years after sentinel lymph node biopsy or axillary dissection: objective measurements.* J Clin Oncol; 26:5213-9.

McLaughlin SA, Wright MJ, Morris KT, Sampson MR, Brockway JP, Hurley KE, Riedel ER & Van Zee KJ. (2008). *Prevalence of lymphedema in women with breast cancer 5 years after sentinel lymph node biopsy or axillary dissection: patient perceptions and precautionary behaviors.* J Clin Oncol; 26:5220-6.

McMasters KM, Wong SL, Martin II RCG, Chao C, Tuttle TM, Noyes RD, Carlson DJ, Laidley AL, McGlothin TQ, Ley PB, Brown CM, Glaser RL, Pennington RE, Turk PS, Simpson D, Cerrito PB & Edwards MJ. (2001). *Dermal injection of radioactive colloid is superior to peritumoral injection for breast cancer sentinel lymph node biopsy : results of a multiinstitutional study.* Ann Surg; 233 : 676-87.

Nieweg OE, Estourgie SH, van Rijk MC & Kroon BB. (2004). *Rationale for superficial injection techniques in lymphatic mapping in breast cancer patients.* J Surg Oncol; 87(4):153-6.

Noguchi M. (2004). *Current controversies concerning sentinel lymph node biopsy for breast cancer.* Breast Cancer Res Treat.; 84(3):261-71.

Paredes P, Vidal-Sicart S, Zanón G, Roé N, Rubí S, Lafuente S, Pavía J & Pons F. (2008). *Radioguided occult lesion localisation in breast cancer using an intraoperative portable gamma camera: first results.* Eur J Nucl Med Mol Imaging; 35(2):230-5.

Patel A, Pain SJ, Britton P, Sinnatamby R, Warren R, Bobrow L, Barber RW, Peters AM & Purushotham AD. (2004). *Radioguided occult lesion localisation (ROLL) and sentinel node biopsy for impalpable invasive breast cancer.* Eur J Surg Oncol.30(9):918-23.

Perry NM. (2001). *EUSOMA Working Party. Quality assurance in the diagnosis of breast disease. EUSOMA Working Party.* Eur J Cancer; 37:159-72.

Querci della Rovere G, Ahmad I, Singh P, Ashley S, Daniels IR & Mortimer P. (2003). *An audit of the incidence of arm lymphoedema after prophylactic level I/II axillary dissection without division of the pectoralis minor muscle.* Ann R Coll Surg Engl; 85:158–61.

Rampaul RS, Dudley NJ, Thompson JZ, Burrell H, Evans AJ, Wilson AR & Macmillan RD. (2003). *Radioisotope for occult lesion localisation (ROLL) of the breast does not require extra radiation protection procedures.* Breast. 12(2):150-2

Schrenk P, Rieger R, Shamiyeh A & Wayand W. (2000). *Morbidity following sentinel lymph node biopsy versus axillary lymph node dissection for patients with breast carcinoma.* Cancer; 88: 608–14.

Somasundaram SK, Chicken DW & Keshtgar MR. (2007). *Detection of the sentinel lymph node in breast cancer.* Br Med Bull; 84:117–31.

Straver ME, Aukema TS, Olmos RA, Rutgers EJ, Gilhuijs KG, Schot ME, Vogel WV & Peeters MJ. (2010). *Feasibility of FDG PET/CT to monitor the response of axillary lymph node metastases to neoadjuvant chemotherapy in breast cancer patients.* Eur J Nucl Med Mol Imaging; 37(6):1069-76.

Straver ME, Loo CE, Alderliesten T, Rutgers EJ & Vrancken Peeters MT. (2010). *Marking the axilla with radioactive iodine seeds (MARI procedure) may reduce the need for axillary dissection after neoadjuvant chemotherapy for breast cancer.* Br J Surg; 97(8):1226-31.

Strnad P, Rob L, Halaska MG, Chod J, Zuntova A & Moravcova Z. (2006*) Radioguided occult lesion localisation in combination with detection of the sentinel lymph node in non-palpable breast cancer tumours.*Eur J Gynaecol Oncol.27(3):236-8.

Tafra L, Lannin DR, Swanson MS, Van Eyk JJ, Verbanac KM, Chua AN, Ng PC, Edwards MS, Halliday BE, Henry CA, Sommers LM, Carman CM, Molin MR, Yurko JE, Perry RR, Williams R. (2001). *Multicenter trial of sentinel node biopsy for breast cancer using both technetium sulfur colloid and isosulfan blue dye.* Ann Surg; 233(1):51–9.

van der Ploeg IM, Hobbelink M, van den Bosch MA, Mali WP, Borel Rinkes IH & van Hillegersberg R. (2008). *'Radioguided occult lesion localisation' (ROLL) for non-palpable breast lesions: a review of the relevant literature.* Eur J Surg Oncol. 34(1):1-5.

van Riet YE, Maaskant AJ, Creemers GJ, van Warmerdam LJ, Jansen FH, van de Velde CJ, Rutten HJ & Nieuwenhuijzen GA. (2010). *Identification of residual breast tumour localization after neo-adjuvant chemotherapy using a radioactive 125Iodine seed.* Eur J Surg Oncol; 36(2):164-9.

Veronesi U, Paganelli G, Viale G, Luini A, Zurrida S, Galimberti V, Intra M, Veronesi P, Robertson C, Maisonneuve P, Renne G, De Cicco C, De Lucia F & Gennari R.. (2003). *A randomized comparison of sentinel-node biopsy with routine axillary dissection in breast cancer.* N Engl J Med.; 349:546–53.

White V, Harvey JR, Griffith CDM, Youssef M & Carr M. (2011). *Sentinel lymph node biopsy in early breast cancer surgery: Working with the risks of vital blue dye to reap the benefits.* EJSO 37 101-108

13

3-Dimensional CT Lymphography in Identifying the Sentinel Node in Breast Cancer

Junko Honda[1], Chieko Hirose[2], Masako Takahashi[3], Sonoka Hisaoka[2],
Miyuki Kanematsu[1], Yoshimi Bando[4] and Mitsunori Sasa[5]

*[1]Department of Surgery, National Hospital Organization Higashitokushima
Medical Center, 1-1, Ohmukai-kita, Ootera, Itano, Tokushima,
[2]Department of Radiology, National Hospital Organization Higashitokushima
Medical Center, 1-1, Ohmukai-kita, Ootera, Itano, Tokushima,
[3]Department of Radiology, Tokushima Breast Care Clinic,
Nakashimada-Cho, Tokushima,
[4]Department of Molecular and Environmental Pathology,
Institute of Health Biosciences, The University of Tokushima Graduate
School, Kuramoto-Cho, Tokushima,
[5]Department of Surgery, Tokushima Breast Care Clinic,
Nakashimada-Cho, Tokushima,
Japan*

1. Introduction

Sentinel node biopsy (SNB) has become a standard surgical procedure for patients with early-stage breast cancer. The sentinel nodes (SN) can be identified by a double-mapping procedure based on a gamma probe-guided method and a dye-guided method using a radioisotope (RI), or by a triple-mapping procedure that includes lymphoscintigraphy and is even more effective. However, the RI method can be performed only at institutions that are trained and licensed to use RI, and other institutions must rely on dye methods alone for SN identification. On the other hand, to obtain images of the lymph vessels and nodes, indirect lymphography seems to be a more convenient than direct intralymphatic administration of a contrast medium. Several studies of indirect lymphography were reported in the 1980s, and Suga *et al.* reported successfully identifying SN by 3-dimensional computed tomographic lymphography (CTLG) using a nonionic contrast medium. We have also been identifying SN in breast cancer patients by CTLG and a dye-guided method since February 2003 as a clinical trial. Here, we report our findings to date regarding the clinical efficacy and problems associated with SN identification by the combination of CTLG and a dye-guided method.

2. Patients and methods

All studies in this paper were approved by the ethics committee of National Hospital Organization Higashitokushima Medical Center. After presenting a detailed explanation of this clinical trial, written informed consent was obtained from all patients.

2.1 Patients

During the period from February 2003 through March 2007, 218 Japanese patients with T1N0M0 or T2N0M0 primary breast cancer were treated at the Tokushima Breast Care Clinic. SN identification was performed by combined application of CTLG and a dye-guided method. In two of the patients the CTLG was performed before and after excisional biopsy, and in one patient with bilateral disease CTLG was performed on both sides. Thus, CTLG was performed 221 times in total, while the dye-guided method was performed a total of 219 times. In principle, backup dissection was performed for patients found to be SN metastasis-positive provided that informed consent was granted. For metastasis-negative patients, axillary dissection was omitted on the basis of informed consent.

2.2 Methods

CTLG was performed as previously described. Each patient was placed in the supine position, with the arms positioned upward but bent at the elbow with the hands at the side of the cranium. This position is similar to the surgical position. First, plain CT scanning of the affected axilla was carried out using a high-speed FX/i single-detector helical CT scanner (GE Yokogawa Medical Systems, Tokyo, Japan). The X-ray beam thickness was 3 mm and the pitch was 1.5. Then, one mL of iopamidol (Iopamiron 300; 300 mgI/ml, Schering, Osaka, Japan) was injected subcutaneously (or intradermally) to the areola or both subcutaneously (or intradermally) to the areola and subcutaneously (or intradermally) above the tumor for a few minutes with a 26-gauge 5/8-inch hypodermic needle attached to a tuberculin syringe (Figure 1a). One minute after the injection, axillary CT scanning (CT lymphography) was done with the same thickness and pitch as described above. In the case that the enhancement of the lymph vessels and lymph nodes was poor, the site(s) where the contrast medium had been injected was gently massaged, and after 3-5 min the CT was repeated. Finally, the remaining medium (about 99ml) was intravenously administered from the unaffected forearm to examine distal metastases. A CT scanning was done in the thoracic and upper abdominal areas with 7mm beam thickness and 1.5 pitch in the axillary region with 3mm beam thickness and 1.5 pitch. 2-dimensional images of axilla (Figure 1b) were reconstructed at intervals of 1 mm, and then 3-dimensional images were created by the volume-rendered method to correlate with the surgical view (Figure 1c). SNs were predicted from CT images by identification of enhanced lymph vessels and/or lymph nodes and assessment of the CT values. The CT values of lymph nodes were measured at their maximally enhanced point. If an SN(s) was identified, a mark was made on the skin immediately above the SN by using an oil painting pen (Figure 1d).

The dye-guided method was also performed as previously described. Briefly, 2~3 mL of indigo carmine blue was injected subcutaneously (or intradermally) to the areola, followed by massage of the injection site for a few minutes. Just before performing an operation, we used ultrasonography to confirm the location of SN(s) that has been identified by CTLG. Five to 15 min later a skin incision was made at the site(s) that were marked in the CTLG, and any blue-stained lymph vessels and lymph nodes were identified. If no SN(s) was identified by either the dye-guided method or CTLG, axillary lymph node dissection was performed.

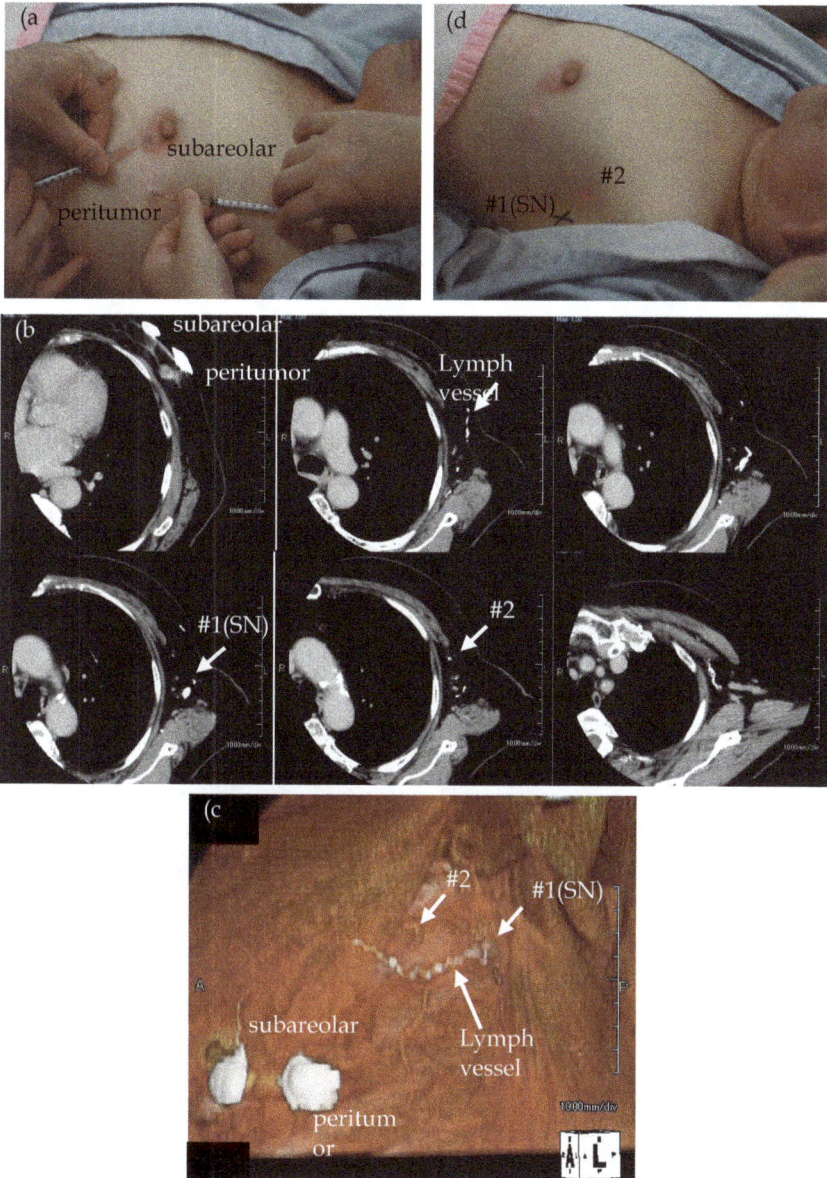

Fig. 1. Method for identification of the SNs performed by CTLG. Images in patient with left breast cancer. (a) One ml of iopamidol was injected both subcutaneously (or intradermally) to the areola and subcutaneously (or intradermally) above the tumor. (b) Axial image demonstrating a tortuous, enhanced lymph vessel that reaches SN. (c) 3-dimensional CT lymphography image clearly showing an enhanced lymh vessel and an opacified SN (#1). (d) The mark was made on the skin immediately above the identificated SN by using an oil painting pen.

2.3 Definition of SNs

CTLG: A lymph node was defined as an SN if visual inspection or the CT value (the increase of Hounsfield units after subcutaneous injection) confirmed it to be enhanced or if it was confirmed to connect to an enhanced lymph vessel.

Dye-guided method: A lymph node was defined as an SN if it was dyed blue or if it was confirmed to connect to a blue lymph vessel.

2.4 SN identification rates and the clinicopathological findings

The results were analyzed in relation to the success/failure of SN identification and various clinicopathological parameters, including the patient age, menopausal status, body mass index (BMI), tumor diameter, presence/absence of excisional biopsy, histological type, presence/absence of lymph node metastasis and presence/absence of vascular invasion.

2.5 Statistical analysis

The relationships between the success/failure of SN identification and the clinicopathological findings were analyzed for statistical significance using the chi-square test. A p value of <0.05 was considered to indicate statistical significance.

3. Results

3.1 SN identification rates by CTLG and the dye-guided method

Identification of the SN(s) was achieved in 212 (96%) of the total 221 performances of CTLG. The number of detected SNs ranged from 1 to 3, with a mean of 1.2 SNs. With the dye-guided method the identification rate was 92% (202/219 tests). The combined identification rate was 99% (Table 1).

	Dye	CTLG	Combination
Lymph vessels	212 (97%)	202 (91%)	215 (98%)
Sentinel nodes	202 (92%)	212 (96%)	216 (99%)

Dye : Dye-guided method, CTLG : CT Lymphography, Combination : Combination of CTLG and Dye

Table 1. Identification rates of lymph vessels and sentinel nodes

With CTLG, both the lymph vessels and lymph nodes were clearly enhanced and the SNs could be identified in 189 patients (86%). In 5 patients (2%) the lymph vessels were not enhanced but the lymph nodes were, and the SNs could be identified. In 7 other patients (3%), the lymph vessels were enhanced while the lymph nodes were not clearly enhanced, but the SNs could be identified from the CT value. In 5 patients (2%) neither the lymph vessels nor the lymph nodes were enhanced, but the SNs could be identified from the CT value. In 6 patients (3%) the lymph vessels were enhanced but the lymph nodes were not, but the SNs could be identified on the basis of confirmation of their connection to the enhanced lymph vessels. Finally, in 9 patients (4%) neither the lymph vessels nor the lymph nodes were enhanced, and SNs could not be demonstrated (Table 2).

Lymph vessels	Node	
Enhanced	Cleary enhanced	189 (86%)
Unenhanced	Cleary enhanced	5 (2%)
Enhanced	Assessment of CT values	7 (3%)
Unenhanced	Assessment of CT values	5 (2%)
Enhanced	Unenhanced	6 (3%)
Unenhanced	Unenhanced	9 (4%)

Table 2. Findings of CT lymphography (n=221)

3.2 SN identification rates and the clinicopathological findings

Analysis of the relationships between the SN identification rates and the clinicopathological findings showed that the SN identification rates with CTLG, the dye-guided method and the combined method showed no differences as a function of the age, menopausal status, tumor diameter or histopathological type. However, with the dye-guided method the SN identification rate was significantly lower in patients with a BMI of 25 or higher, whereas with CTLG and the combined method the SN identification rate was not influenced by the BMI. With CTLG and the combined method, the SN identification rate was significantly lower in the node-positive patients, and with the combined method it was significantly lower in patients with vascular invasion. In addition, in the patients who had undergone lateral-upper region excisional biopsy the SN identification rate with CTLG was lower than in the patients who had undergone excisional biopsy in a region other than the lateral-upper region, although the difference did not reach statistical significance. Moreover, with the dye-guided method the SN identification rate was lower in the patients who had undergone excisional biopsy, regardless of the region, compared with the patients who had not undergone excisional biopsy (Table 3).

	Dye (+), n=202	CTLG (+), n=212	Combination (+), n=216
Age (years)			
≤35	6 (100%)	6 (100%)	6 (100%)
36 - 50	87 (90%)	92 (95%)	96 (98%)
≥51	109 (94%)	114 (97%)	114 (98%)
Menopausal state			
Pre	103 (92%)	108 (96%)	105 (98%)
Post	99 (93%)	104 (95%)	111 (99%)
BMI			
<25	160 (95%) #	164 (96%)	166 (98%)
≥25	42 (84%)	48 (96%)	50 (100%)
Tumor size (cm)			
≤2.0	183 (92%)	193 (96%)	197 (99%)
>2.0	19 (100%)	19 (100%)	19 (100%)

	Dye (+), n=202	CTLG (+), n=212	Combination (+), n=216
With excisionary biopsy *			
In lateral-upper region	17 (89%)	18 (95%)	19 (100%)
In other regions	46 (88%)	52 (100%)	52 (100%)
Histological type			
IIa1	71 (88%)	80 (99%)	80 (99%)
IIa2	22 (100%)	23 (100%)	22 (100%)
IIa3	64 (96%)	63 (94%)	65 (97%)
IIb3	8 (89%)	8 (89%)	9 (100%)
DCIS	13 (100%)	12 (92%)	13 (100%)
Others	24 (89%)	26 (93%)	27 (100%)
Nodal status			
n (-)	177 (93%)	189 (98%) #	191 (100%) #
n (+)	25 (89%)	23 (82%)	25 (89%)
Vascular invasion			
v (-)	160 (91%)	172 (97%)	174 (99%) #
v (+)	42 (95%)	40 (91%)	42 (95%)

BMI : Body mass index, Dye : Dye-guided method, CTLG : CT Lymphography, Combination : Combination of CTLG and Dye, # ; p<0.05, * ; cases with excisional biopsy (n=71)

Table 3. Sentinel node(s) identification rate and clinicopathological findings (all cases) (Dye n=219, CTLG n=221, Combination n=219)

3.3 Size of metastasis and the SN identification rate in node-positive patients

The patients found to be positive for metastasis were classified and analyzed on the basis of the size of metastasis: <0.2 mm, 0.2mm~2mm and >2mm. The results showed that the SN identification rate by both CTLG and the dye-guided method decreased as the size of metastasis increased, although the difference did not reach statistical significance (Table 4).

Metastatic lesion size	Dye (+)	CTLG (+)
Isolated tumor cells (<0.2 mm)	2 (100%)	2 (100%)
Micrometastasis (0.2 - 2 mm)	10 (91%)	9 (82%)
Macrometastasis (≥2.0 mm)	13 (87%)	12 (80%)

Dye : Dye-guided method, CTLG : CT Lymphography

Table 4. Size of the metastatic lesion in the node positive cases (n=28)

3.4 Number of lymph vessels leading to SNs and the number of SNs

In this study, we were able to analyze the data on the lymph vessels leading to SNs and the number of SNs of 200 patients. A single route with a single SN was the most common pattern, seen in 68% of the patients. Multiple routes with a single SN were detected in 4% of the patients, while a single route with multiple SNs was seen in 10% and multiple routes with multiple SNs were seen in 8%. Overall, multiple SNs were identified in 18% of the patients (Table 5).

Single rout and single node	151 (68%)
Multi routs and single node	8 (4%)
Single rout and multi nodes	23 (10%)
Multi routs and multi nodes	18 (8%)

Table 5. Number of lymph vessels leading to sentinel nodes and the number of sentinel nodes

4. Discussion

In breast cancer surgery, dissection of the axillary lymph nodes is considered effective for the objectives of performing staging and achieving local control. Therefore, in recent years SNB has become a standard procedure for patients with no metastasis of the axillary lymph nodes since it avoids unnecessary dissection. Combined use of a RI (gamma probe-guided method and lymphoscintigraphy) and a dye is currently considered to be more efficient for identification of the SN(s) than the single use of either of these methods. In Japan, there are many institutions that do not have the necessary facilities for using RI and thus must use the dye-guided method alone to identify SNs. On the other hand, Suga et al. proposed using a method called CTLG, which employs a nonionic contrast medium, and we have applied that method for SN identification since February 2003. In this paper we have reported our findings regarding the usefulness and problems associated with combined application of CTLG and the dye-guided method for SN identification.

Our experimental subjects were 219 Japanese patients with primary breast cancer that was thought to be clinically free of axillary lymph node metastasis. CTLG and the dye-guided method were employed in an attempt to identify the SN(s) in each of the patients. The SN identification rates were 96% with CTLG alone, 92% with the dye-guided method alone and 99% when the findings with the two methods were combined. Those results are good when compared with the published data for combined use of an RI and the dye-guided method. In particular, our SN identification rate with the dye-guided method is better than the rates that have been reported to date. We think that this is because the location of the SN had already been determined by the CTLG, making it easy to identify the lymph vessels and lymph nodes. The breakdown of the identification by CTLG showed clear enhancing of the lymph nodes in 88% of the patients, while in 5% of the patients identification of the SN was possible on the basis of the CT value in spite of the fact that the lymph nodes were not clearly enhanced. Because the observations were macroscopic, when the dye-guided method was used alone, SN identification was difficult unless the lymph node was stained to a sufficient degree. With CTLG, on the other hand, even if the enhancement is not very striking it can be surmised that the SN identification rate will be improved since the CT value can be taken into consideration.

In 14 patients, the SN was identified by CTLG, but not by dye-guided method. The location of SN was confirmed by ultrasonography just before the operation in all of these patients. In these 14 patients, not only SN but also neighboring nodes were sampled at the same time. Therefore, it seemed that the SN biopsies had been performed accurately.

The sensitivity of CTLG could not be investigated in our patient population because backup axillary dissection was not performed for patients who were metastasis-negative in the SNB. In addition, none of the patients experienced postoperative recurrence in the axillary lymph nodes, but that does not serve as a basis for claiming that there were few false-negatives. However, Tangoku et al. reported the sensitivity of CTLG to be 98%, and for that reason it can be thought that the sensitivity of CTLG is not inferior to that of the RI method.

With regard to correlations between the clinicopathological findings and the SN identification rate, the patient age, the location of the tumor, whether or not surgical biopsy was performed, the tumor size and the presence/absence of lymph node metastasis have been reported to influence the identification of the SN. In our patient series, the results showed no differences as a function of the age, menopausal status, tumor diameter or histopathological type. However, with the dye-guided method the SN identification rate was significantly lower in patients with a BMI of 25 or higher, whereas with CTLG and combination method the SN identification rate was not influenced by the BMI. As the reason for this difference, it is noted that with the dye-guided method it can be difficult to discern staining of lymph vessels and nodes with the naked eye if considerable subcutaneous fat is present, whereas with CTLG the subcutaneous fat plays no role since the observation is done by CT. In addition to the BMI, discrepancies between the SN identification rates with these three methods were observed as a function of the presence/absence of lymph node metastasis, the site of excisional biopsy (i.e., the lateral upper region and regions other than the lateral upper region), and the presence/absence of vascular invasion. Especially, the identification rates with CTLG and the combined method were significantly lower in node-positive patients compared to node-negative patients, and significantly lower with the combined method in vascular invasion-positive patients compared to negative patients. The size of injected particles, the injected volume, the injection site have been reported as examination factors that can influence SN identification. The nonionic contrast media that are used in CTLG have a larger particle size than dyes. For that reason it can be hypothesized that, in patients with lymph node metastasis or vascular invasion and in patients who had undergone lateral upper region excisional biopsy, who can be predicted to have lymph vessel occlusion, the movement of a nonionic contrast medium would be impeded compared with that of a dye, thus resulting in a lower SN identification rate. On the other hand, the SN identification rate with the dye-guided method was lower in the patients with excisional biopsy in any region than in patients without excisional biopsy. It can be hypothesized that this is probably because excisional biopsy leads to edema of the connective tissue, which makes it difficult to distinguish the dye.

We also investigated the number of SNs and the number of lymph vessels leading to them. These evaluations are difficult to achieve by the gamma probe-guided method and the dye-guided method, but CTLG permits detailed investigation. Our results showed that there was a single route leading to a single SN in 68% of the patients, multiple routes leading to a single SN in 4%, a single route leading to multiple SNs in 10% and multiple routes leading to multiple SNs in 8%. These results are in agreement with those reported by Tangoku et al. Thus, approximately 18% of the patients in our series had multiple SNs, and it can be hypothesized that it would be difficult to biopsy all of them if only the dye-guided method were employed. We think that combined use of CTLG with the dye-guided method would permit accurate biopsy, and for that reason we anticipate that combined performance of CTLG will prove useful.

CTLG is a diagnostic test that is performed prior to surgery for breast cancer. Studies are warranted to determine whether the CT findings or performance of fine-needle aspiration cytology of the SN will make it possible to achieve diagnosis of SN metastasis preoperatively and then decide whether or not SNB should be performed. Such diagnosis leading to avoidance of unnecessary surgical procedures would represent a great clinical advantage by reducing the burden on the patient. Moreover, in the future it will be necessary to compare the usefulness of CTLG with the RI method.

5. Conclusion

Combined performance of CTLG with the dye-guided method permits better elucidation of the location of the SN(s) in breast cancer. This makes it easier to identify the SN(s) and results in a higher SN identification rate compared with application of the dye-guided method alone. Combination of CTLG and the dye-guided method was especially useful in obese patients. In addition, in patients with multiple SNs, it is advantageous to be able to perform SNB accurately. However, in patients with occlusion of the lymph vessels due to lymph node metastasis, having undergone lateral upper region lumpectomy or the presence of vascular invasion, there is a possibility that false-negative diagnostic results will be generated with CTLG.

6. Acknowledgement

Written consent was obtained from the patients or their relative for publication of this article.

7. References

Minato M, Hirose C, Sasa M, Nishitani H, Hirose Y & Morimoto T. 3-dimensional computed tomography lymphography-guided identification of sentinel lymph nodes in breast cancer patients using subcutaneous injection of nonionic contrast medium: a clinical trial. *Jornal of Computer Assisted Tomog*raphy, Vol.28, No.1, (January-February 2004), pp. 46-51

Suga K, Ogasawara N, Okada M & Matsunaga N. Interstitial CT lymphography-guided localization of breast sentinel lymph node: preliminary results. *Surgery*, Vol.133, No.2, (February 2003), pp. 170-179

Suga K, Ogasawara N, Yuan Y, Okada M, Matsunaga N & Tangoku A. Visualization of breast lymphatic pathways with an indirect computed tomography lymphography using a nonionic monometric contrast medium iopamidol: preliminary results. *Investigative Radiology,*Vol.38, No.2, (February 2003), pp. 73-84

Takahasi M, Sasa M, Hirose C, Hisaoka S, Taki M, Hirose T & Bando Y. Clinical efficacy and problems with CT lymphography in identifying the sentinel node in breast cancer. *World Journal of Surgical Oncology*, Vol.6, No.57, (June 2008)

Tangoku A, Yamamoto S, Suga K, Ueda K, Nagashima Y, Hida M, Sato T, Sakamoto K
& Oka M. Sentinel lymph node biopsy using computed tomography-
lymphography in patients with breast cancer. *Surgery*, Vol.135, No.3, (March
2004), pp. 258-265

Tangoku A, Seike J, Nakano K, Nagao T, Honda J, Yoshida T, Yamai H, Matsuoka H,
Uyama K, Goto M, Miyoshi T & Morimoto T. Current status of sentinel lymph node
navigation surgery in breast and gastrointestinal tract. *The Journal of Medical
Investigation*, Vol.54, No.1-2, (February 2007), pp. 1-18

14

Axillary Reverse Mapping in Breast Cancer

Masakuni Noguchi, Miki Yokoi,
Yasuharu Nakano, Yukako Ohno and Takeo Kosaka
Kanazawa Medical University Hospital
Japan

1. Introduction

Axillary lymph node dissection represents the standard surgical treatment for breast cancer patients with clinically or histologically involved axillary lymph nodes. However, it is associated with significant morbidity, including postoperative arm lymphedema and neuropathy of the involved extremity, and seroma formation in the axilla (Noguchi et al., 1997). Particularly, arm lymphedema develops in 7%- 77% of patients who undergo axillary lymph node dissection (Blanchard et al., 2003; Leidenius et al., 2005; Haid et al., 2002; Mansel et al., 2006; Ronka et al., 2005; Schijven et al., 2003; Schrenk et al., 2000; Swenson et al., 2002). At present, sentinel lymph node biopsy is accepted as the standard method of surgical staging for axillary lymph nodes in breast cancer. It can avoid unnecessary axillary lymph node dissection in node-negative patients, thereby minimizing arm lymphedema. Nevertheless, node-positive patients who undergo axillary lymph node dissection do not benefit from sentinel lymph node biopsy. Moreover, sentinel lymph node biopsy does not completely eliminate arm lymphedema. Several cooperative group trials have shown lymphedema rates in range of approximately 7% with sentinel lymph node biopsy alone (Sakorafas et al., 2006; Wilke et al., 2006).

Recently, the axillary reverse mapping technique has been developed to map and preserve arm lymphatic drainage during axillary lymph node dissection and/or sentinel lymph node biopsy (Nos et al., 2007; Thompson et al., 2007). This technique is based on the hypothesis that the lymphatic pathway from the arm cannot be involved by metastasis of the primary breast cancer. The assumption is that the lymphatic drainage from the upper arm is different from that of the breast, allowing safe removal of only the lymphatics of the breast and protection of the lymphatic channels draining the upper extremity during axillary lymph node dissection or sentinel lymph node biopsy, thereby preventing arm lymphedema. However, several studies have shown that there are limits to the principle of non-overlap between breast and arm nodes, including: (a) the axillary reverse mapping nodes may be involved with metastatic foci in patients with extensive axillary lymph node metastases (Bedrosian et al., 2010; Kang et al., 2009; Noguchi et al., 2010b; Nos et al., 2008; Ponzone et al., 2009), and (b) the sentinel lymph node draining the breast may be the same as the axillary reverse mapping node draining the upper extremity in some patients (Boneti et al., 2009; Britton et al., 2009; Kang et al., 2009; Noguchi et al., 2010b). Therefore, the oncological safety of this procedure has not yet determined. This article presents a review of current knowledge regarding in the axillary reverse mapping procedure, and discusses its practical applicability and relevance.

2. Lymph nodes and lymphatics from the breast and the upper extremity

Knowledge of the lymphatic pathway from the breast tissue is essential for the diagnosis and treatment of axillary lymph node metastases in breast cancer. Sappey's classic studies of the lymphatic anatomy of the chest wall are familiar to most physicians (Sappey, 1874) **(Fig. 1)**. Sappey distinguished a superficial group of lymphatics originating in the skin over the breast (subcutaneous lymphatics) and a deep group draining the mammary gland itself (intramammary lymphatics). The superficial and intramammary lymphatics anastomose extensively in the breast, and flow from the two lymphatic groups moves centrifugally toward the axillary and internal mammary nodes. Particularly, the axillary lymph nodes constitute the major regional drainage site for breast cancer. Sappey reported that the lymphatics of the breast collected in a subareolar plexus and then drained toward the axilla *via* lymph collecting vessels. Rouviere (1932), Grant et al. (1953), and Borgstein et al. (2000) supported Sappey's concept of the subareolar plexus for the breast lymph drainage. However, Turner-Warwick (1959), Spratt (1979), Tanis et al. (2001), and Suami et al. (2008) observed that lymphatic pathways from the breast drained directly into the axilla without first passing through the subareolar plexus. Thereby, two potential routes of lymphatic connections from the breast parenchyma to the axilla have been suggested: (a) direct lymphatic connections from the breast parenchyma to the axilla and (b) drainage of parenchyma *via* the subareolar complex into the axillary lymph nodes **(Fig. 2)** (Noguchi, 2009). This knowledge is important with regard to the optimal injection site for identifying sentinel lymph node in breast cancer, because subareolar injection may not always identify the same sentinel lymph node as peritumoral injection (Noguchi et al., 2009).

On the other hand, superficial and deep lymphatics from the upper extremity always flow into the axillary lymph node (called the "sentry node") **(Fig. 3)** (Suami et al., 2007a, 2007c).In the upper extremity, there is usually no communication between the superficial and deep lymphatics except in the epitrochlear region (Suami et al., 2007b). The superficial and deep lymphatics differ in that the former go straight to the axillary lymph nodes, whereas the

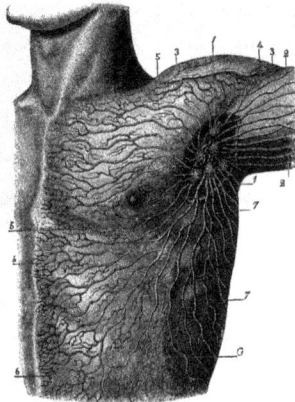

Fig. 1. **Sappey's drawing of the superficial lymphatics of the upper torso.** [Reprinted from Sappey, M.P.C. In A. Delahaye and E. Lecrosnier (Eds.), Anatomie, Physiologie, Pathologie des Vaisseaux Lymphatiques Consideres Chez l'Homme et Les Vertebres. Paris: Adrien Delahaye, 1874].

latter first pass through several interval lymph nodes before reaching the axilla. Thereafter, the lymphatics pass through several lymph nodes before merging into one vessel to reach the subclavian vein (Suami et al., 2007b). In the axilla, however, there are lymphatic interconnections between lymph nodes draining the upper extremity and nodes draining the breast (Suami et al., 2007a).

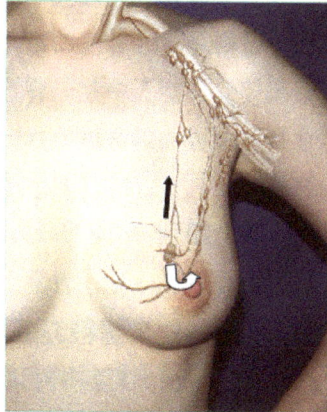

Fig. 2. **Lymphatic flow from the breast tumor to the axilla.** The black arrow shows a direct lymphatic connection from the breast parenchyma to the axilla and the white arrow shows drainage of the parenchyma *via* the subareolar complex into the axillary lymph nodes.

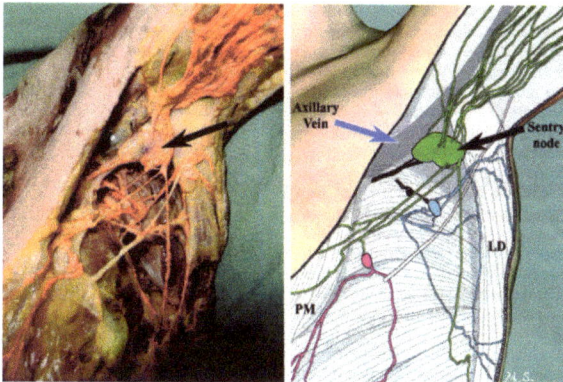

Fig. 3. **Lymphatic anatomy in the axilla.** Left: The left axilla region of the dissected male specimen; Right: a schematic diagram of the same area. The sentry node (black arrow) was connected with the lymphatics from both the upper limb and the upper torso (green). (Reprinted from Suami H: Plast Reconstr Surg 122: 1231-1239, 2008).

3. Axillary reverse mapping

The concept of axillary reverse mapping involves mapping the drainage of the arm with blue dye to determine the anatomical variation in these lymphatics and thus to provide a roadmap for their preservation (Klimberg, 2008). Five variations of axillary reverse mapping lymphatics have been identified: (1) above or below the axillary vein; (2) a sling pattern that may come as much as 4 cm below the axillary vein; (3) a lateral apron; (4) a medial apron; and (5) a twine of cord-like pattern of multiple small nodes. All of these usually emanate from the arm just lateral to the thoracodorsal vessels just under the axillary vein—the so-called axillary ring (Klimberg, 2010). Variations in arm lymphatic drainage put the arm lymphatics at risk for disruption during axillary lymph node dissection or sentinel lymph node biopsy. If arm lymphedema is caused by cutting axillary lymphatics, then being able to see and identify them would allow their preservation. In effect, "axillary reverse mapping is the reverse of sentinel lymph node mapping that serves to map and then remove the lymph nodes draining the breast; axillary reverse mapping involves mapping the arm drainage to allow its preservation" (Klimberg, 2008) **(Fig. 4)**. This procedure is based on the hypothesis that the lymphatic pathway from the arm is not involved by metastasis of the primary breast cancer (Ponzone et al., 2008). However, the preservation of axillary reverse mapping nodes and/or lymphatics is not always possible, because oncological radicality with complete lymphatic preservation may be difficult (Ponzone et al., 2008).

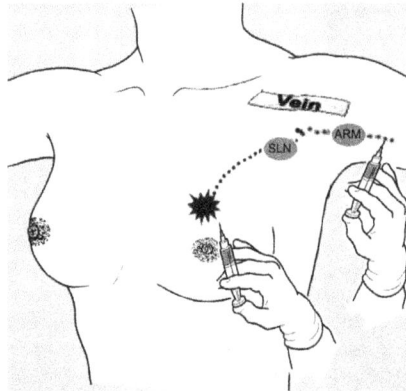

Fig. 4. **Concept of axillary reverse mapping.** Drainage from the breast sentinel lymph node and other lymph nodes from the breast rarely overlap with the lymphatics draining the arm. (Reprinted from Noguchi M : Breast Cancer Res Treat 119:529-535, 2010)

On the other hand, the extent of lymphatic channel disruption required to cause clinically significant lymphedema is unknown. However, it has been suggested that identification and preservation of axillary reverse mapping nodes and/or lymphatics are essential for decreasing postoperative lymphedema rates (Boneti et al., 2008). Boneti et al. (2008) observed the development of lymphedema in 2 of 12 patients in whom the axillary reverse mapping node and/or lymphatics were sacrificed, whereas no lymphedema occurred in patients in whom the axillary reverse mapping node was spared regardless of whether sentinel lymph node biopsy alone or axillary lymph node dissection was performed. In our previous study (Yokoi et al., submitted), 5 of 100 patients developed lymphedema after a mean follow-up of 8 months: 3 patients had undergone axillary lymph node dissection with removal of axillary reverse mapping nodes and lymphatics, and 2 patients had undergone sentinel lymph node biopsy with removal of axillary reverse mapping nodes because the SLN was the same as the axillary reverse mapping node. Although the follow-up has been short in this study, no lymphedema occurred in the remaining patients who had undergone sentinel lymph node biopsy without removal of axillary reverse mapping nodes. As prevention is the key to avoiding lymphedema (Boneti et al., 2008), preservation of axillary reverse mapping nodes and/or lymphatics is worthwhile. If axillary reverse mapping can be confirmed to be both safe and effective in preventing lymphedema, this technique will become the most important technological advancement since sentinel lymph node biopsy.

4. Mapping of axillary reverse mapping node and lymphatics

4.1 Mapping by blue dye injection

Several investigators have reported feasibility studies of the axillary reverse mapping procedure using blue dye **(Fig. 5)** (Boneti et al., 2008; Casabona et al., 2008, 2009; Nos et al., 2007; Ponzone et al., 2008; Thompson et al., 2007). Thompson et al. (2007) injected 2.5 mL of blue dye intradermally or subdermally into the upper inner arm along the medial intramuscular groove of the ipsilateral arm. After injection, the site was massaged and the arm elevated for 5 min to enhance arm lymphatic drainage. Consequently, blue lymphatics and/or nodes in relation to axillary reverse mapping were identified in 11 of 18 (61%) patients, although no blue lymphatics or nodes were identified in the remaining 7 patients. In the same year, Nos et al. (2007) identified axillary reverse mapping nodes in 15 of 21 patients (71%) using a similar technique. Subsequently, several investigators identified axillary reverse mapping nodes using blue dye (Boneti et al., 2008, Casabona et al., 2008; Casabona et al., 2009; Ponzone et al., 2008).

However, identification rates of axillary reverse mapping nodes using blue dye alone were insufficient, ranging from 61% to 86% (Boneti et al., 2008; Casabona et al., 2009; Nos et al., 2007; Ponzone et al., 2008; Thompson et al., 2007) **(Table 1),** and the blue staining at the injection site may persist for up to 6 months after injection (Thompson et al., 2007). To use axillary reverse mapping, moreover, surgeons must use only isotope in the breast to use blue dye in the arm (Klimberg, 2008). This would not be acceptable for those surgeons who believe that the blue dye and radioisotope techniques are complementary for identifying sentinel lymph nodes (Albertini et al., 1996; Noguchi et al., 2000; Noguchi et al., 2009).

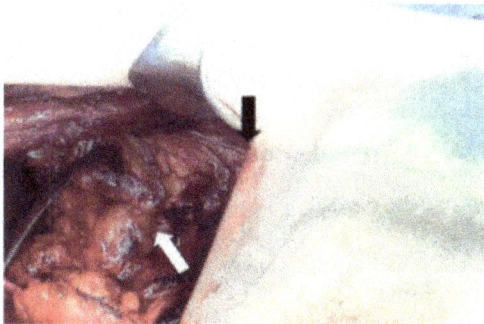

Fig. 5. **Blue axillary reverse mapping lymphatic.** The blue dye tattoo is shown on the inner aspect of the left upper arm in this patient. The white arrow shows a "lateral apron" of blue nodes well below the axillary vein, which is at the level of the black arrow (Reprinted from Klimberg VS: J Surg Oncol 97:563-564, 2008).

Method of ARM procedure Authors/Years	No. of ARM procedures	Identification rates of ARM (ALND field)	Identification rates of ARM (SLN biopsy field)
(a) Blue dye			
Thompson et al. (2007)	18	Nodes/lymphatics: 61% (11/18)	/
Nos et al. (2007)	21	Nodes: 71% (15/21)	/
Ponzone et al.(2007)	4	Nodes/lymphatics: 50% (2/4)	/
Boneti et al.(2008)	131	/	Lymphatics: 43% (56/131)
Casabona et al.(2009)	72	Lymphatics: 89% (8/9)	Lymphatics: 38% (27/72)
(b) Isotope			
Nos et al. (2008)	23	Nodes: 91% (21/23)	/
Britton et al.(2009)	15	Nodes: 100% (15/15)	/
(c) Fluorescence			
Noguchi et al. (2010b)	20	Nodes/lymphatics: 88% (7/8)	Nodes/lymphatics: 75% (9/12)
Yokoi et al. (submitted)	100	Nodes: 93% (25/27)	Nodes: 37% (27/73)

ALND: Axillary lymph node dissection; ARM: Axillary reverse mapping; SLN: sentinel lymph node.

Table 1. Results of Axillary Reverse Mapping Procedure

4.2 Mapping by isotope injection

To improve the identification rate of the axillary reverse mapping nodes and to prevent a persistent blue stain at the site of injection, Nos et al. (2008) injected an isotope into the web space of the ipsilateral hand. During axillary lymph node dissection, the radioactive axillary reverse mapping node was localized above the second intercostal brachial nerve, and then blue dye was injected directly into the node to visualize the efferent ducts constituting the lymphatic axillary reverse mapping chain. Consequently, the axillary reverse mapping nodes were identified in 21 of 23 patients (91%) **(Table 1)**. However, this procedure may be somewhat cumbersome and result in longer operating time. Moreover, direct injection into the axillary reverse mapping node using a syringe with high pressure may cause backflow into the sentinel lymph node, thereby increasing the rate of sentinel lymph node and axillary reverse mapping node confluence. On the other hand, Britton et al. (2009) injected ^{99m}Tc-human polyclonal immunoglobulin G into the breast to identify the sentinel lymph node and injected ^{111}In-human polyclonal immunoglobulin G into the hand to identify axillary reverse mapping nodes. In the specimen of axillary lymph node dissection, the axillary reverse mapping nodes were identified postoperatively in all of 15 patients (100%) using a well scintillation counter **(Table 1)**. Thus, radioisotope labeling seems to be more sensitive for detecting axillary reverse mapping nodes than use of blue dye alone. Identification rates of axillary reverse mapping nodes were improved by using radioisotope with or without blue dye (Britton et al., 2009; Nos et al., 2008). However, radioisotope alone does not permit the visual mapping of axillary reverse mapping lymphatics (Nos et al., 2008).

4.3 Mapping by fluorescent imaging

We have used an invisible near-infrared fluorescence imaging system (PhotoDynamic Eye; Hamamatsu Photonics, Hamamatsu, Japan) for identifying the axillary reverse mapping nodes and/or lymphatics **(Figs. 6 & 7)** (Noguchi et al., 2010b; Yokoi et al., submitted). Before the surgical prep, a smaller volume of indocyanine green (0.1 mL, 0.25 mg) (Diagnogreen; Daiichi Pharmaceutical, Tokyo, Japan) was injected into the forearm to decrease the risk of long-term tattooing, and the injection site was massaged until fluorescent axillary reverse mapping lymphatics were observed in the upper inner arm. During axillary lymph node dissection or sentinel lymph node biopsy, the light was occasionally switched off in the operating room and fluorescent axillary reverse mapping nodes and/or lymphatics were observed in the axilla using the fluorescence imaging system. Consequently, the axillary reverse mapping nodes were identified in 25 (93%) of 27 patients who underwent axillary lymph node dissection alone (Yokoi et al., submitted) **(Table 1).** Although the fluorescent axillary reverse mapping nodes and/or lymphatics were not observed in 2 patients, they were early cases in the study (Noguchi et al., 2010b). Thus, the fluorescence imaging technique is useful for detecting lymphatic drainage from the upper extremity, and it also permits differentiation of fluorescent axillary reverse mapping nodes and/or lymphatics from blue and/or hot sentinel lymph nodes (Noguchi et al., 2009).

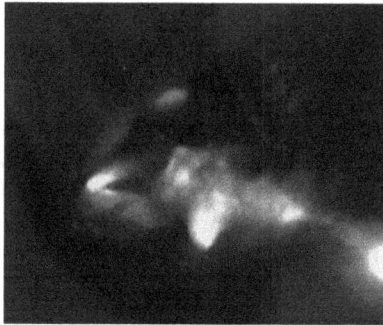

Fig. 6. Fluorescence imaging in the upper extremity.

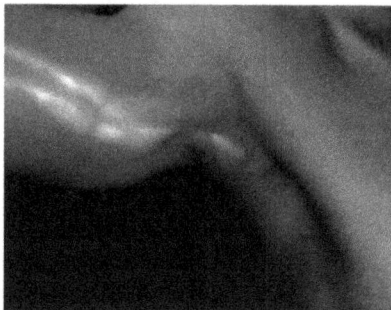

Fig. 7. Fluorescence imaging in the axilla.

5. Mapping of axillary reverse mapping node and lymphatics in the fields of axillary lymph node dissection and sentinel lymph node biopsy

Generally, identification of the axillary reverse mapping nodes is not sufficient in patients undergoing sentinel lymph node biopsy alone (Boneti et al., 2008; Casabona et al., 2009;

Noguchi et al., 2010b). Boneti et al. (2008) reported that blue lymphatics draining the arm were visible from the sentinel lymph node biopsy incision, and so were located near or within the sentinel lymph node field in 56 (42.7%) of 131 patients. Casabona *et al.* (2009) also reported the blue lymphatics draining from the arm in the sentinel lymph node biopsy field in 27 of 72 patients (37.5%). Similarly, in our recent study, axillary reverse mapping nodes were identified in 27 (37%) of 73 patients who underwent sentinel lymph node biopsy. In the remaining 46 patients, however, the axillary reverse mapping node was not observed in the sentinel lymph node field (Yokoi et al., submitted), suggesting that the ARM nodes were located in different fields with respect to the sentinel lymph node area. In these studies (Boneti et al., 2008; Casabona et al., 2009; Noguchi et al., 2010b; Yokoi et al., submitted), a difference was observed in the identification rate of axillary reverse mapping nodes and/or lymphatics in the axillary lymph node dissection field and in the sentinel lymph node field **(Table 1)**. This may be because the majority of lymphatics draining the arm are anatomically located deeper than the sentinel lymph node (Casabona et al., 2009). This hypothesis is consistent with the higher incidence of lymphedema after axillary lymph node dissection than after sentinel lymph node biopsy (Celebioglu et al., 2007; Lucci et al., 2007; McLaughlin et al., 2008; Schrenk et al., 2000).

6. Metastases in axillary reverse mapping nodes

The concept of axillary reverse mapping is based on the hypothesis that the lymphatic pathway from the arm is not involved by metastasis of the primary breast cancer (Ponzone et al., 2008). Initial studies showed that no cancer cells were found in the ARM nodes even when the patients had many positive axillary nodes (Boneti et al., 2008; Casabona et al., 2009; Thompson et al., 2007) **(Table 2)**. Subsequently, they preserved the axillary reverse mapping nodes in patients in the later series (Boneti et al., 2008; Thompson et al., 2007).

Authors/references	No. of patients	No. of patients with ARM involvement	% of ARM involvement
Thompson et al. (2007)	7	0	0%
Nos et al. (2007)	10	0	0%
Boneti et al. (2008)	7	0	0%
Nos et al. (2008)	21	3	14%
Ponzone et al. (2009)	27	3	11%
Kang et al. (2009)	101	9	8.9%
Bedrosian et al. (2010)	11	2	18%
Noguchi et al. (2010)	7	3	43%
Yokoi et al. (submitted)	25	11	44%

ARM: axillary reverse mapping

Table 2. The Involvement of Axillary Reverse Mapping Nodes in Patients who underwent Axillary Lymph Node Dissection with Removal of Axillary Reverse Mapping Nodes

However, recent studies have demonstrated involvement rates of axillary reverse mapping node ranging from 8.9% to 44% **(Table 2)**. In a study performed in France, Nos et al. (2008) reported that the axillary reverse mapping nodes showed metastatic involvement in 3 of 21 patients with N0-3 (14%). Ponzone et al. (2009) also found that three patients with extensive nodal metastatic involvement (*i.e.*, pN2a and pN3a) showed metastatic cells in the axillary reverse mapping nodes, although the ARM was clear of metastases in the remaining 24 (89%) of 27 patients. In a large study, Kang et al. (2009) reported that ARM node metastases were found in 9 (8.9%) of 101 patients who underwent axillary reverse mapping procedure. In our recent study (Yokoi et al., submitted), on the other hand, the axillary reverse mapping nodes were positive in 11 (44%) of 25 patients with a clinically positive node who underwent axillary lymph node dissection without sentinel lymph node biopsy. The fluorescence imaging system was highly sensitive for identification of the fluorescent axillary reverse mapping nodes, and the excised axillary reverse mapping nodes were cut into serial sections at 2 mm intervals for the histological examination (Noguchi et al., 2010b). This may increase the detection of metastatic axillary reverse mapping nodes. It is not surprising that the axillary reverse mapping nodes could be involved by metastasis of the primary breast cancer, because of anatomical interconnections between lymphatics draining from the upper extremity and lymphatics draining from the breast (Suami et al., 2007a). It has been suggested that effacement of nodes by the gross tumor may alter the pattern of lymph flow in these patients, allowing metastasis to the axillary reverse mapping nodes.

7. Convergence of axillary reverse mapping node and sentinel lymph node

The sentinel lymph node is most commonly located in the central nodal group, and it is possible that the axillary reverse mapping nodes are located in the central nodal group. If the sentinel lymph node draining the breast were the same node as the axillary reverse mapping node draining the upper extremity, it would be impossible to preserve the axillary reverse mapping node at sentinel lymph node biopsy. Boneti et al. (2009) reported that crossover (axillary reverse mapping node = sentinel lymph node) occurred in only 6 (2.2%) of 220 patients, although axillary reverse mapping lymphatics were near or within the sentinel lymph node biopsy field in 40.6% of patients. However, Britton et al. (2009) reported that sentinel lymph node from the breast was the same as the axillary reverse mapping node from the upper extremity in 2 (13%) of 15 patients, indicating convergence of the two drainage pathways through the same node. Kang et al. (2009) also reported a concordance rate of 18.9% (19/96) between axillary reverse mapping node and sentinel lymph node. Noguchi et al. (2010b) reported that the sentinel lymph node from the breast was the same as the axillary reverse mapping node from the upper extremity in 3 (21%) of 14 patients **(Fig. 8)**. In our recent study (Yokoi et al., submitted), the sentinel lymph node from the breast was the same as the axillary reverse mapping node from the upper extremity in 20 of 27 patients in whom axillary reverse mapping nodes were identified, but the axillary reverse mapping node was not observed in the sentinel lymph node field in the remaining 46 patients, yielding a concordance rate of 27% (20/73) between axillary reverse mapping node and sentinel lymph node **(Table 3)**. These findings were supported by a recent anatomical description of the lymphatic territories of the upper extremity (Suami et al., 2008). Removal of this common lymph node at sentinel lymph node biopsy will result in disruption of lymphatic drainage of the upper extremity and an increased risk of lymphedema, explaining why sentinel lymph node biopsy does not correct the problem of lymphedema (Schrenk et al., 2000; Sener et al., 2001; Wilke et al., 2006). This is an important limitation of the axillary reverse mapping procedure (Khan & Lurie, 2009), although the axillary reverse mapping lymphatics vary significantly from patient to patient.

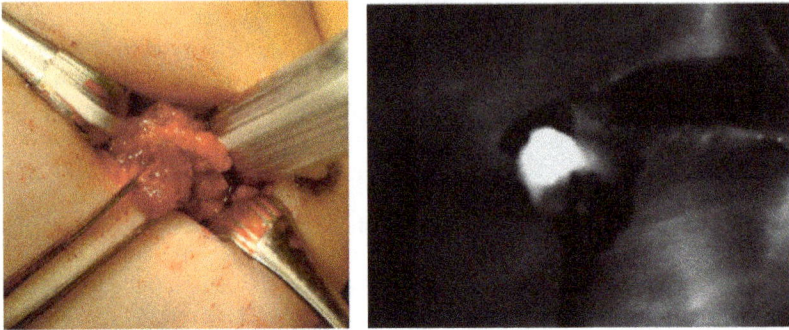

Fig. 8. **Sentinel lymph node and axillary reverse mapping node.** Left: sentinel lymph node biopsy: A hot and blue node was identified. Right: Fluorescence imaging by Photodynamic Eye: A fluorescent ARM node in the axilla, which is the same as the SLN.

Authors/references	No. of patients	No. of patients with converged sentinel lymph node – axillary reverse mapping node	Rates of convergence (%)
Boneti et al. (2009)	220	6	2.2%
Britton et al. (2009)	15	2	13%
Kang et al. (2009)	96	19	18.9%
Noguchi et al. (2010)	14	2	14%
Yokoi et al. (submitted)*	73	20	27%

*: This study included patients from the previous study (Noguchi et al., 2010).

Table 3. Concordance of sentinel lymph node and axillary reverse mapping nodes in patients who underwent sentinel lymph node biopsy and axillary reverse mapping procedure.

8. Preservation of axillary reverse mapping node and lymphatics

Several studies have demonstrated that the axillary reverse mapping nodes are involved with metastatic foci in some patients with extensive axillary lymph node metastasis (Bedrosian et al., 2010; Kang et al., 2009; Noguchi et al., 2010; Nos et al., 2008; Ponzone et al., 2009; Yokoi et al., submitted). Therefore, patients with suspected extensive nodal disease at clinical examination, ultrasonography of the axilla, or intraoperative pathologic assessment should not be candidates for preservation of axillary reverse mapping nodes and lymphatics (Ponzone et al., 2009). On the other hand, the SLN draining the breast is the same node as the axillary reverse mapping node draining the upper extremity in some patients (Boneti et al., 2009; Britton et al., 2009; Kang et al., 2009; Noguchi et al., 2010; Yokoi et al., submitted). It is impossible to preserve converged sentinel lymph node – axillary reverse mapping node, although the excision of one converged node does not always translate into lymphedema, because multiple lymphatic channels drain the arm. Thus, there is no reliable separation of arm and breast lymphatic pathways, because there are lymphatic interconnections between lymph nodes draining the upper extremity and nodes draining the breast.

However, it has been suggested that patients with clinically uninvolved nodes might derive the most benefit from the axillary reverse mapping procedure (Ponzone et al., 2009). In a previous study by Boneti et al. (2009), 220 patients with clinically uninvolved nodes underwent sentinel lymph node biopsy and axillary reverse mapping procedure. Forty of these patients subsequently underwent axillary lymph node dissection because of positive sentinel lymph nodes. Consequently, axillary reverse mapping nodes were negative for malignancy. In our recent study (Yokoi et al., submitted), axillary reverse mapping node was identified in 27 of 73 patients with clinically uninvolved nodes who underwent sentinel lymph node biopsy, and it was the same as the sentinel lymph node in 20 patients. In 11 patients with a positive sentinel lymph node who subsequently underwent axillary lymph node dissection, however, axillary reverse mapping nodes were tumor-free as far as it was not the same as the positive sentinel lymph node. Therefore, it may be possible to spare the remaining axillary reverse mapping nodes during axillary lymph node dissection in patients with clinically negative node but positive sentinel lymph node.

To minimize prolonged seroma and prevent arm lymphedema, on the other hand, Kodama routinely performed lower axillary dissection without using either axillary reverse mapping or sentinel lymph node biopsy. The lower axillary dissection is defined as dissecting axillary lymph nodes only below the second intercostal brachial nerve. Consequently, they found that the 5-year overall and relapse-free survival rates were 95.3% and 88.3%, respectively, in 1043 clinically node-negative patients from 2001 to 2008 (Kodama et al., 2010). They were not significantly different from those of 1084 clinically node-negative patients who underwent total or partial axillary lymph node dissection from 1994 to 2000 (94.9% and 88.4%, respectively). Only 6 (0.6%) of 1043 patients developed axillary recurrence with a median follow-up of 72 months, while no lymphedema occurred. Although axillary reverse mapping nodes are usually localized above the second intercostal brachial nerve, therefore, it may be oncologically safe to spare axillary lymph nodes and lymphatics above the second intercostal brachial nerve in patients with clinically negative nodes. However, sentinel lymph node is not infrequently located above the second intercostal brachial nerve, and the axillary reverse mapping procedure is a more accurate means of preserving lymphatics from the upper extremity than partial axillary lymph node dissection. Therefore, sentinel lymph node biopsy followed by axillary lymph node dissection with axillary reverse mapping is oncologically more safe than the lower axillary dissection in patients with clinically negative nodes.

9. Microsurgical lymphatic-venous anastomosis

It is not always possible to preserve ARM nodes and/or lymphatics during axillary lymph node dissection or sentinel lymph node biopsy. Therefore, Casabona et al. (2008, 2009) performed microsurgical lymphatic-venous anastomosis using lymphatic collectors coming from the arm and one of the collateral branches of the axillary vein. Lymphatic collectors were introduced inside the vein and the inferior edge of the lymphatics introduced into the vein lumen acted as valves to avoid backflow of blood into the lymphatics (Campisi et al., 2007), thus preventing the occurrence of thrombosis. In fact, lymphatic microsurgery techniques have been shown to be effective in the treatment of peripheral lymphedema (Campisi et al., 2007). To perform lymphatic-venous microanastomosis, however, the axillary reverse mapping lymphatics must be visible and preserved as much as possible during axillary lymph node dissection, while maintaining oncological radicality.

On the other hand, postmastectomy radiotherapy is currently accepted as a standard adjuvant treatment in patients with more than 4 positive axillary nodes (Noguchi et al., 2002; Recht et al., 2001). According to the Guidelines of the American Society of Clinical Oncology (Recht et al., 2001), the chest wall as well as the supraclavicular region should be irradiated, whereas full axillary radiotherapy should not be performed routinely in patients undergoing axillary lymph node dissection. In postmastectomy radiotherapy, the lymphatic-venous microanastomosis in the axilla can be exposed to non-negligible irradiation as tangents with chest wall irradiation. It is well known that arm lymphedema can be caused by scar formation from surgery and/or radiation therapy. Therefore, long-term follow-up studies are required before we can conclude that this microsurgical technique is effective for prevention of arm lymphedema even after postmastectomy radiotherapy.

10. Conclusions

The axillary reverse mapping procedure is not completely accurate in differentiating between the arm and breast lymphatic pathways. The ARM node is involved with metastatic foci in some patients with extensive axillary lymph node metastasis. Moreover, the sentinel lymph node draining the breast is the same node as the axillary reverse mapping node draining the upper extremity in a minority of patients. It is oncologically unacceptable to preserve a metastatic axillary reverse mapping node in axillary lymph node dissection or converged sentinel lymph node – axillary reverse mapping node in sentinel lymph node biopsy. In patients with a positive sentinel lymph node who subsequently underwent axillary lymph node dissection, however, remaining axillary reverse mapping nodes are tumor-free as far as it is not the same as positive sentinel lymph node. Therefore, it may be oncologically safe to spare the axillary reverse mapping nodes during axillary lymph node dissection only in patients with clinically uninvolved nodes. Further studies are needed before this technology can be accepted as a standard procedure in the surgical management of breast cancer. On the other hand, it is not always possible to preserve ARM nodes and/or lymphatics during axillary lymph node dissection. Therefore, microsurgical lymphatic-venous anastomosis may be effective for prevention of arm lymphedema in patients who underwent axillary lymph node dissection with removal of axillary reverse mapping nodes. However, long-term follow-up studies are required before we can conclude that it is effective for prevention of arm lymphedema even after postmastectomy radiotherapy.

11. References

Albertini JJ. Lyman GH, Cox C, Yeatman T, Balducci L, Ku N, Shivers S, Berman C, Wells K, Rapaport D, Shons A, Horton J, Greenberg H, Nicosia S, Clark R, Cantor A, &

Reintgen DS. (1996). Lymphatic mapping and sentinel node biopsy in the patients with breast cancer. *J Am Med Assoc* Vol. 276, No. 22, pp. 1818-1822, ISSN 0098-7484.

Bedrosian I, Babiera GV, Mittendorf EA, Kuerer HM, Pantoja L, Hunt KK, Krishnamurthy S, & Meric-Bernstam FM (2010). A phase I study to assess the feasibility and oncologic safety of axillary reverse mapping in breast cancer patients. *Cancer Vol.* 116, No. 11, pp. 2543-2548, ISSN 0008-543X.

Blanchard DK, Donohue JH, Reynolds C, & Grant CS (2003). Relapse and morbidity in patients undergoing sentinel lymph node biopsy alone or with axillary dissection for breast cancer, *Arch Surg* Vol. 138, No. 5, pp. 482-488, ISSN 0344-8444.

Boneti C, Korourian S, Bland K, Cox K, Adkins LL, Henry-Tillman RS, & Klimberg VS (2008). Axillary reverse mapping: mapping and preserving arm lymphatics may be important in preventing lymphedema during sentinel lymph node biopsy. *J Am Coll Surg* Vol. 206, No. 5, pp. 1038-1044, ISSN 1072-7515 .

Boneti C, Korourian S, Diaz Z, Santiago C, Mumford S, Adkins L, & Klimberg VS (2009). Scientific Impact Award: Axillary reverse mapping (ARM) to identify and protect lymphatics draining the arm during axillary lymphadenectomy. *Am J Surg* Vol. 198, No. 4, pp. 482-487, ISSN 0002-9610.

Borgstein PJ, Meijer S, Pijers RJ, & van Diest PJ (2000). Functional lymphatic anatomy for sentinel node biopsy in breast cancer: echoes from the past and the periareolar blue method. *Ann Surg* Vol. 232, No. 1, pp. 81-89, ISSN 0003-4932.

Britton TB, Solanki CK, Pinder SE, Mortimer PS, Peters AM, & Purushotham AD (2009). Lymphatic drainage pathways of the breast and the upper limb. *Nucl Med Comm* Vol. 30, No. 6, pp. 427-430, ISSN 0143-3636.

Campisi C, Eretta C, Pertile D, Da Rin E, Campisi C, Maccio A, Campisi M, Accogli S, Bellini C, Boniolo E, & Boccardo F (2007). Microsurgery for treatment of peripheral lymphedema: long-term outcome and future perspectives. Microsurgery Vol. 27, No. 4, pp. 333-338, ISSN 0738-1085.

Casabona F, Bogliolo S, Ferrero S, Boccardo F, & Campisi C (2008). Axillary reverse mapping in breast cancer: A new microsurgical lymphatic-venous procedure in the prevention of arm lymphedema. *Ann Surg Oncol* Vol. 15, No. 11, pp. 3318-3319, ISSN 1068-9265.

Casabona F, Bogliolo S, Valenzano Menada M, Sala P, Villa G, & Ferrero S (2009). Feasibility of axillary reverse mapping during sentinel lymph node biopsy in breast cancer patients. *Ann Surg Oncol Vol.* 16, No. 9, pp. 2459-2463, ISSN 1068-9265.

Celebioglu F, Perbeck L, Frisell J, Grondal E, Svensson L, & Danielsson R (2007). Lymph drainage studied by lymphoscintigraphy in the arms after sentinel node biopsy compared with axillary lymph node dissection following conservative breast cancer surgery. *Acta Radiol* Vol. 48, No. 5, pp.488-495, ISSN 0001-6926.

Grant RN, Tabah EJ, & Adair FF (1953). The surgical significance of the subareolar lymph plexus in cancer of the breast. *Surgery* Vol. 33, No. 1, pp. 71-78, ISSN 0039-6060.

Haid A, Koberle-Wuhrer R, Knauer M, Burtscher J, Fritzsche H, Peschina W, Jasarevic Z, Ammann M, Hergan K, Sturn H, & Zimmermann G (2002). Morbidity of breast cancer patients following complete axillary dissection or sentinel node biopsy only: a comparative evaluation. *Breast Cancer Res Treat* Vol. 73, No. 1, pp. 31-36, ISSN 0167-6806.

Kang SH, Choi JE, Jeon YS, Lee SJ, & Bae YK (2009). Preservation of lymphatic drainage from arm in breast cancer surgery: Is it safe? *Cancer Research* Vol. 69, No. Suppl 2, pp. 87s, ISSN 0576-6656.

Khan SA & Lurie RH (2009). Axillary reverse maping to prevent lymphedema after breast cancer surgery: Defining the limits of the concept. *J Clin Oncol* Vol. 27, No. 33, pp.5494-5496, ISSN 0732-183X.

Klimberg VS (2008). A new concept toward the prevention of lymphedema: Axillary reverse maping. *J Surg Oncol* Vol. 97, No. 7, pp.563-564, ISSN 0022-4790.

Klimberg VS (2010). Chapter 14 Axillary reverse mapping, *Atlas of Breast Surgical Techniques*, Klimberg (ed), pp174-181, Surgical Techniques Atlas Series , Townsend Jr CM, Evers BM, eds., Saunders Co., ISBN 978-1-4160-4691-2, Philadelphia, PA, USA.

Kodama H, Mise K & Kan N (2010). Lower axillary dissection for early breast cancer. *J Jap Surg Assoc Vol.* 71, No. 12. pp. 3031-3038 (in Japanese with English abstract), ISSN 1345-2843.

Leidenius M, Leivonen M, Vironen J, & von Smitten K (2005). The consequences of long-time arm morbidity in node-negative breast cancer patients with sentinel node biopsy or axillary clearance. *J Surg Oncol Vol.* 92, No. 1, pp 23-31, ISSN 0022-4790

Lucci A, McCall LM, Beitsch PD, Whitworth PW, Reintgen DS, Blumencranz PW, Leitch AM, Saha S, Hunt KK, & Giuliano AE (2007). Surgical complications associated with sentinel lymph node dissection (SLND) plus axillary lymph node dissection compared with SLND alone in the American College of Surgeons Oncology Group Trial Z0011. *J Clin Oncol* Vol. 25, No. 24, pp.3657-3663, ISSN 0732-183X.

Mansel RE, Fallowfield L, Kissin M, Goyal A, Newcombe RG, Dixon JM, Yiangou C, Horgan K, Bundred N, Monypenny I, England D, Sibbering M, Abdullah TI, Barr L, Chetty U, Sinnett DH, Fleissig A, Clarke D, & Ell PJ (2006). Randomized multicenter trial of sentinel node biopsy versus standard axillary treatment in operable breast cancer: The ALMANAC Trial. *JNCI* Vol. 98, No. 9, pp. 599-609, ISSN 0027-8874.

McLaughlin SA, Wright MJ, Morris KT, Giron GL, Sampson MR, Brockway JP, Hurley KE, Riedel ER, & Van Zee KJ (2008). Prevalence of lymphedema in women with breast cancer 5 years after sentinel lymph node biopsy or axillary dissection: objective measurements. *J Clin Oncol Vol.* 26, No. 32, pp. 5313-5319, ISSN 0732-183X.

Noguchi M, Miwa K, Michigishi T, Yokoyama K, Nishijima H, Takanaka T, Kawashima H, Nakamura S, & Nonomura A (1997). The role of axillary lymph node dissection in breast cancer management. *Breast Cancer Vol.* 4, No. 3, pp. 143-153, ISSN 1340-6868.

Noguchi M, Motomura K, Imoto S, Miyauchi M, Sato K, Iwata H, Ohta M, Kurosumi M, & Tsugawa K (2000). A multicenter validation study of sentinel lymph node biopsy by the Japanese Breast Cancer Society. *Breast Cancer Res Treat* Vol. 63, No. 1, pp. 31-40, ISSN 0167-6806.

Noguchi M (2002). Does regional treatment improve the survival in patients with operable breast cancer? *Breast Cancer Res Treat Vol.* 76, No. 3, pp. 269-282, ISSN 0167-6806.

Noguchi M, Inokuchi M, & Zen Y (2009). Complement of peritumoral and subareolar injection in breast cancer sentinel lymph node biopsy. *J Surg Oncol* Vol. 100, No. 2, pp.100-105, ISSN 0022-4790.

Noguchi M (2010a). Axillary reverse mapping for preventing lymphedema in the axillary lymph node dissection and/or sentinel lymph node biopsy (Editorial). Breast Cancer Vol. 17, No. 3, pp. 155-157. ISSN 1340-6868.

Noguchi M, Yokoi M, & Nakano Y (2010b). Axillary reverse mapping with indocyanine fluorescence imaging in patients with breast cancer. *J Surg Oncol* Vol. 101, No. 3, pp.217-221, ISSN 0022-4790.

Noguchi M (2010c). Axillary reverse mapping for breast cancer. *Breast Cancer Res Treat* Vol. 119, No. 3, pp. 529-535, ISSN 0167-6806.

Nos C, Lesieur B, Clough KB, & Lecuru F (2007). Blue dye injection in the arm in order to conserve the lymphatic drainage of the arm in breast cancer patients requiring an axillary dissection. *Ann Surg Oncol* Vol. 14, No. 9, pp. 2490-2496, ISSN 1068-9265.

Nos C, Kaufmann G, Clough KB, Collignon M-A, Zerbib E, Cusumano P, & Lecuru F (2008). Combined axillary mapping (ARM) technique for breast cancer patients requiring axillary dissection. *Ann Surg Oncol* Vol. 15, No. 9, pp. 2550-2555, ISSN 1068-9265.

Ponzone R, Mininanni P, Cassina E, & Sismondi P (2008). Axillary reverse mapping in breast cancer: can we spare what we find? *Ann Surg Oncol* Vol. 15, No. 1, pp.390-391, ISSN 1068-9265.

Ponzone R, Cont NT, Maggiorotto F, Cassina E, Mininanni P, Biglia N, & Sismondi P (2009) Extensive nodal disease may impair axillary reverse mapping in patients with breast cancer. *J Clin Oncol* Vol. 27, No. 33, pp.5547-5551, ISSN 0732-183X.

Recht A, Edge SB, Solin LJ, Robinson DS, Estabrook A, Fine RE, Fleming GF, Formenti S, Hudis C, Kirshner JJ, Krause DA, Kuske RR, Langer AS, Sledge GW Jr, Whelan TJ, & Pfister DG (2001). Postmastectomy radiotherapy: clinical practice guidelines of the American Society of Clinical Oncology. *J Clin Oncol* Vol. 19, No. 5, pp. 1539-1569, ISSN 0732-183X.

Ronka R, von Smitten K, Tasmuth T, & Leidenius M (2005). One-year morbidity after sentinel node biopsy and breast surgery. Breast Vol. 14, No. 1, pp. :28-36, ISSN 0960-9776.

Rouviere H (1932) *Anatomie des Lymphatiques de l'Homme.* Paris: Masson.

Sakorafas GH, Peros G, Cataliotti L, & Vlastos G (2006). Lymphedema following axillary lymph node dissection for breast cancer. *Surg Oncol Vol.* 15, No. 3, pp. 153-165, ISSN 0960-7404.

Sappey MPC (1874). Anatomie, Physiologie, Pathologie des vaisseaux Lymphatiques consideres chez L' homme at les Vertebres. Paris: A. Delahaye and E. Lecrosnier.

Schijven MP, Vingerhoets AJ, Rutten HJ, Nieuwenhuijzen GA, Roumen RM, van Bussel ME, & Voogd AC (2003). Comparison of morbidity between axillary lymph node dissection and sentinel node biopsy. *Eur J Surg Oncol* Vol. 29, No. 4, pp. 341-350, ISSN 0748-7983.

Schrenk P, Rieger R, Shamiyeh A, & Wayand W (2000). Morbidity following sentinel lymph node biopsy versus axillary lymph node dissection for patients with breast carcinoma. *Cancer Vol.* 88, No. 3, pp. 608-614, ISSN 0008-543X.

Sener SF, Winchester DJ, Martz CH, Feldman JL, Cavanaugh JA, Winchester DP, Weigel B, Bonnefoi K, Kirby K, & Morehead C (2001). Lymphedema after sentinel lymphadenectomy for breast carcinoma. *Cancer Vol.* 92, No. 4, pp. 748-752, ISSN 0008-543X.

Spratt JS (1979). Anatomy of the breast. *Major Prob Clin* Vol. 5, pp. 1-13, ISSN 0025-1062.

Suami H, Taylor GI, & Pan WR (2007a). The lymphatic territories of the upper limb: anatomical study and clinical implications. *Plast Reconstr Surg* Vol. 119, No. 6, pp. 1813-1822, ISSN 0032-1052.

Suami H, Pan WR, & Taylor GI (2007b). Changes in the lymph structure of the upper limb after axillary dissection: radiographic and anatomical study in a human cadaver. *Plast Reconstr Surg* Vol. 120, No. 4, pp. 982-991, ISSN 0032-1052.

Suami H, O' Neill JK, Pan WR, & Taylor GI (2008) Superficial lymphatic system of the upper torso: preliminary radiographic results in human cadavers. *Plast Reconstr Surg* Vol. 121, No. 4, pp. 1231-1239, ISSN 0032-1052.

Swenson KK, Nissen MJ, Ceronsky C, Swenson L, Lee MW, & Tuttle TM (2002). Comparison of side effects between sentinel lymph node and axillary lymph node dissection for breast cancer. *Ann Surg Oncol* Vol. 9, No. 8, pp. 745-753, ISSN 1068-9265.

Tanis PJ, Nieweg OE, Valdes Olmos RA, & Kroon BB (2001). Anatomy and physiology of lymphatic drainage of the breast from the perspective of sentinel node biopsy. *J Am Coll Surg* Vol. 192, No. 3, pp. 399-409, ISSN 1072-7515 .

Thompson M, Korourian S, Henry-Tillman R, Adkins L, Mumford S, Westbrook KC, & Klimberg VS (2007). Axillary reverse mapping (ARM): a new concept to identify and enhance lymphatic preservation. *Ann Surg Oncol* Vol. 14, No. 6, pp. 1890-1895, ISSN 1068-9265.

Turner-Warwick RT (1959). The lymphatics of the breast. Br J Surg Vol. 46, pp. 574-582, ISSN 0007-1323.

Wilke LG, McCall LM, Posther KE, Whitworth PW, Reintgen DS, Leitch AM, Gabram SG, Lucci A, Cox CE, Hunt KK, Herndon JE II, & Giuliano AE (2006). Surgical complications associated with sentinel lymph node biopsy: results from a prospective international cooperative group trial. *Ann Surg Oncol* Vol. 13, No. 4, pp. 491-500, ISSN 1068-9265.

Yokoi M, Noguchi M, Nakano Y, Ohno Y, & Kosaka T (n.d.). Axillary reverse mapping using fluorescence imaging system in breast cancer. *Breast Cancer Res Treat* (submitted)., ISSN 0167-6806.

15

Lymphedema: Clinical Picture, Diagnosis and Management

Tanja Planinšek Ručigaj and Vesna Tlaker Žunter

Department of Dermatovenereology, University Clinical Centre Ljubljana

Slovenia

1. Introduction

Lymphedema is swelling due to excess accumulation of lymph in the tissues caused by inadequate lymph drainage (Mortimer, 2010). The protein-rich lymph accumulates in tissue when reabsorption is hindered or when the lymphatics are absent or blocked (Gaber, 2009). Lymphedema is most commonly seen on lower limbs (Figure 1), rarely on the upper limbs, and extremely rarely on the head or the trunk. The development of this process may take several years (Cohen et al, 2001).

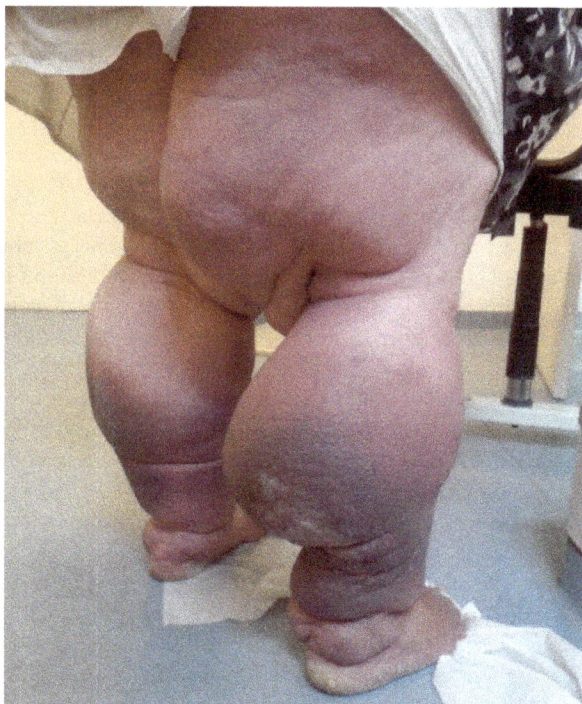

Fig. 1. Bilateral leg lymphedema.

The leading causes of lymphedema include cancer and therapy for malignant diseases, infections, chronic venous insufficiency, and congenital malformations (Planinšek Ručigaj et al, 2005).

2. Etiology and pathogenesis of lymphedema

Within the interstitial space – the space between tissue cells – fluid is always present. The amount of fluid depends on two factors: the amount introduced into the interstitial space, and the amount removed from it. Fluid enters the space from arterioles and venules; some returns to the venules, and the remainder is taken up by the lymphatics. In the normal physiologic state, entrance and exit are approximately equal, so that tissues retain their usual morphologic appearance and function. Edema (swelling) develops when the volume of interstitial fluid increases, either from increased inflow or decreased outflow, or both (Tretbar et al, 2008).

An imbalance between lymphatic flow and the capacity of lymphatic circulation (capillary filtration exceeding the capacity of lymphatic circulation) bring up to edema (Mortimer, 1998). The protein constituents of static lymph fluid may cause inflammation and subsequent tissue fibrosis. In patients with lymphedema, the overlying skin may develop cobblestoning, as well as a verrucous or mossy appearance (Philips, 2008).

Based on pathogenesis, one distinguishes two types of lymphedema: primary and secondary lymphedema (Gaber, 2009).

2.1 Primary lymphedema

Lymphedema arising from an intrinsic abnormality of the lymph-conducting pathways is referred to as primary lymphedema (Browse et al, 1985). The deficient lymphatics do not or can not propel the lymph in adequate amounts, and the fluid sequesters within the interstitial or lymphatic spaces (Tretbar et al, 2008). There is a simple classification by age of onset without reference to etiology or other clinical features into three types based on age at first appearance of disease: congenital lymphedema, which typically develops within two years of life, lymphedema praecox (present <35 years of age), and lymphedema tarda (first present >35 years of age). The problem can be unilateral or bilateral and may be familial or sporadic (Gaber 2009, Mortimer 2010, Moneta 2009). Primary lymphedema often occurs in the lower extremities, and affects women more often than men (Browse et al, 1985).

An increasing understanding of the genetic basis of primary lymphedema has further changed classification and to some extent made the praecox and tarda categories redundant. For example in lymphoedema–distichiasis syndrome (LDS), the onset of lower-limb oedema can range from puberty to 40 years old despite the same cause, namely a mutation in the FOXC2 gene (Brice G et al, 2002). Conversely, LDS and Meige forms of primary lymphoedema can both present at puberty with similar lower limb lymphedema but have different genotypes and different mechanisms; LDS is associated with hyperplasia and valve reflux of collecting lymphatics whereas Meige's disease is associated with hypoplasia of collecting vessels. In the future, classification of lymphoedema is likely to be based on phenotype, unless the genotype is known (Mortimer 2010).

The development of lymphangiography in the 1950s resulted in a radiological classification: aplasia (no formed lymph pathways found), hypoplasia (lymphatics smaller of fewer than normal) and hyperplasia (lymphatics larger and more numerous). Aplasia, hypoplasia and hyperplasia refer to abnormalities in the main (leg) conducting lymph vessels as opacified on lymphangiography, and not to the initial lymphatics, which are not imaged with this method. Further investigation revealed types of lymphoedema where few, if any, lymph conducting vessels could be identified in the foot, but vessels were found to be normal further up the limb (Browse 1996).

The causes of primary lymphedema are summarized in Table 1 (Mortimer 2010).

Congenital onset		Postpubertal onset	
Familial	Sporadic	Familial	Sporadic
Milroy's disease	Turner's syndrome	Distichiasis-lymphedema	Distal hypoplasia
	Noonan's syndrome	Meige's disease	Lymph reflux
	Neurofibromatosis		Ilioinguinal node sclerosis
	Proteus syndrome		Yellow nail syndrome
	Pure or mixed vascular lymphatic malformations		
	Lymphangiomatosis		
	Klippel-Trenaunay syndrome		
	Mafucci's syndrome		
	Amniotic bands		

Table 1. Causes of primary lymphedema (Mortimer 2010)

2.2 Secondary lymphedema

Secondary lymphedema is swelling that follows some other incident or event, such as infection or injury (Tretbar 2008). The dysfunction of lymphatics in secondary lymphedema may be due to surgery, post-radiation fibrosis, infections, and primary or metastatic tumors, all causing lymphatic obstruction. Secondary lymphedema is much more frequent than primary (Planinšek Ručigaj T, 2010). The causes are summarized in Table 2 (Gaber 2009, Mortimer 2010).

In the Western world, surgery and radiation therapy for cancer (eg. breast and pelvic carcinoma, melanoma, head/neck cancer, Caposi sarcoma) are the leading cause of secondary lymphedema (Figures 2 and 3.) Lymphedema may appear as late as 30 years after the intervention. Due to increased cancer survival, the incidence of lymphedema is on the rise (Rockson 2001, Carpentier, 2002). Lymphatic tumors (e.g. lymphoma, sarcoma) may also cause lymphedema, however infrequently (Planinšek Ručigaj 2010).

Worldwide, venous insufficiency is a highly prevalent condition and is commonly accompanied by edema. It is assumed that venous edema is the sole consequence of increased capillary filtration from venous hypertension. As lymph drainage is the main buffer against edema, it is in fact the failure of local lymphatics to compensate for the

Tumour	Cancer surgery
	Radiotherapy
	Kaposi's sarcoma
	Infiltrative cancer
	Lymphoma
	Relapsed tumour
Infection	Filariasis
	Erysipelas, cellulitis, tuberculosis, lymphogranuloma inguinale, lice
Inflammation	Lymphatic occlusion: podoconiosis, pretibial myxedema, dermatitis (eg. hand eczema), rosacea
	Granulomatous disease: orofacial granulomatosis, Crohn disease, sarcoidosis
Vascular	Venous disease: postthrombotic syndrome, chronic venous insufficiency
Trauma	Surgery: lymphadenectomy, vein harvesting
	Self-harm: tourniquet application, intravenous drug abuse, Secretan syndrome
	Accident: degloving injury, burns
Functional	Static (prolonged standing), premenstrual, pregnancy
Medications	Oral contraceptives, psoralens, corticosteroids, antihypertensives, diuretics

Table 2. Causes of secondary lymphedema (Gaber 2009, Mortimer 2010)

increased lymph load from filtration that leads to edema. The small and precollecting lymphatics of the skin and subcutaneous tissues of the lower leg are damaged by prolonged venous hypertension. Lymphedema develops with chronic lipodermatosclerosis, with or without venous ulceration. Lymphedema associated with venous disease can give rise to the most gross swelling and skin changes owing to the combined effect of impaired lymph drainage and the increased lymph load (capillary filtration) (Mortimer 2010).

The leading cause of secondary lymphedema in the third world is filariasis, a worm (nematode) infection. Lymphatic filariasis is caused by *Wuchereria bancrofti*, *Brugia malayi*, or *Brugia timori*. The infection is transmitted by mosquitoes and other arthropodes. The adult worms reside in lymphatic channels or lymph nodes, where they cause lymphatic dilatation and thickening of the vessel walls with subsequent fibrosis, lymphatic obstruction and lymphedema (elephantiasis tropica) (Rockson 2001, Neese 2000).

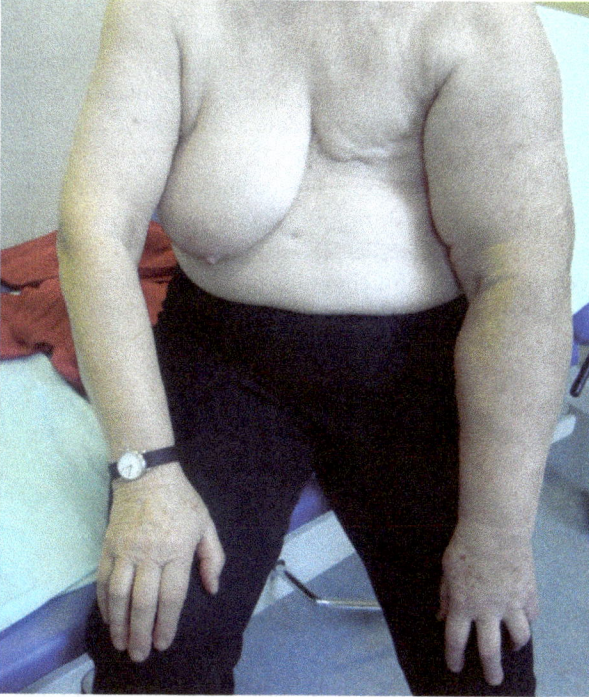

Fig. 2. Secondary lymphedema of the left arm after breast cancer surgery, including axillary lymphadenectomy.

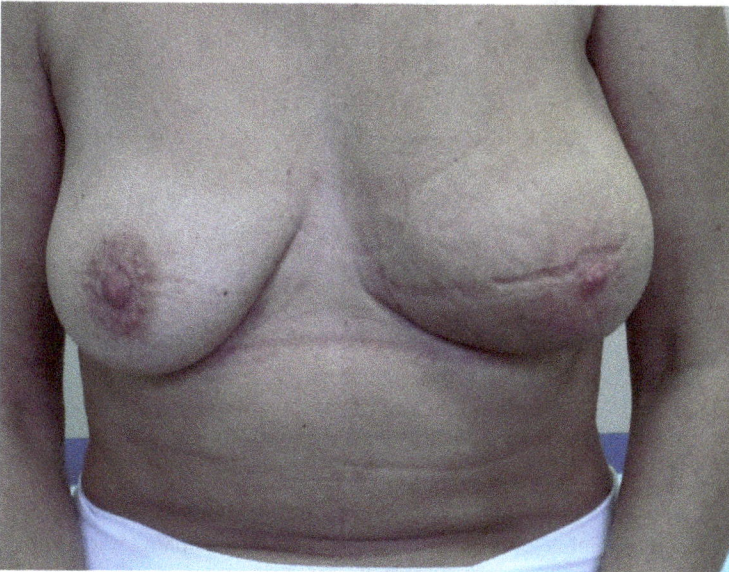

Fig. 3. Secondary lymphedema of the left breast after axillary lymphadenectomy.

3. Clinical picture and staging of lymphedema

Lymphedema differs from all other edemas (in which increased capillary filtration is the major factor) in that cells, proteins, lipids and debris accumulate in addition to water. This results in a 'solid' as well as a 'fluid' component to the swelling, so giving rise to the brawny nature of the edema which does not readily pit (Mortimer 1995). However, during the early stages, the swelling of lymphedema may be soft and pitting, therefore indistinguishable from other causes of oedema. Only during the later stages it develops the characteristics of 'true' lymphedema (Planinšek Ručigaj 2005).

The Stemmer-Kaposi sign is a helpful clinical sign in diagnosing lymphedema. It refers to the fact that, with lymphedema, the skin over the metatarsal-phalangeal joint of the second toe cannot be pinched up into a fold (Figure 4) (Stemmer 1976).

In lymphedema, skin creases become enhanced and hyperkeratosis develops. Dilatation of upper dermal lymphatics with consequent organization and fibrosis gives rise to papillomatosis. As dermal lymph stasis progresses, these skin changes become more marked and are referred to as elephantiasis. Occasionally, the tissue fibrosis and thickening may become so marked in the later stages of lymphedema that pitting is absent.

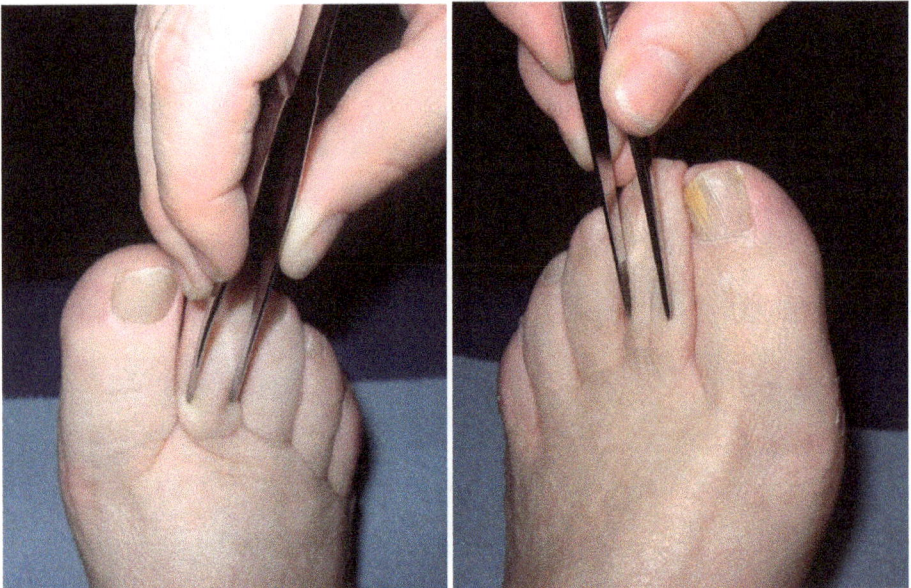

Fig. 4. The Stemmer-Kaposi sign. In lymphedema, the skin over metatarsal-phalangeal joint of the second toe can not be pinched into a fold (left, positive sign). In the absence of lymphedema (right), the skin can be readily pinched into a fold (negative sign).

Limb swelling leads to discomfort, limb heaviness, pain, reduced mobility and, on occasions, impaired function. The size and weight of affected limbs can result in secondary musculoskeletal complications such as back pain and joint problems. Thickening of the skin causes pseudoscleroderma and consequently impairs small-joint mobility. The difficulty in finding clothes or shoes to fit create social problems. Poor footwear will further compound the swelling by discouraging a normal gait or enough exercise. Leakage of lymph through the skin (lymphorrhoea) may occur from engorged dermal lymphatics (Mortimer 2010).

3.1 Staging of lymphedema

Regardless of etiology, lymphedema is clinically staged by the extent of visible tissue degradation (Moneta 2009).

Latent or subclinical (stage 0) lymphedema may persist for months to years without any clinical evidence of lymphatic disturbance. Trigger events, e.g. insect sting, physical exertion, injuries or surgery, inflammation or warming of the limb may cause some minor foot edema or even severe limb edema, which is either reversible or may, with additional lymphatic overload, proceed to the following stage. In stage I, the edema is reversible, soft, disappearing spontaneously overnight or, with compression therapy, during the day. The skin is smooth, with small pits. Stage I may persist for several years. However, if left untreated, it sooner or later proceeds to the chronical stage II.

During the stage II, edema persists despite limb elevation. In the early stage, edema is still pittng, later the edema is non-pitting, elastic. The skin feels harder, fibrotic. This is the phase can not be reversed spontaneously without therapy.

During the stage III, which is also called elephanthiasis, the edema is enormous. The skin shows trophic changes (fibrosis, hyperkeratoses, papillomatosis, hyperpigmentations, lymphorea, ulcerations) and is prone to bacterial and fungal infections. The condition may only partly improve with appropriate therapy. (International Society of Lymphology 2003, Cohen 2001, Rockson 2001).

Lymphedema may be further classified by the volume of edema (minimal, moderate, or severe); speed of progression (benign, malignant type); and localization (distal, proximal type).

3.2 Complications of lymphedema

Chronic lymphedema is often complicated by recurrent lymphangitis/cellulitis and in the long-standing disease, numerous neoplastic complications have been described (Szuba & Rockson, 1998).

The accumulated fluid and proteins serve as a perfect culture medium for bacterial growth. Impaired lymphatic drainage impedes the local immune response, which, in turn, promotes bacterial and fungal invasion (Mallon & Ryan, 1994). The infection further impairs lymphatic drainage and the aggravation of the edema usually persists after the infection resolves. With recurrent infections, there is progressive damage of the lymphatic capillaries (Bollinger 1993). The clinical picture may vary from acute attacks of a rapidly progressive infection (high fever, chills and general malaise, with localized edema, erythema and characteristic changes of *peau d'orange*) to a subclinical course with, at best, subtle skin changes and normal body temperature. Recurrent attacks of cellulitis damage existing cutaneous lymphatics, worsen skin changes and further aggravate existing edema. Acute attacks of cellulitis usually resolve quickly after antibiotic therapy but tend to reoccur, becoming more resistant to antibiotic therapy when they do. Prophylaxis against cellulitis includes meticulous skin care, avoidance of minor trauma and the prophylactic use of antibiotics (Olszewski & Jamal, 1996, VanScoy & Wilkowske 1992).

In rare cases, chronic lymphedema may be complicated by the development of malignant tumors within the involved limb. Malignant tumors of a lymphedematous extremity can evolve from lymphedema of any etiology: postsurgical, traumatic, filarial and primary, but the phenomenon is most often observed in postmastectomy edema of the arm, with a described frequency of 0.45% (Szuba & Rockson 1998). The best-known associated malignancy is lymphangiosarcoma, however, other tumours have been recorded and include basal cell carcinoma, lymphoma, melanoma, malignant fibrous histiocytoma and Kaposi's sarcoma (Mortimer 2010).

4. Diagnosis of lymphedema

Lymphedema can be surprisingly difficult to diagnose, especially in its early stages. Without a proper diagnosis, therapy is often delayed, allowing secondary fibrosis and lipid deposition to take place. Early treatment often results in rapid clinical improvement and prevents progression to the chronic phase of the disease (Szuba et al 2003).

The diagnosis of lymphedema requires careful history and clinical assessment, and exclusion of other causes of oedema.

History should include:

- Duration (acute vs. chronic) and previous therapy of edema;
- Pain, fatigue, paresthesias, mobility disturbances of the affected limb;
- Trigger events,
- Concurrent systemic signs/symptoms;
- Personal history of tumours, tumour therapy, postoperative complications;
- Personal history of injuries, surgery, infections, chronic venous insufficiency, deep venous thrombosis;
- Personal history of malabsorption syndrome, thyroid disease, cardiovascular disease, arterial hypertension, rheumatoid arthritis, transient obesity and lipedema, cerebrovascular disease, peripheral arterial disease, diabetes;
- Past and current medications, allergies;
- Occupation, hobbies;
- Family history.

Clinical assessment should include:

- Appearance of the skin (dry, flaky skin, fibrosis, papillomatosis, hyperkeratosis, hyperpigmentation, lymphorrhea, ulceration, warmth, erythema, pain of the affected area);
- Localization of edema;
- Assessment of the oedema (soft/hard or elastic/nonelastic);
- Comparison of limb diameters at several points (at the ankle, 10 cm above the upper and 10 cm below the lower margin of the patella);
- Stemmer-Kaposi sign – a diagnostic test that involves pinching of the skin on the upper surface of the second toe. In a negative result, it is possible to grasp a thin fold of tissue. In a positive result, which is characteristic of lymphedema, a plateau or a small pit is formed on the upper surface of the second toe when the tissue is pinched;
- Lymph node palpation;
- Mobility of the edematous limb;

- Presence of peripheral pulses, varicose veins;
- Inspection of feet and interdigital spaces, ingrowing nails,
- Systemic signs/symptoms (eg. fever, dyspnea etc.)

4.1 Diagnostic tests in lymphedema

In many cases of advanced sustained disease, a typical history and characteristic clinical presentation establish the diagnosis of lymphedema with near certainty. Nevertheless, additional tests are sometimes necessary to confirm the presence of impaired lymphatic flow and/or the typical pattern of abnormal fluid distribution within the tissues. The diagnosis is more difficult to ascertain in the early stages, particularly when edema is mild or intermittent (Szuba & Rockson 1998).

Lymphoscintigraphy is the golden standard of imaging in diagnosing lymphedema. It offers an objective and reliable approach to diagnose and characterize the severity of lymphedema (Szuba et al 2003). Other available tests include indirect and direct lymphography, lymphatic capillaroscopy, magnetic resonance imaging (MRI), axial tomography, and ultrasonography (Szuba & Rockson 1998). Magnetic resonance lymphangiography is a new, and also promising, imaging modality of the lymphatic system (Peller et al. 2009).

Lymphoscintigraphy, broadly described as an assessment of the lymphatic clearance of injected radioactive particles, was initially developed in the 1950s (Moneta 2009). Particles smaller than a few nanometers usually leak into blood capillaries, whereas larger particles (up to about 100 nm) can enter the lymphatic capillaries and be transported to lymph nodes (Moghimi & Bonnemain 1999). The optimal colloidal size for lymphoscintigraphy is believed to be approximately 50–70 nm (Strand & Bergqvist 1989). Technetium ($^{99}Tc^m$) labelled tracers such as sulphur colloid, microaggregated albumin and antimony sulphide are the most commonly used agents. In this minimally invasive procedure, intradermal or subcutaneous injection of the chosen radiolabeled tracer is performed, followed by subsequent gamma camera monitoring. Subfascial injection of the radiotracer may also be performed to analyse deep lymphatic system. The protocol for lymphoscintigraphy is not standardized and differs among diagnostic centers. Differences include the choice of radiotracer, the type and site of injection, the use of dynamic and static acquisition, and the acquisition times themselves. (Szuba 2003). A normal lymphoscintigram of the lower extremities is shown in Fig. 5, whereas Fig. 6 features a lymphoscintigraphic picture of lymphedema.

In extremity lymphedema assessment, a small volume of tracer (approximately 0.2 ml) is injected subcutaneously into the first to third web spaces on each hand or foot. Both limbs are examined, even if one appears normal, to detect sub-clinical abnormalities, provide internal comparison and monitor injection and camera technique. Images are acquired using a high-resolution, parallel hole, collimator. The arrival of tracer at the knees and groin (or elbows and axillae) is timed. A transmission scan using a flood source may be useful for anatomical localization. Quantitative parameters derived from tracer clearance data can be used to detect incipient lymphedema. After 30 min, if no groin (or axillary) activity is demonstrated, the patient is encouraged to mobilize/stress their limbs briefly. Lower extremity stress manoeuvres include walking, limb massage or bicycle exercise. For upper limb stressing, repetitive squeezing of a rubber ball, use of a hand-grip or massage can be used. They are then re-imaged to see if the proximal lymphatic system has been demonstrated. If there is still no activity, patients are encouraged to mobilize for several hours and delayed imaging is performed after 3–4 h (Scarsbrook et al 2007).

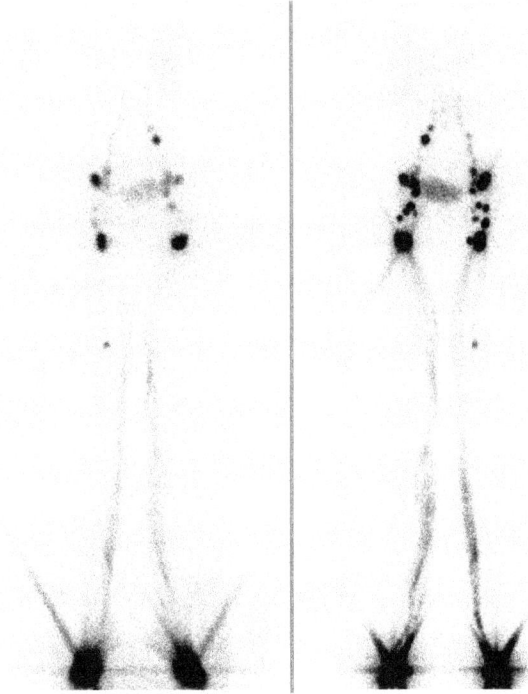

Fig. 5. Radioisotopic lymphography of the lower extremities in a patient with normally developed lymphatic system and normal lymphatic circulation (left, PA projection, right, AP projection). The radioindicator application site is seen as a large star-shaped activity in the lower part of the picture. Activity within the lymphatic vessels of the lower extremities can be observed, from the application site of the radioindicator on the dorsa of the feet, to the inguinal region, including several ingunal and abdominal lymph nodes. The lymphatic system of both extremities is virtually symmetrical. Image and interpretation courtesy of Marko Grmek.

5. Therapy of lymphedema

For decades, lymphedema has been thought by many clinicians to be non-curable and also largely untreatable. There now exist data supporting many types of successful treatment interventions, including medical and physical therapy (Gamble et al 2009). Surgery is not a preferred treatment modality, but may be useful in selected cases. However, at the moment, there is no consensus on the therapy of lymphedema. Therapy should be tailored individually to the patient's clinical situation, history, and any coexisting illnesses, e.g. cancer. The aims of therapy are to reduce edema, to prevent its further accumulation, and to prevent any complications (e.g. infections). Patient compliance is of crucial importance, making continuous patient education and encouragement essential parts of management. (Planinšek Ručigaj et al 2010) Therefore, it is central to the management of lymphedema that patients get to understand their condition and know what they can do for themselves. Only then can a high level of motivation and compliance with treatment be generated.

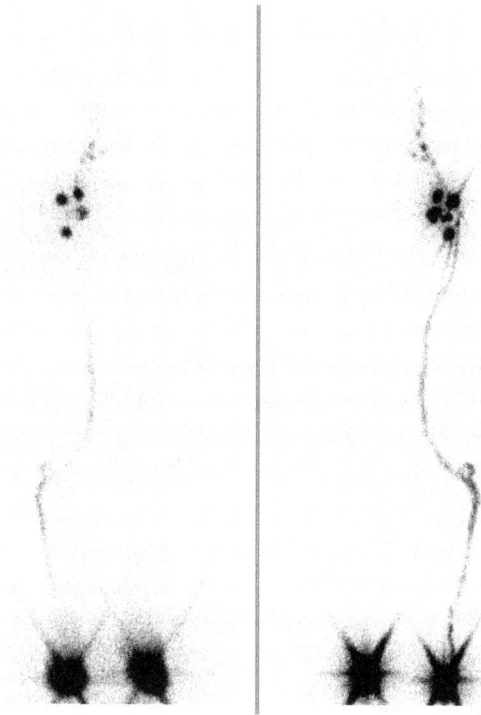

Fig. 6. Radioisotopic lymphography in a patient with absent lymphatic circulation in the right lower extremity (left, PA projection, right, AP projection). A star-shaped activity at the radioisotope application site is seen on both feet, and the activity in lymphatic vessels and nodes on the left side. There is no activity in the lymphatic vessels or lymph nodes of the right lower extremity. Image and interpretation courtesy of Marko Grmek.

5.1 Lymphedema risk factor management and risk reduction measures

Management of lymphedema should always include management of lymphedema risk factors and risk reduction measures (EWMA 2005, Planinšek Ručigaj 2005, Gamble et al 2009).

Other (accompanying) causes of edema must continue to be in the forefront of the clinician's thougts in order not to neglect potentially dangerous and reversible causes. Deep venous thrombosis, recurrent malignancy, and active infection must be considered and ruled out as potential causes prior to initial treatment, and periodically thereafter. Cardiopulmonary, renal, or hepatic dysfunction must also be considered (Gamble et al 2009).

Meticulous skin care is of utmost importance to avoid scaling and cracking, and minimizing the risk for secondary bacterial of fungal infection. The skin should be daily washed with soap and water, and treated with an emollient cream to keep the protective skin barrier. An individually tailored exercise regimen should be recommended. Patients after cancer surgery should be monitored for the development of lymphedema, and treated as soon as possible (Planinšek Ručigaj et al 2010).

The patients with lymphedema should avoid injury to the affected extremity, including injections and insect stings, and temperature extremes, whether hot or cold, which can create vascular challenge and increase lymphatic load. Tight consctriction and excessive overuse should also be avoided (Gamble et al 2009).

Increased body mass index (BMI) is a statistically significant predisposing factor for patients at risk for lymphedema as well (Soran et al 2010, Kizer et al 2011). Thus, a diet which maintains weight, or even promotes weight reduction, is encouraged to minimize development of edema (EWMA 2005, Gamble et al 2009).

5.2 Compression and physical therapy of lymphedema

The key treatment programme for lymphedema, known as decongestive lymphatic therapy, involves a combination of skin care, manual lymphatic drainage, inelastic multi-layer compression bandaging and exercise. Although this programme of care is well established, there is little understanding of how the different treatment components work or how to optimise the effects. Much of the evidence on how compression works is based on research into venous disease, which has been extrapolated to lymphedema (EWMA 2005).

Several studies have shown an impressive reduction in swelling as a result of compression (Badger et al 2004, McNeely 2004) but few have tried to elucidate the mechanism of action for this improvement. The following mechanisms may explain how compression reduces volume in a lymphedematous limb:

- reduction in capillary filtration
- shift of fluid into non-compressed parts of the body
- increase in lymphatic reabsorption and stimulation of lymphatic transport
- improvement in the venous pump in patients with veno-lymphatic dysfunction
- breakdown of fibrosclerotic tissue (EWMA 2005).

Compression therapy for lymphedema is based mainly on the use of elastic, short-stretch bandages, which has been clinically endorsed by only one randomised controlled trial that showed a consistently greater volume reduction in lymphedematous arms with short-stretch multi-layer bandages compared to hosiery (Badger 2000). The summary of therapeutic recommendations for lymphedema by stages is shown in Table 3.

Stage I	1. Evacuation of edema: long-stretch bandages/adhesive or nonadhesive short-stretch bandages and/or manual lymphatic drainage 2. Maintenance: medical compression stockings class II
Stage II	1. Evacuation of edema: adhesive or nonadhesive short-stretch bandages and/or manual lymphatic drainage 2. Maintenance: medical compression stockings class III/IV
Stage III	1. Evacuation of edema: adhesive or nonadhesive short-stretch bandages and/or manual lymphatic drainage 2. Maintenance: medical compression stockings class III/IV

Table 3. Recommended therapeutic interventions in lymphedema by stages (EWMA 2005).

6. Conclusion

Lymphedema is a chronic, progressive swelling of a part of the body, usually a limb. It may appear secondary to other diseases or as an isolated process. Lymphedema is most commonly seen on lower limbs, rarely on the upper limbs, and extremely rarely on the head or the trunk.

Ethiopathogenetically, lymphedema is classified into primary and secondary forms. Primary lymphedema may be hereditary or sporadic. Edema builds up due to lymphovascular or lymphonodular anomalies. In secondary lymphedema, which is much more common than the primary forms, the dysfunction of lymphatics is secondary due to other causes, such as infections, surgery, post-radiation fibrosis, and primary or metastatic tumors, all causing lymphatic obstruction. The leading cause of secondary lymphedema in the third world is filariasis, a worm (nematode) infection. Surgery and radiation therapy for cancer are the leading cause of secondary lymphedema in the Western world.

In some cases of lymphedema, particularly during the early stages, diagnostic tests may be necessary to confirm the lymphatic disturbance or point to the differential diagnosis. Lymphoscintigraphy (isotope lymphography) is the golden standard of imaging in diagnosing lymphedema. Lymphoscintigraphy involves the interstitial (dermis or subcutis) injection of a radio-labelled protein or colloid. Radioactivity, measured using a wide field-of-view gamma-camera, is determined over the injection site depot and at regions of interest over vessels or nodes. Measurement of transit times and time activity curves permit quantitative analysis of lymph drainage. Measurement of tracer uptake within axillary or inguinofemoral lymph nodes at a specified tme following a standardized exercise routine will discriminate lymphedema from edema of non-lymphatic origin.

At the moment, there is no consensus on the therapy of lymphedema. Therapy should be tailored individually to the patient's clinical situation, history, and any coexisting illnesses, e.g. cancer. The aims of therapy are to reduce edema, to prevent its further accumulation, and to prevent any complications (e.g. infections). Patient compliance is of crucial importance, making continuous patient education and encouragement essential parts of management. Edema should be reduced as early as possible, using compression therapy and/by manual lymph drainage. During improvement, compression stockings are required to maintain the improved condition.

7. Acknowledgment

The authors thank to Marko Grmek, MD, PhD, Department of Nuclear Medicine, University Medical Center Ljubljana, for kindly providing the images and interpretation of normal and abnormal lymphoscintigraphy of the lower extremities (Figures 5 and 6).

8. References

Badger CM, Peacock JL, Mortimer PS. A randomized, controlled, parallelgroup clinical trial comparing multilayer bandaging followed by hosiery versus hosiery alone in the treatment of patients with lymphedema of the limb. *Cancer* 2000; 88(12): 2832-37

Badger C, Preston N, Seers K, Mortimer P. Physical therapies for reducing and controlling lymphoedema of the limbs. Cochrane Database Syst Rev 2004; 4: CD003141

Bollinger A. Microlymphatics of human skin. *Int J Microcirc Clin Exp* 1993; 12(1): 1–15

Brice G, Mansour S, Bell R *et al.* Analysis of the phenotypic abnormalities in lymphoedema distichiasis syndrome in 74 patients with FOXC-2 mutations or linkage to 16q.24. *J Med Genet* 2002; (39) : 478–83

Browse NL, Stewart G. Lymphoedema: pathophysiology and classification. *J Cardiovasc Surg* 1985; (6) : 91–106

Browse NL. The diagnosis and management of primary lymphedema. *J Vasc Surg* 1996; (3): 181-4

Carpentier PH. Physiopathology of lymphedema. *Rev Med Interne* 2002; 23 (3): 371-374

Cohen SR et al. Lymphedema - Strategies for Management. Cancer 2001; 92 (4):980-987

European Wound Management Association (EWMA). Focus Document: *Lymphoedema bandaging in practice.* London: MEP Ltd, 2005. Available from: http://ewma.org/fileadmin/user_upload/EWMA/pdf/Position_Documents/200 5_Lymphoedema/English_focus_doc_05.pdf

Gaber Y. Diseases of Lymphatics. In: Burgdorf WHC, Plewig G, Wolff HH, Landthaler M, eds. *Braun-Falco's Dermatology.* Heidelberg: Springer, 2009: 930-5

Gamble GL, Cheville A, Strick D. Lymphedema: medical and physical therapy. . In: Gloviczki P, ed. *Handbook of venous disorders.* Guidelines of the American Venous Forum. London: Edward Arnold, 2009: 649-57

International Society of Lymphology. The diagnosis and Treatment of Peripheral Lymphedema. *Lymphology* 2003; 36: 84-91

Kizer NT, Thaker PH, Gao F, Zighelboim I, Powell MA, Rader JS, Mutch DG, Grigsby PW. The effects of body mass index on complications and survival outcomes in patients with cervical carcinoma undergoing curative chemoradiation therapy. *Cancer* 2011 Mar 1;117(5) :948-56

Mallon EC, Ryan TJ. Lymphedema and wound healing. *Clin Dermatol* 1994; 12(1): 89–93

McNeely ML, Magee DJ, Lees AW, et al. The addition of manual lymph drainage to compression therapy for breast cancer related lymphoedema: a randomized controlled trial. *Breast Cancer Res Treat* 2004; 86(2): 95-106.

Moghimi SM, Bonnemain B. Subcutaneous and intravenous delivery of diagnostic agents to the lymphatic system: applications in lymphoscintigraphy and indirect lymphography. *Adv Drug Deliv Rev* 1999;37:295–312

Moneta G, ed. Lymphedema. In: Gloviczki P, ed. *Handbook of venous disorders. Guidelines of the American Venous Forum.* London: Edward Arnold, 2009: 629-634

Mortimer PS. Pathophysiology of Lymphoedema. *Lymphplogy* 1998; 31: 3-6

Mortimer PS. Managing lymphoedema. *Clin Exp Dermatol* 1995; (20): 98–106

Mortimer PS. Disorders of Lymphatic vessels. In: Burns T, Breathnach S, Cox N, Griffiths C, editors. *Rook's Textbook of Dermatology.* Oxford: Wiley-Blackwell; 2010. p. 48.1-25

Neese PY,. Management of Lymphedema. *Lippincotts Prim Care Pract* 2000; 4(4): 390-399

Olszewski W, Jamal S. The role of antibiotics in control of progression of lymphedema. In: Network NL, ed. *The Second National Lymphedema Network Conference.* Burlingame, California, 1996

Peller PJ, Bender CE, Gloviczki P. Lymphoscintigraphy and lymphangiography. In: Gloviczki P, ed. *Handbook of venous disorders. Guidelines of the American Venous Forum.* London: Edward Arnold, 2009: 635-48

Philips T. Ulcers. In: Bolognia JL, Jorizzo LJ, Rapini RP, eds. *Dermatology.* St. Louis: Mosby/Elsevier, 2008: 1601-2

Planinšek Ručigaj T, Košiček M, Kozak M, Grmek M. Obravnava bolnikov z limfedemom. *Slikovne metode v odkrivanju in zdravljenju žilnih bolezni.* Združenje za žilne bolezni Slovenskega zdravniškega društva. Ljubljana 2005: 168-83

Planinšek Ručigaj T, Tlaker Žunter V, Miljković J. Compression therapy for lymphedema: our experience. *Acta Med Croatica* 2010 Jul;64(3):167-73

Rockson SG; Lymphedema. *Am J Med.* 2001; 110: 288-29

Scarsbrook AF, Ganeshan A, Bradley KM. Pearls and pitfalls of radionuclide imaging of the lymphatic system. Part 2: evaluation of extremity lymphoedema. *Br J Radiol* 2007 Mar; 80(951): 219-26

Soran A, Wu WC, Dirican A, Johnson R, Andacoglu O, Wilson J. Estimating the Probability of Lymphedema After Breast Cancer Surgery. *Am J Clin Oncol.* 2010 Nov 30

Stemmer R. Ein klinisches Zeichen zur Früh- und Differentialdiagnose des Lymphödems. *Vasa* 1976; (5) : 261–2

Strand SE, Bergqvist L. Radiolabeled colloids and macromolecules in the lymphatic system. *Crit Rev Ther Drug Carrier Syst.* 1989;6:211–238

Szuba A, Rockson G. Lymphedema: classification, diagnosis and therapy. *Vasc Med* 1998 3: 145

Szuba A, Shin WS, Strauss HW, Rockson S. The Third Circulation: Radionuclide Lymphoscintigraphy in the Evaluation of Lymphedema. *J Nucl Med* 2003 Jan;44(1):43-57

Tretbar LL, Morgan CL, Byung-Boong L, Simonian SJ, Blondeau B. *Lymphedema.* London: Springer, 2008: 12-31

Van Scoy RE, Wilkowske CJ. Prophylactic use of antimicrobial agents in adult patients. *Mayo Clin Proc* 1992; 67(3): 288–92

16

Targeting the Causes of Intractable Chronic Constipation in Children: The Nuclear Transit Study (NTS)

Yee Ian Yik[1,5,6] et al.[*]
[1]Department of Paediatrics, University of Melbourne,
[5]Murdoch Childrens Research Institute,
[6]Division of Paediatric Surgery, Department of General Surgery,
Faculty of Medicine, University of Malaya
[1,5]Australia
[6]Malaysia

1. Introduction

Chronic constipation is a symptom and not a disease! It is a common and major health problem affecting both adults and children. It is a difficult health issue for the patients and their treating physicians, with major psycho-socio-economic impacts. As the underlying cause varies, there is no standard therapy for chronic constipation. Many investigations have been used in the past, both invasive and non-invasive to study colonic physiology and pathophysiology to identify the underlying cause(s) for chronic constipation, but none of these tests have been able to provide reliable information for satisfactory treatment of this complex problem.

The nuclear transit study (NTS) has begun to gain acceptance in both adults and children in recent years. At our institute (The Royal Children's Hospital, Melbourne, Victoria, Australia), nuclear transit studies were initially used to investigate total colonic transit time and to define the site where slowing occurs in children with intractable chronic constipation (Cook et al. 2005), leading to the description of slow-transit constipation (STC) in children in 1998 (Southwell et al. 2009). In addition to colonic transit, gastric emptying and small bowel transit can be characterised producing a complete picture of gastrointestinal tract dynamics. This study reviews NTS collected over 12 years (1999-2010) at our tertiary children's hospital to identify sites of hold-up or delay in children with chronic constipation not responding to medical treatment.

[*] David J. Cook[2], Duncan M. Veysey[2], Stephen J. Rutkowski[4], Coral F. Tudball[2], Brooke S. King[2], Timothy M. Cain[2], Bridget R. Southwell[5] and John M. Hutson[1,3,5]
[1]Department of Paediatrics, University of Melbourne, Australia
[2]Department of Medical Imaging, Royal Children's Hospital, Melbourne, Australia
[3]Department of General Surgery, Royal Children's Hospital, Melbourne, Australia
[4]Australian Radiation Services Pty. Ltd., Australia
[5]Murdoch Childrens Research Institute, Australia

2. Intractable chronic constipation

We defined intractable chronic constipation in children as chronic constipation with duration of symptoms > 2 years, not responding to maximal laxative therapy, behavioural therapy or a toilet training program. Most children had consulted multiple paediatricians and gastroenterologists and suffered persisting symptoms by the time they entered our program. Some of the children had been subjected to surgical interventions as a last attempt to improve their symptoms and quality of life. The majority of children with chronic constipation with a palpable faecoloma are diagnosed as having anorectal retention (AR) and are treated conservatively (laxatives, diet therapy and/or toilet training program). The most severe form of chronic constipation is slow-transit constipation (STC) and this does not respond to most therapies. We have employed a non-invasive form of therapy using transcutaneous electrical stimulation (TES), and we have successful outcomes in STC (Chase et al. 2005; Clarke et al. 2009; Clarke et al. 2009; Ismail et al. 2009). Also, we have identified a subgroup of children with rapid proximal colonic transit and AR, and found that dietary exclusion could be an important treatment strategy for this form of intractable chronic constipation (Yik et al. in press).

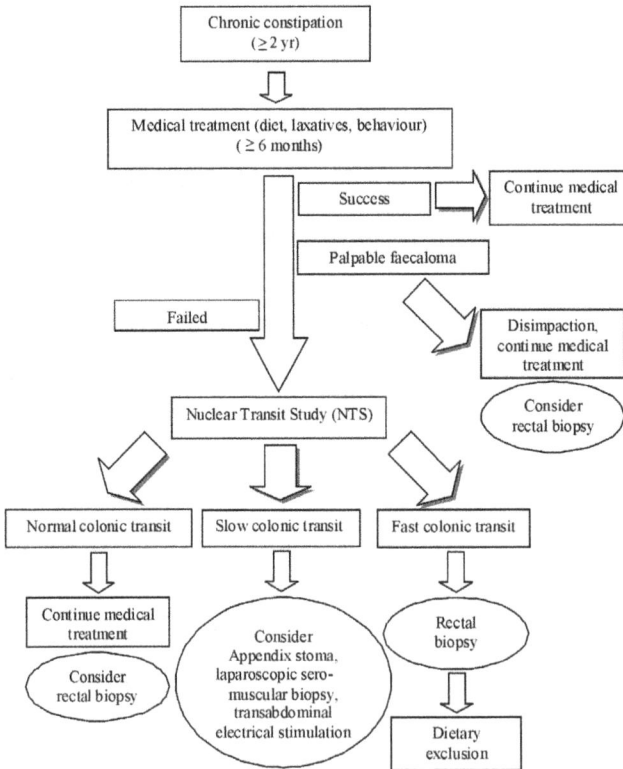

Fig. 1. Our current algorithm for the management of children with treatment-resistant constipation

The algorithm adopted in the management of children with intractable chronic constipation at our hospital is shown in Figure 1.

3. Gastrointestinal transit study

The NTS has evolved from plastic (sitz) marker studies (radio-opaque markers and X-rays), which provide a "snapshot" from which colonic transit time can be inferred. The NTS provides additional information on gastric emptying and small bowel transit, in addition to colonic transit, which permits further characterization of transit through the whole of the gastrointestinal tract. This feature, in addition to its reproducibility (Clarke et al. 2009) and the physiological characteristics of a real (labelled) meal, has led to its increasing clinical acceptance as a means for monitoring colonic dysmotility and its response to treatment. It is especially useful in monitoring the more severe form of proximal colonic dysmotility in children leading to intractable chronic constipation, i.e. slow-transit constipation (STC).

3.1 Test procedure and technique (Royal Children's Hospital, Melbourne, Victoria, Australia)

All patients are instructed to fast for a period of 4 hours prior to the study (3 hours for infants). They are to cease all medications affecting gastric motility and colonic motility (e.g. laxatives) for 5 days prior to, and for the duration of the study. After the gastric emptying phase (2 hours), patients resume their normal diet.

The test involves administration (orally, NG Tube or PEG) of a radiopharmaceutical (gallium-67 citrate) in 10-400mls labelled full cream milk (or standard formula), with the radiopharmaceutical dose calculated according to the patient's weight (dose range of 3 - 10MBq). The dose is a calculated fraction of an adult dose (10MBq), with a minimum of 3MBq (refer Table 1).

Weight (kg)	Fraction of Adult dose
10	0.30
15	0.30
20	0.50
30	0.58
40	0.70
50	0.81
60	0.89
70	0.97
80	1.00

Table 1. Example weight and dose calculation for 67-Ga Citrate for children undergoing gastrointestinal transit study (minimum dose 3 MBq)

Effective dose as a function of colonic transit time for adult patients

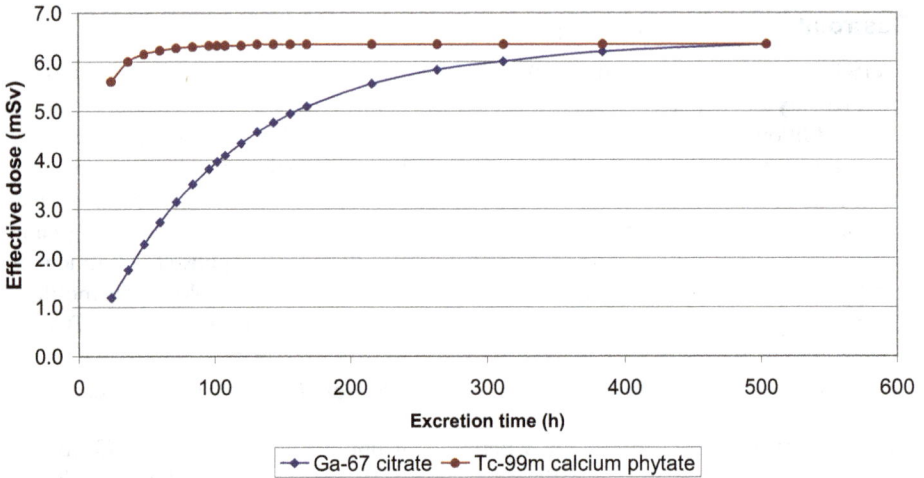

Fig. 2. Effective dose (mSv) as a function of colonic transit time (hours) for an adult-sized patient based on full excretion. The relationship between effective dose and transit is similar for children, although the magnitude differs (refer Table 2)

Gallium-67 citrate effective dose (mSv)								
Age	Weight (kg)	Administration activity (MBq)	Transit time (full excretion) (hr)					
			24	36	48	72	96	120
3	14	3	1.5	2.2	2.8	3.9	4.7	5.4
5	19	5	1.9	2.8	3.6	5.0	6.0	6.9
10	32	6	1.4	2.1	2.7	3.8	4.6	5.2
15	55	8	1.3	1.9	2.5	3.4	4.2	4.7
Adult	>71	10	1.2	1.8	2.3	3.1	3.8	4.3
Technetium-99m calcium phytate colloid effective dose (mSv)								
Age	Weight (kg)	Administration activity (MBq)	Transit time (full excretion) (hr)					
			24	36	48	72	96	120
3	14	111	10.3	11.0	11.3	11.5	11.6	11.6
5	19	111	7.9	8.4	8.6	8.8	8.9	8.9
10	32	150	6.7	7.2	7.4	7.5	7.6	7.6
15	55	211	6.1	6.5	6.7	6.8	6.9	6.9
Adult	>71	250	5.6	6.0	6.2	6.3	6.3	6.3

Table 2. Effective dose (mSv) for children undergoing NTS [effective dose is dependent on dose administered and total bowel transit time (hrs)]

Technetium-99m calcium phytate colloid was used prior to 2000 and gallium-67 citrate used post-2000 due to the preferential radiation dosimetry. The effective radiation dose to the patient has been calculated for both technetium-99m calcium phytate colloid and gallium-67 citrate as a function of patient's weight and transit time (Figure 2 and Table 2). The effective dose was calculated using MIRD methodology with standard phantom sizes (except for the 3 year-old phantom which was interpolated from the 1 and 5 year-old phantoms) with varying transit time through the gastrointestinal tract. Gallium-67 citrate delivers a lower radiation dose than technetium-99m calcium phytate, particularly for short transit times, with less than half the dose if transit takes < 100 hours (Figure 2). Due to the longer half-life of gallium-67 citrate, the effective dose continues to increase for patients with long transit times. For transit times greater than 500 hrs (approximately 21 days), the effective dose is similar for technetium-99m calcium phytate and gallium-67 citrate.

Serial static anterior and posterior images are obtained at T_0, $T_{0+30 \text{ mins}}$, $T_{0+60 \text{mins}}$, $T_{0+120 \text{ mins}}$ for the gastric emptying study, and further static images at $T_{0+6 \text{ hours}}$, $T_{0+24 \text{ hours}}$, $T_{0+30 \text{ hours}}$ and $T_{0+48 \text{hours}}$ for the small bowel and colonic transit study. All images are acquired using opposed dual head detectors (anterior/posterior), with medium energy, general purpose (MEGP) collimation (using a Philips "SKYLight" or Philips "BrightView" dual-head scintillation camera). All imaging acquisition parameters, including distance of the detectors from the patient, and from each other, are recorded and are strictly maintained throughout the study. Bowel motion(s) are recorded during the period of the study to correlate with excreted activity.

Following the completion of imaging, the laxative or other regime in place prior to the investigation is reinstituted. If, following the 48 hour images, there is significant colonic retention of activity, bowel disimpaction[1] can be undertaken in the interests of radiation dose minimisation.

3.2 Gastric emptying time

Gastric half–emptying time is calculated using background and decay-corrected regions of interest (ROI) drawn around the stomach in T_0, $T_{0+30 \text{ mins}}$, $T_{0+60 \text{mins}}$, $T_{0+120 \text{ mins}}$ images. T_0 and $T_{0+120 \text{ mins}}$ are used to determine the two hour stomach retention.

3.3 Geometric centre (GC) measurement

The geometric centre (GC) at each imaging time is determined for six ROI, assigned as: 1=pre-colonic; 2=caecum/ascending colon, 3=transverse colon; 4=descending colon; 5=recto-sigmoid colon and 6=excreted.

ROI are drawn on each image as illustrated in Figure 3 (below), and counts are corrected for decay and background. Geometric mean counts are calculated from the corrected data and activity in each region is expressed as a fraction of total activity [see example below of calculation steps (Figure 4)]:

---

[1] Suggested protocol for bowel disimpaction at the end of NTS using Movicol® (macrogol with electrolytes): 1. 1-5 yrs: 2 sachets on day 1, then 4 sachets for 2 days, then 6 sachets for 2 days and 8 sachets daily thereafter; 2. 5-12 yrs: 4 sachets on day 1, then increase by 2 sachets daily until max of 12 sachets daily; 3. >12 yrs: 8 sachets per day for 3 days; if Movicol® is not tolerated, use Lactulose® and Senna® for disimpaction

Geometric Centre (GC) of activity = $\sum$ (fraction of activity x region number) (Notghi et al. 1994)

Example for ROI are as follows:

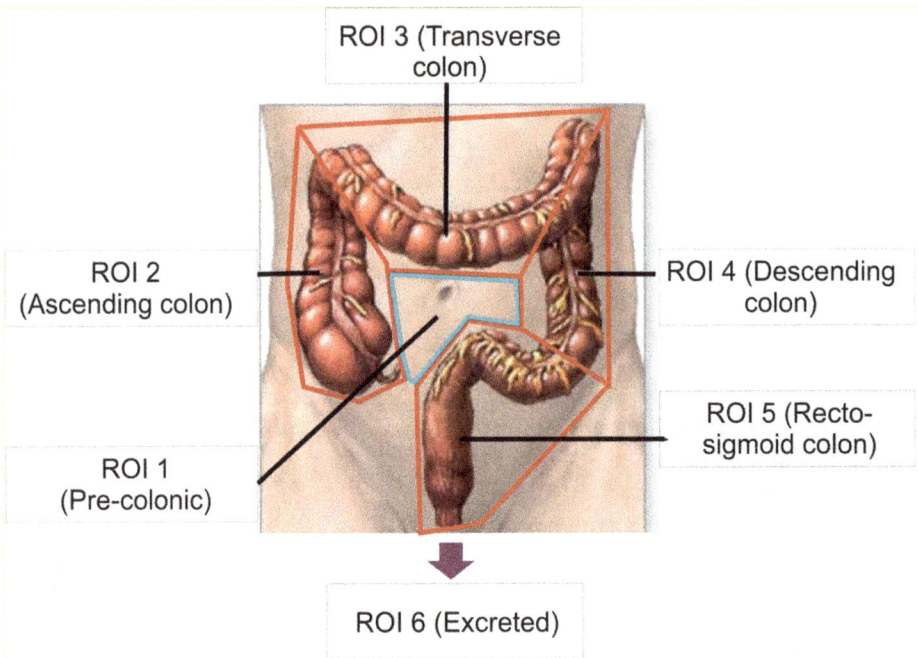

Fig. 3. Regions of interest (ROI) used for GC calculation; 1 = pre-colonic (total image activity less colonic activity after background activity correction), 2 = ascending colon, 3 = transverse colon, 4 = descending colon, 5 = recto-sigmoid colon, 6 = excreted.

A

RAW DATA

			EXPECTED COUNTS(FROM T0):	PIXELS
ANTERIOR COUNTS			ANTERIOR	138350	128
			POSTERIOR	75592	128

HRS + T0	ASC CTS	PIXELS	TRANS CTS	PIXELS	DESC CTS	PIXELS	REC/SIG CTS	PIXELS	BGD CTS	PIXELS	TOTAL CTS
6.00	94457.0	665.0	0.0	0.0	0.0	0.0	0.0	0.0	54.0	25.0	131189.0
24.00	16393.0	600.0	21144.0	679.0	31912.0	859.0	19914.0	613.0	78.0	25.0	101617.0
30.00	17618.0	530.0	14506.0	614.0	6199.0	540.0	40732.0	587.0	45.0	25.0	95856.0
48.00	0.0	0.0	3502.0	839.0	5335.0	680.0	19942.0	944.0	28.0	25.0	40463.0

POSTERIOR COUNTS

HRS + T0	ASC CTS	PIXELS	TRANS CTS	PIXELS	DESC CTS	PIXELS	REC/SIG CTS	PIXELS	BGD CTS	PIXELS	TOTAL CTS
6.00	37870.0	665.0	0.0	0.0	0.0	0.0	0.0	0.0	62.0	25.0	71680.0
24.00	11774.0	600.0	12260.0	679.0	22168.0	859.0	12478.0	613.0	67.0	25.0	74110.0
30.00	11855.0	530.0	7632.0	614.0	4571.0	540.0	25480.0	587.0	41.0	25.0	67455.0
48.00	0.0	0.0	2764.0	839.0	4246.0	680.0	12532.0	944.0	41.0	25.0	31749.0

ANTERIOR PIXEL CORRECTION

HRS + T0	ASC	TRANS	DESC	REC/SIG	BGD	TOT CTS
6.00	142.0	0	0	0	2.2	8.0
24.00	27.3	31.1	37.2	32.5	3.1	6.2
30.00	33.2	23.6	11.5	69.4	1.8	5.9
48.00	0	4.2	7.8	21.1	1.1	2.5

POSTERIOR PIXEL CORRECTION

HRS + T0	ASC	TRANS	DESC	REC/SIG	BGD	TOT CTS
6.00	56.9	0	0	0	2.5	4.4
24.00	19.6	18.1	25.8	20.4	2.7	4.5
30.00	22.4	12.4	8.5	43.4	1.6	4.1
48.00	0	3.3	6.2	13.3	1.6	1.9

ANTERIOR BGD CORRECTION

HRS + T0	ASC	TRANS	DESC	REC/SIG		TOT CTS		EXP CTS
6.00	93020.6	0.0	0.0	0.0		131189.0		138349.5
24.00	14521.0	19025.5	29231.9	18001.4		101617.0		
30.00	16664.0	13400.8	5227.0	39675.4		95856.0		
48.00	0.0	2562.3	4573.4	18884.7		40463.0		

POSTERIOR BGD CORRECTION

HRS + T0	ASC	TRANS	DESC	REC/SIG		TOT CTS		EXP CTS
6.00	36220.8	0.0	0.0	0.0		71680.0		75592.4
24.00	10166.0	10440.3	19865.9	10835.2		74110.0		
30.00	10985.8	6625.0	3685.4	24517.3		67455.0		
48.00	0.0	1388.0	3130.8	10983.8		31749.0		

B

GEOMETRIC MEAN CALCULATION

HRS + T0	ASC	TRANS	DESC	REC/SIG		TOT CTS		EXP CTS
6.00	58045.5	0.0	0.0	0.0		96972.3		102719.1
24.00	12149.9	14093.7	24098.1	13966.0		86780.4		
30.00	13530.2	9422.4	4389.0	31188.7		80411.2		
48.00	0.0	1885.9	3784.0	14402.3		35842.2		

C

DECAY CORRECTION

HRS + T0	ASC	TRANS	DESC	REC/SIG		TOT CTS	EXC CTS		EXP CTS
6.00	61213.7	0.0	0.0	0.0		102265.2	41505.4		102719.1
24.00	15027.8	17431.9	29806.0	17274.0		107335.3	0.0		
30.00	17648.5	12290.3	5724.9	40681.6		104886.1	0.0		
48.00	0.0	2885.1	5788.8	22033.0		54832.2	47886.9		

D

HRS	SMALL BOWEL	ASC	TRANS	DESC	REC/SIG		EXCRETED		CHECK
6.00	40%	60%	0%	0%	0%		0%		100%
24.00	0%	19%	22%	37%	22%		0%		100%
30.00	0%	23%	16%	7%	53%		0%		100%
48.00	0%	0%	4%	7%	28%		61%		100%

E

HRS + T0	GEOMETRIC CENTRE CALCULATION
6.00	1.60
24.00	3.62
30.00	3.91
48.00	5.46
INDEX =	**14.59**

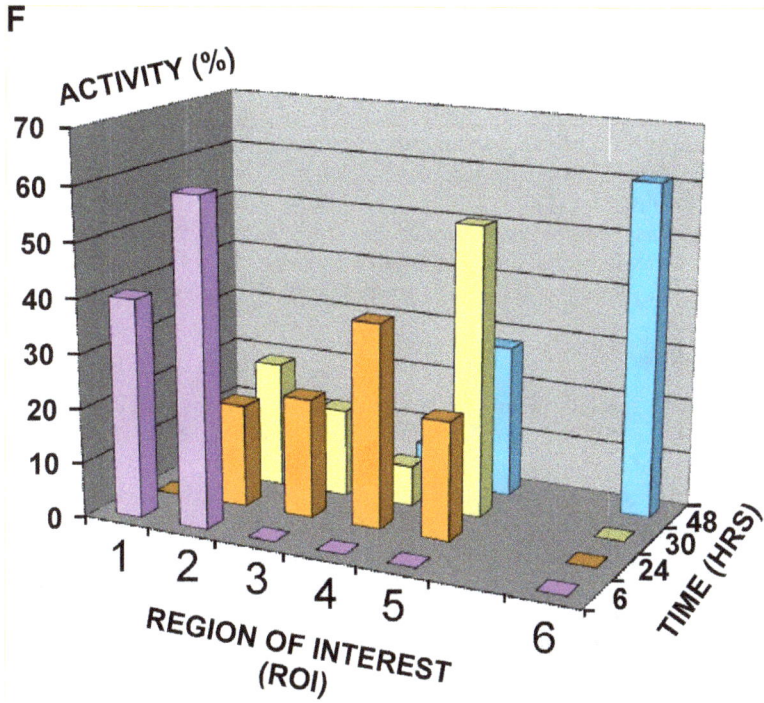

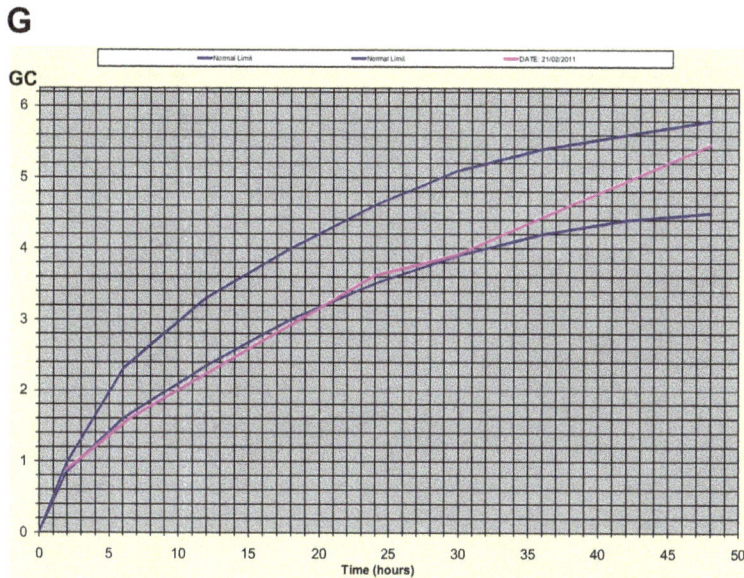

Fig. 4. Example of GC calculation steps - based on ROI at each time interval.

A - Background correction is applied, based on counts/pixel for each ROI

B - Geometric mean calculated from background corrected counts for both anterior and posterior

C - Geometric mean counts corrected for time (decay corrected)

D - Counts expressed as a % of total counts for each time point

E - Geometric centre (GC) calculation for each time point [e.g. GC calculation at 24 hour = $(0.00x1)+(0.19x2)+(0.22x3)+(0.37x4)+(0.22x5)+(0.00x6) = 0.38+0.66+1.48+1.10 = 3.62$] and the GC Index is the sum of all GC's

F - Histograms displaying the percentage of the ingested meal present in the pre-colonic, each colonic segment and excreted at each time of imaging

G - Patient's GC data plotted vs time (with normal range referenced)

4. Groups identified

Since 1998, 770 NTS have been undertaken in 626 children, investigating the underlying colonic dysmotility contributing to their chronic constipation. Four (4) main groups were identified based on rates of segmental colonic transit:

1. Normal colonic transit (>40% excretion at 48 hour) with no evidence of colonic dysmotility (Figure 5A) (Southwell et al. 2009; Sutcliffe et al. 2009)
2. Anorectal retention (AR, Figure 5B) (Southwell et al. 2009; Sutcliffe et al. 2009)
3. Rapid proximal colonic transit (rapid transit through proximal colon at 6 and 24 hr, Figure 5C) (Yik et al. in press)
4. Slow proximal colonic transit (Sutcliffe et al. 2009)
   a. Slow proximal colonic transit (slow-transit constipation, STC, Figure 5D)
   b. Focal hold-up (Figure 5E)
   c. STC with AR (Figure 5F, common in adults, may need additional imaging at 72 hr for confirmation in paediatrics)

Gastric emptying can be normal or delayed (NGE or DGE). Normal gastric emptying time is defined as $T_{50\%} \leq 50$ mins and % retention at 2 hr < 15% (Yik et al. 2011).

Likewise, small bowel transit can be normal (NSBT) or slow (SSBT). Normally the small bowel is empty by 4 hours and slow small bowel transit is defined as > 25% of radiopharmaceutical remaining within the small bowel at 6 hr (Yik et al. 2011).

The subgroup of children with rapid proximal colonic transit are identifiable at 6 hour and 24 hr with > 25% of the ingested meal having passed beyond hepatic flexure and distal descending colon respectively (Figure 5C) (Yik et al. in press).

5. Clinical importance

Treatment strategies differ for different causes of colonic dysmotility. Anorectal retention usually responds to a toilet training program, behavioural therapy and laxative therapy. For children with rapid proximal colonic transit and associated anorectal retention, dietary exclusion is worth considering with a reasonable probability of success (Yik et al. in press). However, this may take time (usually months and not weeks) before improvement becomes apparent because the rectum is chronically distended. The time may be needed to recover receptor function to sense filling with faecal matter. The most difficult to manage is the group with slow-transit constipation (STC). STC does not respond to multiple laxative therapies, and a high-fibre diet is not an appropriate option as it may aggravate symptoms

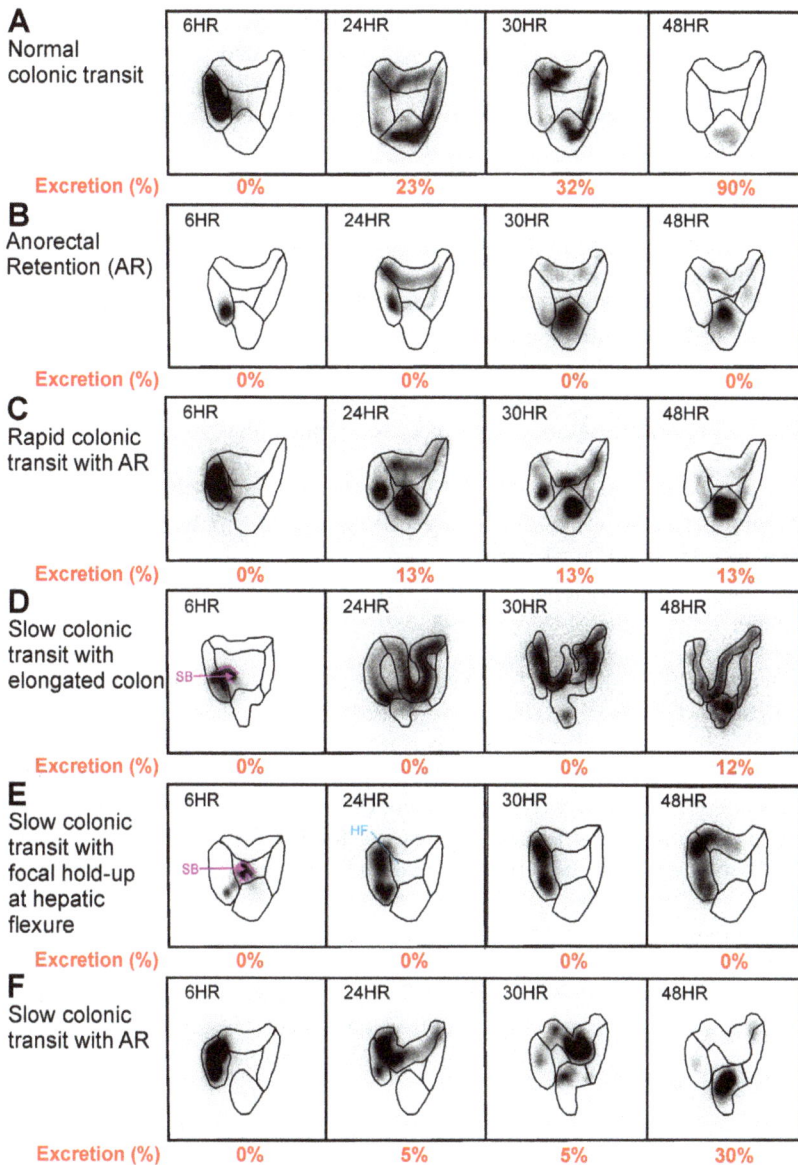

Fig. 5. Colonic transit patterns in children:

A – Normal colonic transit with normal excretion;

B – Anorectal retention (AR) with retention of activity in the rectum at 48 hr;

C – Rapid proximal colonic transit with anorectal retention at 48 hr;

D – Slow colonic transit [with an unusually elongated colon (Southwell 2010)];

E – Slow colonic transit with focal hold-up at hepatic flexure, and

F – Slow colonic transit with AR (may be difficult to identify as this may required imaging at 72 hr to diagnose in adults). [HF = Hepatic flexure; SB = small bowel]

(Voderholzer et al. 1997; Knowles & Martin 2000; Wald 2002; Hutson et al. 2009). Invasive surgical interventions may have complications with associated morbidities and mortality. High rates of symptom recurrence have been reported in this group (Kamm et al. 1988; Yoshioka & Keighley 1989; de Graaf et al. 1996; Lubowski et al. 1996; Mollen et al. 2001; Sample et al. 2005). A less invasive surgical option - antegrade continence enema (ACE) - was introduced in 1990 (Malone et al. 1990). This has helped many children with faecal incontinence and/or impaction. We have employed appendix stoma/ACE in STC children since 1997 (Stanton et al. 2002; King et al. 2005). Since 2006, we have used transcutaneous electrical stimulation (TES) therapy to treat this special group of children and have shown promising results (Chase et al. 2005; Clarke et al. 2009; Clarke et al. 2009; Ismail et al. 2009). TES has not only improved colonic transit in these children (Clarke et al. 2009) but also alleviated their soiling, feelings of chronic fullness and abdominal discomfort which has improved their quality of life (Clarke et al. 2009; Ismail et al. 2009). The use of TES therapy has significantly reduced the need for appendix stoma formation in STC children in our institute (unpublished).

The success of therapy largely depends on accurate identification of the underlying colonic dsymotility, which is aided by the use of NTS at our institute. The NTS helps triage patients into aetiological groups where appropriate treatment strategies can be instituted. The positive psycho-social impact on families previously labelled "dysfunctional", when an explanation for their child's problem is suggested, cannot be overestimated.

Significantly, identifying the hold-up site may guide the site for biopsies to look for underlying pathology (King et al. 2005), as opposed to the standard practice of investigating children with chronic constipation by performing rectal mucosal biopsy to exclude Hirschsprung disease. Colonic seromuscular biopsies (collected laparoscopically) may be more useful for children with dysmotility proximal to the anorectum. However, rectal mucosal biopsy can be useful for the diagnosis of food intolerance/allergy if eosinophilia and mast cells are demonstrated (Carroccio et al. 2000; Daher et al. 2001; Carroccio & Iacono 2006; Scaillon & Cadranel 2006).

Some centres advocate the identification of "the site of slowing in the colon" to decide on the part/segment of the colon to be resected in patients with intractable symptoms, desperate for a cure (Youssef et al. 2004). This has important implications because patients with pancolonic slow transit will not benefit from a segmental resection as the risk of relapse of symptoms is much higher. In this situation, NTS may have a crucial role in surgical planning (CB et al. 2001).

6. Limitations

We are currently performing NTS using a 48-hour protocol which does not encompass the total colonic transit time in some children with a severe form of STC. However, we feel that by 48 hours, we are able to characterise most colonic dysmotility permitting us to treat associated symptoms with appropriate strategies. However, using the 48-hour protocol, we will miss AR if it occurs in these patients.

Another limitation is the need for the child (and parent) to attend the hospital Nuclear Medicine Department over 2 ½ days while the serial images are acquired. It is perhaps a telling indication of the social disruption that results from chronic constipation, that in our series, we have not encountered a single family unwilling to attend for this investigation, despite the time involved and the inconvenience.

7. Future directions

A standardised/nationwide protocol for NTS in children and standardised diagnostic criteria would allow more appropriate treatment strategies to help all children with this debilitating condition.

8. Conclusion

NTS is a useful diagnostic tool for the investigation of colonic dysmotility in children with intractable chronic constipation, and provides objective, reproducible and quantifiable data, to monitor the progression of the condition, and allow clinicians to decide on the "best treatment" strategy and response to therapy.

9. Acknowledgment

We would like to extend our sincere thanks to Mrs. Mita Pedersen and Ms. Pam Grant for all their assistance in tracing the patients' data. Most importantly, we would like to thank all the children with intractable chronic constipation and their parents for encouraging us to work harder to seek the answers for their long-standing, previously untreatable condition and to become well again.

10. References

Carroccio, A.; F. Cavataio; G. Montalto; D. D'Amico; L. Alabrese & G. Iacono (2000). Intolerance to hydrolysed cow's milk proteins in infants: clinical characteristics and dietary treatment. *Clin Exp Allergy*, Vol. 30, No. 11, (Nov 2000), pp. (1597-603),0954-7894

Carroccio, A. & G. Iacono (2006). Review article: Chronic constipation and food hypersensitivity--an intriguing relationship. *Aliment Pharmacol Ther*, Vol. 24, No. 9, (Nov 1 2006), pp. (1295-304),0269-2813

CB, O. S.; J. H. Anderson; R. F. McKee & I. G. Finlay (2001). Strategy for the surgical management of patients with idiopathic megarectum and megacolon. *Br J Surg*, Vol. 88, No. 10, (Oct 2001), pp. (1392-6),0007-1323

Chase, J.; V. J. Robertson; B. Southwell; J. Hutson & S. Gibb (2005). Pilot study using transcutaneous electrical stimulation (interferential current) to treat chronic treatment-resistant constipation and soiling in children. *J Gastroenterol Hepatol*, Vol. 20, No. 7, (Jul 2005), pp. (1054-61),0815-9319

Clarke, M. C.; J. W. Chase; S. Gibb; A. G. Catto-Smith; J. M. Hutson & B. R. Southwell (2009). Standard medical therapies do not alter colonic transit time in children with treatment-resistant slow-transit constipation. *Pediatr Surg Int*, Vol. 25, No. 6, (Jun 2009), pp. (473-8),1437-9813

Clarke, M. C.; J. W. Chase; S. Gibb; J. M. Hutson & B. R. Southwell (2009). Improvement of quality of life in children with slow transit constipation after treatment with transcutaneous electrical stimulation. *J Pediatr Surg*, Vol. 44, No. 6, (Jun 2009), pp. (1268-72; discussion 1272),1531-5037

Clarke, M. C.; J. W. Chase; S. Gibb; V. J. Robertson; A. Catto-Smith; J. M. Hutson & B. R. Southwell (2009). Decreased colonic transit time after transcutaneous interferential electrical stimulation in children with slow transit constipation. *J Pediatr Surg*, Vol. 44, No. 2, (Feb 2009), pp. (408-12),1531-5037

Cook, B. J.; E. Lim; D. Cook; J. Hughes; C. W. Chow; M. P. Stanton; S. S. Bidarkar; B. R. Southwell & J. M. Hutson (2005). Radionuclear transit to assess sites of delay in large bowel transit in children with chronic idiopathic constipation. *J Pediatr Surg*, Vol. 40, No. 3, (Mar 2005), pp. (478-83),1531-5037

Daher, S.; S. Tahan; D. Sole; C. K. Naspitz; F. R. Da Silva Patricio; U. F. Neto & M. B. De Morais (2001). Cow's milk protein intolerance and chronic constipation in children. *Pediatr Allergy Immunol*, Vol. 12, No. 6, (Dec 2001), pp. (339-42),0905-6157

de Graaf, E. J.; E. C. Gilberts & W. R. Schouten (1996). Role of segmental colonic transit time studies to select patients with slow transit constipation for partial left-sided or subtotal colectomy. *Br J Surg*, Vol. 83, No. 5, (May 1996), pp. (648-51),0007-1323

Hutson, J. M.; J. W. Chase; M. C. Clarke; S. K. King; J. Sutcliffe; S. Gibb; A. G. Catto-Smith; V. J. Robertson & B. R. Southwell (2009). Slow-transit constipation in children: our experience. *Pediatr Surg Int*, Vol. 25, No. 5, (May 2009), pp. (403-6),1437-9813

Ismail, K. A.; J. Chase; S. Gibb; M. Clarke; A. G. Catto-Smith; V. J. Robertson; J. M. Hutson & B. R. Southwell (2009). Daily transabdominal electrical stimulation at home increased defecation in children with slow-transit constipation: a pilot study. *J Pediatr Surg*, Vol. 44, No. 12, (Dec 2009), pp. (2388-92),1531-5037

Kamm, M. A.; P. R. Hawley & J. E. Lennard-Jones (1988). Outcome of colectomy for severe idiopathic constipation. *Gut*, Vol. 29, No. 7, (Jul 1988), pp. (969-73),0017-5749

King, S. K.; J. R. Sutcliffe & J. M. Hutson (2005). Laparoscopic seromuscular colonic biopsies: a surgeon's experience. *J Pediatr Surg*, Vol. 40, No. 2, (Feb 2005), pp. (381-4),1531-5037

King, S. K.; J. R. Sutcliffe; B. R. Southwell; P. G. Chait & J. M. Hutson (2005). The antegrade continence enema successfully treats idiopathic slow-transit constipation. *J Pediatr Surg*, Vol. 40, No. 12, (Dec 2005), pp. (1935-40),1531-5037

Knowles, C. H. & J. E. Martin (2000). Slow transit constipation: a model of human gut dysmotility. Review of possible aetiologies. *Neurogastroenterol Motil*, Vol. 12, No. 2, (Apr 2000), pp. (181-96),1350-1925

Lubowski, D. Z.; F. C. Chen; M. L. Kennedy & D. W. King (1996). Results of colectomy for severe slow transit constipation. *Dis Colon Rectum*, Vol. 39, No. 1, (Jan 1996), pp. (23-9),0012-3706

Malone, P. S.; P. G. Ransley & E. M. Kiely (1990). Preliminary report: the antegrade continence enema. *Lancet,* Vol. 336, No. 8725, (Nov 17 1990), pp. (1217-8),0140-6736

Mollen, R. M.; H. C. Kuijpers & A. T. Claassen (2001). Colectomy for slow-transit constipation: preoperative functional evaluation is important but not a guarantee for a successful outcome. *Dis Colon Rectum,* Vol. 44, No. 4, (Apr 2001), pp. (577-80),0012-3706

Notghi, A.; R. Hutchinson; D. Kumar; N. Tulley & L. K. Harding (1994). Use of geometric center and parametric images in scintigraphic colonic transit studies. *Gastroenterology,* Vol. 107, No. 5, (Nov 1994), pp. (1270-7),0016-5085

Sample, C.; R. Gupta; F. Bamehriz & M. Anvari (2005). Laparoscopic subtotal colectomy for colonic inertia. *J Gastrointest Surg,* Vol. 9, No. 6, (Jul-Aug 2005), pp. (803-8),1091-255X

Scaillon, M. & S. Cadranel (2006). Food allergy and constipation in childhood: how functional is it? *Eur J Gastroenterol Hepatol,* Vol. 18, No. 2, (Feb 2006), pp. (125-8),0954-691X

Southwell, B. R. (2010). Colon lengthening slows transit: is this the mechanism underlying redundant colon or slow transit constipation? *J Physiol,* Vol. 588, No. Pt 18, (Sep 15 2010), pp. (3343),1469-7793

Southwell, B. R.; M. C. Clarke; J. Sutcliffe & J. M. Hutson (2009). Colonic transit studies: normal values for adults and children with comparison of radiological and scintigraphic methods. *Pediatr Surg Int,* Vol. 25, No. 7, (Jul 2009), pp. (559-72),1437-9813

Stanton, M. P.; Y. M. Shin & J. M. Hutson (2002). Laparoscopic placement of the Chait cecostomy device via appendicostomy. *J Pediatr Surg,* Vol. 37, No. 12, (Dec 2002), pp. (1766-7),1531-5037

Sutcliffe, J. R.; S. K. King; J. M. Hutson; D. J. Cook & B. R. Southwell (2009). Gastrointestinal transit in children with chronic idiopathic constipation. *Pediatr Surg Int,* Vol. 25, No. 6, (Jun 2009), pp. (465-72),1437-9813

Voderholzer, W. A.; W. Schatke; B. E. Muhldorfer; A. G. Klauser; B. Birkner & S. A. Muller-Lissner (1997). Clinical response to dietary fiber treatment of chronic constipation. *Am J Gastroenterol,* Vol. 92, No. 1, (Jan 1997), pp. (95-8),0002-9270

Wald, A. (2002). Slow Transit Constipation. *Curr Treat Options Gastroenterol,* Vol. 5, No. 4, (Aug 2002), pp. (279-283),1534-309X

Yik, Y. I.; T. M. Cain; C. F. Tudball; D. J. Cook; B. R. Southwell & J. M. Hutson (in press). Nuclear transit studies of patients with intractable chronic constipation reveal a subgroup with rapid proximal colonic transit. *J Pediatr Surg,* Vol. 46, No. 7,(July 2011-in press), pp. (1406-1411), 1531-5037

Yik, Y. I.; M. C. Clarke; A. G. Catto-Smith; V. J. Robertson; J. R. Sutcliffe; J. W. Chase; S. Gibb; T. M. Cain; D. J. Cook; C. F. Tudball; J. M. Hutson & B. R. Southwell (2011). Slow-transit constipation with concurrent upper gastrointestinal dysmotility and its response to transcutaneous electrical stimulation. *Pediatr Surg Int,* Vol. 27, No. 7, (July 2011), pp. (705-711), 1437-9813

Yoshioka, K. & M. R. Keighley (1989). Clinical results of colectomy for severe constipation. *Br J Surg*, Vol. 76, No. 6, (Jun 1989), pp. (600-4),0007-1323

Youssef, N. N.; L. Pensabene; E. Barksdale, Jr. & C. Di Lorenzo (2004). Is there a role for surgery beyond colonic aganglionosis and anorectal malformations in children with intractable constipation? *J Pediatr Surg*, Vol. 39, No. 1, (Jan 2004), pp. (73-7),1531-5037

Permissions

The contributors of this book come from diverse backgrounds, making this book a truly international effort. This book will bring forth new frontiers with its revolutionizing research information and detailed analysis of the nascent developments around the world.

We would like to thank Nirmal Singh, for lending his expertise to make the book truly unique. He has played a crucial role in the development of this book. Without his invaluable contribution this book wouldn't have been possible. He has made vital efforts to compile up to date information on the varied aspects of this subject to make this book a valuable addition to the collection of many professionals and students.

This book was conceptualized with the vision of imparting up-to-date information and advanced data in this field. To ensure the same, a matchless editorial board was set up. Every individual on the board went through rigorous rounds of assessment to prove their worth. After which they invested a large part of their time researching and compiling the most relevant data for our readers. Conferences and sessions were held from time to time between the editorial board and the contributing authors to present the data in the most comprehensible form. The editorial team has worked tirelessly to provide valuable and valid information to help people across the globe.

Every chapter published in this book has been scrutinized by our experts. Their significance has been extensively debated. The topics covered herein carry significant findings which will fuel the growth of the discipline. They may even be implemented as practical applications or may be referred to as a beginning point for another development. Chapters in this book were first published by InTech; hereby published with permission under the Creative Commons Attribution License or equivalent.

The editorial board has been involved in producing this book since its inception. They have spent rigorous hours researching and exploring the diverse topics which have resulted in the successful publishing of this book. They have passed on their knowledge of decades through this book. To expedite this challenging task, the publisher supported the team at every step. A small team of assistant editors was also appointed to further simplify the editing procedure and attain best results for the readers.

Our editorial team has been hand-picked from every corner of the world. Their multi-ethnicity adds dynamic inputs to the discussions which result in innovative outcomes. These outcomes are then further discussed with the researchers and contributors who give their valuable feedback and opinion regarding the same. The feedback is then collaborated with the researches and they are edited in a comprehensive manner to aid the understanding of the subject.

Apart from the editorial board, the designing team has also invested a significant amount of their time in understanding the subject and creating the most relevant covers. They scrutinized every image to scout for the most suitable representation of the subject and create an appropriate cover for the book.

The publishing team has been involved in this book since its early stages. They were actively engaged in every process, be it collecting the data, connecting with the contributors or procuring relevant information. The team has been an ardent support to the editorial, designing and production team. Their endless efforts to recruit the best for this project, has resulted in the accomplishment of this book. They are a veteran in the field of academics and their pool of knowledge is as vast as their experience in printing. Their expertise and guidance has proved useful at every step. Their uncompromising quality standards have made this book an exceptional effort. Their encouragement from time to time has been an inspiration for everyone.

The publisher and the editorial board hope that this book will prove to be a valuable piece of knowledge for researchers, students, practitioners and scholars across the globe.

List of Contributors

Nathan C. Sheets and Andrew Z. Wang
University of North Carolina – Chapel Hill, United States of America

Thiago Mastrangelo and Julio Walder
Centre for Nuclear Energy in Agriculture (CENA/USP), Brazil

Sosuke Miyoshi, Keisuke Mitsuoka and Shintaro Nishimura
Bioimaging Research Laboratories, Drug Discovery Research, Astellas Pharma Inc., Tsukuba, Japan

Stephan A. Veltkamp
Global Clinical Pharmacology & Exploratory Development, Astellas Pharma Europe BV, Leiderdorp, Netherlands

A. A. Alharbi
Faculty of Sciences, Physics Department, Princess Nora University Riyadh, Saudi Arabia
Cyclotron institute, Texas A&M University, College Station, TX, USA

M. McCleskey, B. Roeder, A. Spiridon, E. Simmons, V.Z. Goldberg, A. Banu, L. Trache and R. E. Tribble
Cyclotron institute, Texas A&M University, College Station, TX, USA

A. Azzam
Faculty of Sciences, Physics Department, Princess Nora University Riyadh, Saudi Arabia
Nuclear Physics Department, Nuclear Research Center, AEA, Cairo, Egypt

Emilia A. Kimura, Gerhard Wunderlich, Fabiana M. Jordão, Renata Tonhosolo, Heloisa B. Gabriel,
Rodrigo A. C. Sussmann, Alexandre Y. Saito and Alejandro M. Katzin
Department of Parasitology, Institute of Biomedical Sciences, University of São Paulo, São Paulo, Brazil

Nahoko Nagasaki-Takeuchi
Graduate School of Bioagricultural Sciences, Nagoya University, Nagoya, Japan
Graduate School of Biosciences, Nara Institute of Science and Technology, Nara, Japan

Mariko Kato and Masayoshi Maeshima
Graduate School of Bioagricultural Sciences, Nagoya University, Nagoya, Japan

Miki Kawachi
Graduate School of Bioagricultural Sciences, Nagoya University, Nagoya, Japan
Lehrstuhl für Pflanzenphysiologie, Ruhr-Universität Bochum, Bochum,, Germany

Anke Hannemann, Urszula Cytlak and John S. Gibson
Department of Veterinary Medicine, University of Cambridge, UK

Robert J. Wilkins and J. Clive Ellory
Department of Physiology, Anatomy and Genetics, University of Oxford, UK

David C. Rees
Department of Molecular Haematology, King's College Hospital, London, UK

Adam Konefał
Institute of Physics, Department of Nuclear Physics and Its Application, University of Silesia in Katowice, Poland

László Sajo-Bohus and Eduardo D. Greaves
Universidad Simón Bolívar, Valle de Sartenejas, Caracas, Venezuela

József K. Pálfalvi
HAS KFKI Atomic Energy Research Institute, Budapest, Hungary

Dimple Chopra
Department of Pharmaceutical Sciences, Punjabi University, India

Suzanne V. Smith
Centres of Excellence in Antimatter Matter Studies (CAMS) at Australian Nuclear Science and Technology Organisation, Australia
The Australian National University, Australia

Marian Jones and Vanessa Holmes
Centres of Excellence in Antimatter Matter Studies (CAMS) at Australian Nuclear Science and Technology Organisation, Australia

Luciano Izzo and Sara Savelli
Sapienza University of Rome, Umberto I hospital, Department of Surgery "P. Valdoni", Italy

Andrea Stagnitti and Mario Marini
Department of Radiology, Italy

Junko Honda and Miyuki Kanematsu
Department of Surgery, National Hospital Organization Higashitokushima, Medical Center, 1-1, Ohmukai-kita, Ootera, Itano, Tokushima, Japan

Mitsunori Sasa
Department of Surgery, Tokushima Breast Care Clinic, Nakashimada-Cho, Tokushima, Japan

Yoshimi Bando
Department of Molecular and Environmental Pathology, Institute of Health Biosciences, The University of Tokushima Graduate School, Kuramoto-Cho, Tokushima, Japan

Masako Takahashi
Department of Radiology, Tokushima Breast Care Clinic, Nakashimada-Cho, Tokushima, Japan

Chieko Hirose and Sonoka Hisaoka
Department of Radiology, National Hospital Organization Higashitokushima, Medical Center, 1-1, Ohmukai-kita, Ootera, Itano, Tokushima, Japan

Masakuni Noguchi, Miki Yokoi, Yasuharu Nakano, Yukako Ohno and Takeo Kosaka
Kanazawa Medical University Hospital, Japan

Tanja Planinšek Ručigaj and Vesna Tlaker Žunter
Department of Dermatovenereology, University Clinical Centre Ljubljana, Slovenia

Yee Ian Yik
Department of Paediatrics, University of Melbourne, Australia
Murdoch Childrens Research Institute, Australia
Division of Paediatric Surgery, Department of General Surgery, Faculty of Medicine, University of Malaya, Malaysia

David J. Cook, Duncan M. Veysey, Coral F. Tudball, Brooke S. King and Timothy M. Cain
Department of Medical Imaging, Royal Children's Hospital, Melbourne, Australia

Stephen J. Rutkowski
Australian Radiation Services Pty. Ltd., Australia

Bridget R. Southwell
Murdoch Childrens Research Institute, Australia

John M. Hutson
Department of Paediatrics, University of Melbourne, Australia
Department of General Surgery, Royal Children's Hospital, Melbourne, Australia
Murdoch Childrens Research Institute, Australia